Graduate Texts in Mathematics 120

Graduate Texts in Mathematics

continued after Index

William P. Ziemer

Weakly Differentiable Functions

Sobolev Spaces and Functions of Bounded Variation

Springer Science+Business Media, LLC

William P. Ziemer
Department of Mathematics
Indiana University
Bloomington, IN 47405
USA

With 1 illustration.

Mathematics Subject Classifications (1980): 46-E35, 26-B30, 31-B15

Library of Congress Cataloging-in-Publication Data
Ziemer, William P.
Weakly differentiable functions: Sobolev spaces and functions of bounded variation / William P. Ziemer.
p. cm.—(Graduate texts in mathematics; 120)
Bibliography: p.
Includes index.
ISBN 978-1-4612-6985-4 ISBN 978-1-4612-1015-3 (eBook)
DOI 10.1007/978-1-4612-1015-3
1. Sobolev spaces. 2. Functions of bounded variation. I. Title.
II. Series.
QA323.Z53 1989
515'.73—dc20 89-10072

Printed on acid-free paper.

Camera-ready copy prepared using LaTeX.

9 8 7 6 5 4 3 2 1

To Suzanne

Preface

The term "weakly differentiable functions" in the title refers to those integrable functions defined on an open subset of R^n whose partial derivatives in the sense of distributions are either L^p functions or (signed) measures with finite total variation. The former class of functions comprises what is now known as Sobolev spaces, though its origin, traceable to the early 1900s, predates the contributions by Sobolev. Both classes of functions, Sobolev spaces and the space of functions of bounded variation (BV functions), have undergone considerable development during the past 20 years. From this development a rather complete theory has emerged and thus has provided the main impetus for the writing of this book. Since these classes of functions play a significant role in many fields, such as approximation theory, calculus of variations, partial differential equations, and non-linear potential theory, it is hoped that this monograph will be of assistance to a wide range of graduate students and researchers in these and perhaps other related areas. Some of the material in Chapters 1–4 has been presented in a graduate course at Indiana University during the 1987–88 academic year, and I am indebted to the students and colleagues in attendance for their helpful comments and suggestions.

The major thrust of this book is the analysis of pointwise behavior of Sobolev and BV functions. I have not attempted to develop Sobolev spaces of fractional order which can be described in terms of Bessel potentials, since this would require an effort beyond the scope of this book. Instead, I concentrate on the analysis of spaces of integer order which is largely accessible through real variable techniques, but does not totally exclude the use of Bessel potentials. Indeed, the investigation of pointwise behavior requires an analysis of certain exceptional sets and they can be conveniently described in terms of elementary aspects of Bessel capacity.

The only prerequisite for the present volume is a standard graduate course in real analysis, drawing especially from Lebesgue point theory and measure theory. The material is organized in the following manner. Chapter 1 is devoted to a review of those topics in real analysis that are needed in the sequel. Included here is a brief overview of Lebesgue measure, L^p spaces, Hausdorff measure, and Schwartz distributions. Also included are sections on covering theorems and Lorentz spaces—the latter being necessary for a treatment of Sobolev inequalities in the case of critical indices. Chapter 2 develops the basic properties of Sobolev spaces such as equivalent formulations of Sobolev functions and their behavior under the operations of truncation, composition, and change of variables. Also included is a proof of the Sobolev inequality in its simplest form and the related Rellich-Kondrachov Compactness Theorem. Alternate proofs of the Sobolev inequality are given, including the one which relates it to the isoperimetric inequality and provides the best constant. Limiting cases of the Sobolev inequality are discussed in the context of Lorentz spaces.

The remaining chapters are central to the book. Chapter 3 develops the analysis of pointwise behavior of Sobolev functions. This includes a discussion of the continuity properties of functions with first derivatives in L^p in terms of Lebesgue points, approximate continuity, and fine continuity, as well as an analysis of differentiability properties of higher order Sobolev functions by means of L^p-derivatives. Here lies the foundation for more delicate results, such as the comparison of L^p-derivatives and distributional derivatives, and a result which provides an approximation for Sobolev functions by smooth functions (in norm) that agree with the given function everywhere except on sets whose complements have small capacity.

Chapter 4 develops an idea due to Norman Meyers. He observed that the usual indirect proof of the Poincaré inequality could be used to establish a Poincaré-type inequality in an abstract setting. By appropriately interpreting this inequality in various contexts, it yields virtually all known inequalities of this genre. This general inequality contains a term which involves an element of the dual of a Sobolev space. For many applications, this term is taken as a measure; it therefore is of interest to know precisely the class of measures contained in the dual of a given Sobolev space. Fortunately, the Hedberg–Wolff theorem provides a characterization of such measures.

The last chapter provides an analysis of the pointwise behavior of BV functions in a manner that runs parallel to the development of Lebesgue point theory for Sobolev functions in Chapter 3. While the Lebesgue point theory for Sobolev functions is relatively easy to penetrate, the corresponding development for BV functions is much more demanding. The intricate nature of BV functions requires a more involved exposition than does Sobolev functions, but at the same time reveals a rich and beautiful structure which has its foundations in geometric measure theory. After the structure of BV functions has been developed, Chapter 5 returns to the analysis of Poincaré inequalities for BV functions in the spirit developed for Sobolev functions, which includes a characterization of measures that belong to the dual of BV.

In order to place the text in better perspective, each chapter is concluded with a section on historical notes which includes references to all important and relatively new results. In addition to cited works, the Bibliography contains many other references related to the material in the text. Bibliographical references are abbreviated in square brackets, such as [DL]. Equation numbers appear in parentheses; theorems, lemmas, corollaries,and remarks are numbered as $a.b.c$ where b refers to section b in chapter

a, and section $a.b$ refers to section b in chapter a.

I wish to thank David Adams, Robert Glassey, Tero Kilpeläinen, Christoph Neugebauer, Edward Stredulinsky, Tevan Trent, and William K. Ziemer for having critically read parts of the manuscript and supplied many helpful suggestions and corrections.

WILLIAM P. ZIEMER

Contents

3.9 The Implications of Pointwise Differentiability 146
Comparison of L^p-derivatives and distributional derivatives
If $u \in t^{k,p}(x)$ for every x, and if the L^p-derivatives are in L^p, then $u \in W^{k,p}$
3.10 A Lusin-Type Approximation for Sobolev Functions 153
Integral averages of Sobolev functions are uniformly close to their limits on the complement of sets of small capacity
Existence of smooth functions that agree with Sobolev functions on the complement of sets of small capacity
3.11 The Main Approximation 159
Existence of smooth functions that agree with Sobolev functions on the complement of sets of small capacity and are close in norm
Exercises 168
Historical Notes 175

4 Poincaré Inequalities—A Unified Approach 177

4.1 Inequalities in a General Setting 178
An abstract version of the Poincaré inequality
4.2 Applications to Sobolev Spaces 182
An interpolation inequality
4.3 The Dual of $W^{m,p}(\Omega)$ 185
The representation of $(W_0^{m,p}(\Omega))^*$
4.4 Some Measures in $(W_0^{m,p}(\Omega))^*$ 188
Poincaré inequalities derived from the abstract version by identifying Lebesgue and Hausdorff measure with elements in $(W^{m,p}(\Omega))^*$
The trace of Sobolev functions on the boundary of Lipschitz domains
Poincaré inequalities involving the trace of a Sobolev function
4.5 Poincaré Inequalities 193
Inequalities involving the capacity of the set on which a function vanishes
4.6 Another Version of Poincaré's Inequality 196
An inequality involving dependence on the set on which the function vanishes, not merely on its capacity
4.7 More Measures in $(W^{m,p}(\Omega))^*$ 198
Sobolev's inequality for Riesz potentials involving

1

Preliminaries

Beyond the topics usually found in basic real analysis, virtually all of the material found in this work is self-contained. In particular, most of the information contained in this chapter will be well-known by the reader and therefore no attempt has been made to make a complete and thorough presentation. Rather, we merely introduce notation and develop a few concepts that will be needed in the sequel.

1.1 Notation

Throughout, the symbol Ω will generally denote an open set in Euclidean space R^n and $\emptyset$ will designate the empty set. Points in R^n are denoted by $x = (x_1, \ldots, x_n)$, where $x_1 \in R^1$, $1 \leq i \leq n$. If $x, y \in R^n$, the *inner product* of x and y is

$$x \cdot y = \sum_{i=1}^{n} x_i y_i$$

and the *norm* of x is

$$|x| = (x \cdot x)^{1/2}.$$

If $u: \Omega \to R^1$ is a function defined on Ω, the *support* of u is defined by

$$\operatorname{spt} u = \Omega \cap \overline{\{x: u(x) \neq 0\}},$$

where the closure of a set $S \subset R^n$ is denoted by $\overline{S}$. If $S \subset \Omega$, $\overline{S}$ compact and also $\overline{S} \subset \Omega$, we shall write $S \subset\subset \Omega$. The *boundary* of a set S is defined by

$$\partial S = \overline{S} \cap \overline{(R^n - S)}.$$

For $E \subset R^n$ and $x \in R^n$, the distance from x to E is

$$d(x, E) = \inf\{|x - y| : y \in E\}.$$

It is a simple exercise (see Exercise 1.1) to show that

$$|d(x, E) - d(y, E)| \leq |x - y|$$

whenever $x, y \in R^n$. The diameter of a set $E \subset R^n$ is defined by

$$\operatorname{diam}(E) = \sup\{|x - y| : x, y \in E\},$$

and the characteristic function E is denoted by χ_E. The symbol

$$B(x,r) = \{y : |x-y| < r\}$$

denotes the open ball with center x, radius r and

$$\overline{B}(x,r) = \{y : |x-y| \le r\}$$

will stand for the closed ball. We will use $\alpha(n)$ to denote the volume of the ball of radius 1 in R^n. If $\alpha = (\alpha_1, \ldots, \alpha_n)$ is an n-tuple of non-negative integers, α is called a *multi-index* and the *length* of α is

$$|\alpha| = \sum_{i=1}^{n} \alpha_i.$$

If $x = (x_1, \ldots, x_n) \in R^n$, we will let

$$x^\alpha = x_1^{\alpha_1} \cdot x_2^{\alpha_2} \cdots x_n^{\alpha_n}$$

and $\alpha! = \alpha_1!\alpha_2! \cdots \alpha_n!$. The *partial derivative operators* are denoted by $D_i = \partial/\partial x_i$ for $1 \le i \le n$, and the higher order derivatives by

$$D^\alpha = D_1^{\alpha_1} \cdots D_n^{\alpha_n} = \frac{\partial^{|\alpha|}}{\partial x_1^{\alpha_1} \cdots \partial x_n^{\alpha_n}}.$$

The *gradient* of a real-valued function u is denoted by

$$Du(x) = (D_1 u(x), \ldots, D_n u(x)).$$

If k is a non-negative integer, we will sometimes use $D^k u$ to denote the vector $D^k u = \{D^\alpha u\}_{|\alpha|=k}$.

We denote by $C^0(\Omega)$ the space of continuous functions on Ω. More generally, if k is a non-negative integer, possibly ∞, let

$$C^k(\Omega) = \{u : u\colon \Omega \to R^1, D^\alpha u \in C^0(\Omega), \quad 0 \le |\alpha| \le k\},$$

$$C_0^k(\Omega) = C^k(\Omega) \cap \{u : \text{spt } u \text{ compact, spt } u \subset \Omega\},$$

and

$$C^k(\overline{\Omega}) = C^k(\Omega) \cap \{u : D^\alpha u \text{ has a continuous extension to } \overline{\Omega}, 0 \le |\alpha| \le k\}.$$

Since Ω is open, a function $u \in C^k(\Omega)$ need not be bounded on Ω. However, if u is bounded and uniformly continuous on Ω, then u can be uniquely extended to a continuous function on $\overline{\Omega}$. We will use $C^k(\Omega; R^m)$ to denote the class of functions $u\colon \Omega \to R^m$ defined on Ω whose coordinate functions belong to $C^k(\Omega)$. Similar notation is used for other function spaces whose elements are vector-valued.

If $0 < \alpha \leq 1$, we say that u is *Hölder continuous on Ω with exponent α* if there is a constant C such that

$$|u(x) - u(y)| \leq C|x - y|^\alpha, \quad x, y \in \Omega.$$

We designate by $C^{0,\alpha}(\overline{\Omega})$ the space of all functions u satisfying this condition on Ω. In case $\alpha = 1$, the functions are called Lipschitz and the constant C is denoted by $\text{Lip}(u)$. For functions that possess some differentiability, we let

$$C^{k,\alpha}(\overline{\Omega}) = C^{0,\alpha}(\overline{\Omega}) \cap \{u : D^\beta u \in C^{0,\alpha}(\overline{\Omega}),\ 0 \leq |\beta| \leq k\}.$$

Note that $C^{k,\alpha}(\overline{\Omega})$ is a Banach space when provided with the norm

$$\sup_{|\beta|=k} \sup_{\substack{x,y\in\Omega \\ x\neq y}} \frac{|D^\beta u(x) - D^\beta u(y)|}{|x-y|^\alpha} + \max_{0\leq|\beta|\leq k} \sup_{x\in\Omega} |D^\beta u(x)|.$$

1.2 Measures on R^n

For the definition of Lebesgue outer measure, we consider closed n-dimensional intervals

$$I = \{x : a_i \leq x_i \leq b_i,\ i = 1, \ldots, n\}$$

and their volumes

$$v(I) = \prod_{i=1}^n (b_i - a_i).$$

The Lebesgue outer measure of an arbitrary set $E \subset R^n$ is defined by

$$|E| = \inf\left\{\sum_{k=1}^\infty v(I_k) : E \subset \bigcup_{k=1}^\infty I_k,\ I_k \text{ an interval}\right\} \tag{1.2.1}$$

A set E is said to be *Lebesgue measurable* if

$$|A| = |A \cap E| + |A \cap (R^n - E)| \tag{1.2.2}$$

whenever $A \subset R^n$.

The reader may consult a standard text on measure theory to find that the Lebesgue measurable sets form a *σ-algebra,* which we denote by $\mathcal{A}$; that is

(i) $\emptyset, R^n \in \mathcal{A}$.

(ii) If $E_1, E_2, \ldots \in \mathcal{A}$, then

$$\bigcup_{i=1}^\infty E_i \in \mathcal{A}. \tag{1.2.3}$$

(iii) If $E \in \mathcal{A}$, then $R^n - E \in \mathcal{A}$.

Observe that these conditions also imply that $\mathcal{A}$ is also closed under countable intersections. It follows immediately from (1.2.2) that sets of measure zero are measurable. Also recall that if $E_1, E_2, \ldots$ are pairwise disjoint measurable sets, then

$$\left| \bigcup_{i=1}^{\infty} E_i \right| = \sum_{i=1}^{\infty} |E_i|. \tag{1.2.4}$$

Moreover, if $E_1 \subset E_2 \subset \ldots$ are measurable, then

$$\left| \bigcup_{i=1}^{\infty} E_i \right| = \lim_{i \to \infty} |E_i| \tag{1.2.5}$$

and if $E_1 \supset E_2 \supset \ldots$, then

$$\left| \bigcap_{i=1}^{\infty} E_i \right| = \lim_{i \to \infty} |E_i| \tag{1.2.6}$$

provided that $|E_k| < \infty$ for some k.

Up to this point, we find that Lebesgue measure possesses many of the continuity properties that are essential for fruitful applications in analysis. However, at this stage we do not yet know whether the σ-algebra, $\mathcal{A}$, contains a sufficiently rich supply of sets to be useful. This possible objection is met by the following result.

1.2.1. Theorem. *Each closed set $C \subset R^n$ is Lebesgue measurable.*

In view of the fact that the Borel subsets of R^n form the smallest σ-algebra that contains the closed sets, we have

1.2.2. Corollary. *The Borel sets of R^n are Lebesgue measurable.*

Proof of Theorem 1.2.1. Because of the subadditivity of Lebesgue measure, it suffices to show that for a closed set $C \subset R^n$,

$$|A| \geq |A \cap C| + |A \cap (R^n - C)| \tag{1.2.7}$$

whenever $A \subset R^n$. This will follow from the following property of Lebesgue outer measure, which follows easily from (1.2.1):

$$|A \cup B| = |A| + |B| \tag{1.2.8}$$

whenver $A, B \in R^n$ with $d(A, B) = \inf\{|x - y| : x \in A, y \in B\} > 0$. Indeed, it is sufficient to establish that $|A \cup B| \geq |A| + |B|$. For this purpose, choose $\varepsilon > 0$ and let

$$A \cup B \subset \bigcup_{k=1}^{\infty} I_k \quad \text{where}$$

$$\sum_{i=1}^{\infty} v(I_k) < |A \cup B| + \varepsilon. \tag{1.2.9}$$

Because $d(A, B) > 0$, there exists disjoint open sets U and V such that

$$A \subset U, \quad B \subset V. \tag{1.2.10}$$

Clearly, the covering of $A \cup B$ by $\{I_k\}$ can be modified so that, for each k,

$$I_k \subset U \cup V \tag{1.2.11}$$

and that (1.2.9) still remains valid. However, (1.2.10) and (1.2.11) imply

$$\sum_{i=1}^{\infty} v(I_k) \geq |A| + |B|.$$

In order to prove (1.2.7), consider $A \subset R^n$ with $|A| < \infty$ and let $C_i = \{x : d(x, C) \leq 1/i\}$. Note that

$$d(A - C_i, A \cap C) > 0$$

and therefore, from (1.2.8),

$$|A| \geq |(A - C_i) \cup (A \cap C)| \geq |A - C_i| + |A \cap C|. \tag{1.2.12}$$

The proof of (1.2.7) will be concluded if we can show that

$$\lim_{i \to \infty} |A - C_i| = |A - C|.$$

Note that we cannot invoke (1.2.5) because it is not known that $A - C_i$ is measurable since A is an arbitrary set, perhaps non-measurable. Let

$$T_i = A \cap \left\{ x : \frac{1}{i+1} < d(x, C) \leq \frac{1}{i} \right\} \tag{1.2.13}$$

and note that since C is closed,

$$A - C = (A - C_j) \cup \left(\bigcup_{i=j}^{\infty} T_i \right) \tag{1.2.14}$$

which in turn, implies

$$|A - C| \leq |A - C_j| + \sum_{i=j}^{\infty} |T_i|. \tag{1.2.15}$$

Hence, the desired conclusion will follow if it can be shown that

$$\sum_{i=1}^{\infty} |T_i| < \infty. \tag{1.2.16}$$

To establish this, first observe that $d(T_i, T_j) > 0$ if $|i-j| \geq 2$. Thus, we obtain from (1.2.8) that for each positive integer m,

$$\left|\bigcup_{i=1}^{m} T_{2i}\right| = \sum_{i=1}^{m} |T_{2i}| \leq |A| < \infty,$$

$$\left|\sum_{i=1}^{m} T_{2i-1}\right| = \left|\bigcup_{i=1}^{m} T_{2i-1}\right| \leq |A| < \infty.$$

This establishes (1.2.16) and thus concludes the proof. □

1.2.3. Remark. Lebesgue measure and Hausdorff measure (which will be introduced in Section 1.4) will meet most of the applications that occur in this book, although in Chapter 5, it will be necessary to consider more general measures. We say that μ is a *measure* on R^n if μ assigns a non-negative (possibly infinite) number to each subset of R^n and $\mu(\emptyset) = 0$. It is also accepted terminology to call such a set function an *outer measure.* Following (1.2.2), a set E is called *μ-measurable* if

$$\mu(A) = \mu(A \cap E) + \mu(A \cap (R^n - E))$$

whenever $A \subset R^n$. A measure μ on R^n is called a *Borel measure* if every Borel set is μ-measurable. A Borel measure μ with the properties that each subset of R^n is contained within a Borel set of equal μ measure and that $\mu(K) < \infty$ for each compact set $K \subset R^n$ is called a *Radon measure.*

Many outer measures defined on R^n have the property that the Borel sets are measurable. However, it is sometimes necessary to consider a larger σ-algebra of sets, namely, the *Suslin sets,* (often referred to as analytic sets). They have the property of remaining invariant under continuous mappings on R^n, a property not enjoyed by the Borel sets. The Suslin sets of R^n can be defined in the following manner. Let $\mathcal{N}$ denote the space of all infinite sequences of positive integers topologized by the metric

$$\sum_{i=1}^{\infty} \frac{2^{-i}|a_i - b_i|}{1 + |a_i - b_i|}$$

where $\{a_i\}$ and $\{b_i\}$ are elements of $\mathcal{N}$. Let $R^n \times \mathcal{N}$ be endowed with the product topology. If

$$p : R^n \times \mathcal{N} \to R^n$$

is the projection defined by $p(x, a) = x$, then a Suslin set of R^n can be defined as the image under p of some closed subset of $R^n \times \mathcal{N}$.

The main reason for providing the preceding review of Lebesgue measure is to compare its development with that of Hausdorff measure, which is not as well known as Lebesgue measure but yet is extremely important in geometric analysis and will play a significant role in the development of this monograph.

1.3 Covering Theorems

Before discussing Hausdorff measure, it will be necessary to introduce several important and useful covering theorems, the first of which is based on the following implication of the Axiom of Choice.

Hausdorff Maximal Principle. *If $\mathcal{E}$ is a family of sets (or a collection of families of sets) and if $\{\cup F : F \in \mathcal{F}\} \in \mathcal{E}$ for any subfamily $\mathcal{F}$ of $\mathcal{E}$ with the property that*

$$F \subset G \text{ or } G \subset F \quad \textit{whenever} \quad F, G \in \mathcal{F},$$

then there exists $E \subset \mathcal{E}$ which is maximal in the sense that it is not a subset of any other member of $\mathcal{E}$.

The following notation will be used. If B is a closed ball of radius r, let $\hat{B}$ denote the closed ball concentric with B with radius $5r$.

1.3.1. Theorem. *Let $\mathcal{G}$ be a family of closed balls with*

$$R \equiv \sup\{\operatorname{diam} B : B \in \mathcal{G}\} < \infty.$$

Then there is a subfamily $\mathcal{F} \subset \mathcal{G}$ of pairwise disjoint elements such that

$$\{\cup B : B \in \mathcal{G}\} \subset \{\cup \hat{B} : B \in \mathcal{F}\}.$$

In fact, for each $B \in \mathcal{G}$ there exists $B_1 \in \mathcal{F}$ such that $B \cap B_1 \neq 0$ and $B \subset \hat{B}_1$.

Proof. We determine $\mathcal{F}$ as follows. For $j = 1, 2, \ldots$ let

$$\mathcal{G}_j = \mathcal{G} \cap \left\{ B : \frac{R}{2^j} < \operatorname{diam} B \leq \frac{R}{2^{j-1}} \right\},$$

and observe that $\mathcal{G} = \cup_{j=1}^{\infty} \mathcal{G}_j$. Now proceed to define $\mathcal{F}_j \subset \mathcal{G}_j$ inductively as follows.

Let $\mathcal{F}_1 \subset \mathcal{G}_1$ be an arbitrary maximal subcollection of pairwise disjoint elements. Such a collection exits by the Hausdorff maximal principle. Assuming that $\mathcal{F}_1, \mathcal{F}_2, \ldots, \mathcal{F}_{j-1}$ have been chosen, let $\mathcal{F}_j$ be a maximal pairwise disjoint subcollection of

$$\mathcal{G}_j \cap \left\{ B : B \cap B' = \emptyset \text{ whenever } B' \in \bigcup_{i=1}^{j-1} \mathcal{F}_i \right\}. \tag{1.3.1}$$

Thus, for each $B \in \mathcal{G}_j$, $j \geq 1$, there exists $B_1 \in \cup_{i=1}^{j} \mathcal{F}_i$ such that $B \cap B_1 \neq \emptyset$. For if not, the family $\mathcal{F}_j^*$ consisting of B along with all elements of $\mathcal{F}_j$

would be a pairwise disjoint subcollection of (1.3.1), thus contradicting the maximality of $\mathcal{F}_j$. Moreover,

$$\operatorname{diam} B \leq \frac{R}{2^{j-1}} = 2\frac{R}{2^j} \leq 2 \operatorname{diam} B_1$$

which implies that $B \subset \hat{B}_1$. Thus,

$$\{\cup B : B \in \mathcal{G}_j\} \subset \left\{\cup \hat{B} : B \in \bigcup_{i=1}^{j} \mathcal{F}_i\right\},$$

and the conclusion holds by taking

$$\mathcal{F} = \bigcup_{i=1}^{\infty} \mathcal{F}_i. \qquad \square$$

1.3.2. Definition. A collection $\mathcal{G}$ of closed balls is said to cover a set $E \subset R^n$ finely if for each $x \in E$ and each $\varepsilon > 0$, there exists $B(x,r) \in \mathcal{G}$ and $r < \varepsilon$.

1.3.3. Corollary. *Let $E \subset R^n$ be a set that is covered finely by $\mathcal{G}$, where $\mathcal{G}$ and $\mathcal{F}$ are as in Theorem* 1.3.1. *Then,*

$$E - \{\cup B : B \in \mathcal{F}^*\} \subset \{\cup \hat{B} : B \in \mathcal{F} - \mathcal{F}^*\}$$

for each finite collection $\mathcal{F}^* \subset \mathcal{F}$.

Proof. Since $R^n - \{\cup B : B \in \mathcal{F}^*\}$ is open, for each $x \in E - \{\cup B : B \in \mathcal{F}^*\}$ there exists $B \in \mathcal{G}$ such that $x \in B$ and $B \cap [\{\cup B : B \in \mathcal{F}^*\}] = \emptyset$. From Theorem 1.3.1, there is $B_1 \in \mathcal{F}$ such that $B \cap B_1 \neq \emptyset$ and $\hat{B}_1 \supset B$. Now $B_1 \notin \mathcal{F}^*$ since $B \cap B_1 \neq \emptyset$ and therefore

$$x \in \{\cup \hat{B} : B \in \mathcal{F} - \mathcal{F}^*\}. \qquad \square$$

The next result addresses the question of determining an estimate for the amount of overlap in a given family of closed balls. This will also be considered in Theorem 1.3.5, but in the following we consider closed balls whose radii vary in a Lipschitzian manner. The notation $\operatorname{Lip}(h)$ denotes the Lipschitz constant of the mapping h.

1.3.4. Theorem. *Let $S \subset U \subset R^n$ and suppose $h : U \to (0, \infty)$ is Lipschitz with* $\operatorname{Lip}(h) \leq \lambda$. *Let $\alpha > 0$, $\beta > 0$ with $\lambda\alpha < 1$ and $\lambda\beta < 1$. Suppose the collection of closed n-balls $\{B(s, h(s)) : s \in S\}$ is disjointed. Let*

$$S_x = S \cap \{s : \overline{B}(x, \alpha h(x)) \cap \overline{B}(s, \beta h(s)) \neq 0\}.$$

Then

$$(1-\lambda\beta)/(1+\lambda\alpha) \le h(x)/h(s) \le (1+\lambda\beta)/(1-\lambda\alpha) \qquad (1.3.2)$$

whenever $s \in S_x$ *and*

$$\operatorname{card}(S_x) \le [\alpha + (\beta+1)(1+\lambda\alpha)(1-\lambda\beta)^{-1}]^n[(1+\lambda\beta)/(1-\lambda\alpha)]^n$$

where $\operatorname{card}(S_x)$ *denotes the number of elements in* S_x.

Proof. If $s \in S_x$, then clearly $|x-s| \le \alpha h(x) + \beta h(s)$ and therefore

$$\begin{aligned} |h(x)-h(s)| \le \lambda|x-s| &\le \lambda\alpha h(x) + \lambda\beta h(s), \\ (1-\lambda\beta)h(s) &\le (1+\lambda\alpha)h(x), \\ (1-\lambda\alpha)h(x) &\le (1+\lambda\beta)h(s). \end{aligned} \qquad (1.3.3)$$

Now,

$$\begin{aligned} |x-s| + h(s) &\le \alpha h(x) + (\beta+1)h(s) \\ &\le \alpha h(x) + (\beta+1)[(1+\lambda\alpha)/(1-\lambda\beta)]h(x) \\ &= \gamma h(x) \end{aligned}$$

where $\gamma = \alpha + (\beta+1)(1+\lambda\alpha)/(1-\lambda\beta)$. Hence

$$\overline{B}(s, h(s)) \subset \overline{B}(x, \gamma h(x)) \quad \text{whenever} \quad s \in S_x.$$

Since $\{\overline{B}(s, h(s))\}$ is a disjoint family,

$$\sum_{s \in S_x} |\overline{B}(s, h(s))| \le |\overline{B}(s, \gamma h(x))|$$

or from (1.3.3)

$$\operatorname{card}(S_x)\alpha(n)[(1+\lambda\alpha)(1-\lambda\beta)^{-1}h(x)]^n \le \sum_{s \in S_x} \alpha(n)h(s)^n \le \alpha(n)[\gamma h(x)]^n. \square$$

We now consider an arbitrary collection of closed balls and find a subcover which is perhaps not disjoint, but whose elements have overlap which is controlled.

1.3.5. Theorem. *There is a positive number* $N > 1$ *depending only on* n *so that any family* $\mathcal{B}$ *of closed balls in* R^n *whose cardinality is no less than* N *and* $R = \sup\{r : B(a, r) \in \mathcal{B}\} < \infty$ *contains disjointed subfamilies* $\mathcal{B}_1$, $\mathcal{B}_2, \ldots, \mathcal{B}_N$ *such that if* A *is the set of centers of balls in* $\mathcal{B}$, *then*

$$A \subset \bigcup_{i=1}^{N} \{\cup B : B \in \mathcal{B}_i\}.$$

Proof.

Step I. Assume A is bounded.

Choose $B_1 = B(a_1, r_1)$ with $r_1 > \frac{3}{4}R$. Assuming we have chosen $B_1, \ldots, B_{j-1}$ in $\mathcal{B}$ where $j \geq 2$ choose B_j inductively as follows. If $A_j = A \sim \cup_{i=1}^{j-1} B_i = \emptyset$, then the process stops and we set $J = j$. If $A_j \neq \emptyset$, continue by choosing $B_j = B(a_j, r_j) \in \mathcal{B}$ so that $a_j \in A_j$ and

$$r_j > \frac{3}{4} \sup\{r : B(a, r) \in \mathcal{B}, a \in A_j\}. \tag{1.3.4}$$

If $A_j \neq \emptyset$ for all j, then we set $J = +\infty$. In this case $\lim_{j\to\infty} r_j = 0$ because A is bounded and the inequalities

$$|a_i - a_j| > r_i = \frac{r_i}{3} + \frac{2}{3} r_i > \frac{r_i}{3} + \frac{r_j}{2}, \quad \text{for } i < j,$$

imply that

$$\{B(a_j, r_j/3) : 1 \leq j \leq J\} \text{ is disjointed.} \tag{1.3.5}$$

In case $J < \infty$, we clearly have the inclusion

$$A \subset \{\cup B_j : 1 \leq j \leq J\}. \tag{1.3.6}$$

This is also true in case $J = +\infty$, for otherwise there would exist $B(a, r) \in \mathcal{B}$ with $a \in \cap_{j=1}^{\infty} A_j$ and an integer j with $r_j \leq 3r/4$, contradicting the choice of B_j.

Step II. We now prove there exists an integer M (depending only on n) such that for each k with $1 \leq k < J$, M exceeds the number of balls B_i with $1 \leq i \leq k$ and $B_i \cap B_k \neq \emptyset$.

First note that if $r_i < 10 r_k$, then

$$B(a_i, r_i/3) \subset B(a_k, 15 r_k)$$

because if $x \in B(a_i, r_i/3)$,

$$\begin{aligned} |x - a_k| &\leq |x - a_i| + |a_i - a_k| \\ &\leq 10 r_k/3 + r_i + r_k \\ &\leq 43 r_k/3 < 15 r_k. \end{aligned}$$

Hence, there are at most $(60)^n$ balls B_i with

$$1 \leq i \leq k, \;\; B_i \cap B_k \neq \emptyset, \text{ and } r_i \leq 10 r_k$$

because, for each such i,

$$B(a_i, r_i/3) \subset B(a_k, 15 r_k),$$

and by (1.3.4) and (1.3.5)

$$|B(a_i, r_i/3)| = |B_1| \cdot \left(\frac{r_i}{3}\right)^n > |B_1| \cdot \left(\frac{r_k}{4}\right)^n = \frac{1}{60^n} |B(a_k, 15 r_k)|.$$

To complete Step II, it remains to estimate the number of points in the set

$$I = \{i : 1 \le i \le k, B_i \cap B_k \neq \emptyset, r_i > 10r_k\}.$$

For this we first find an absolute lower bound on the angle between the two vectors

$$a_i - a_k \quad \text{and} \quad a_j - a_k$$

corresponding to $i, j \in I$ with $i < j$. Assuming that this angle $\alpha < \pi/2$, consider the triangle

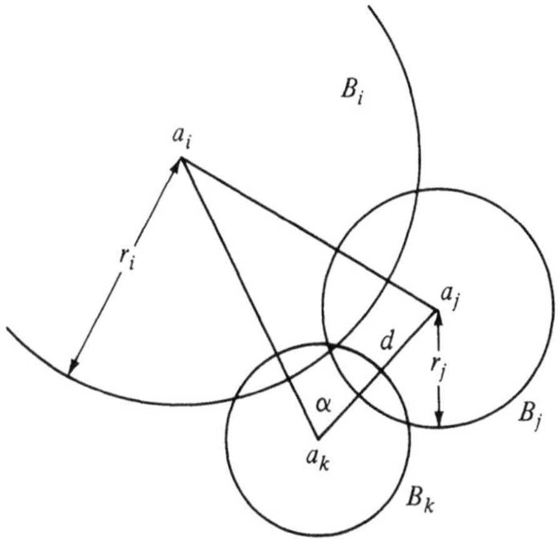

and assume for notational convenience that $r_k = 1$, $d = |a_j - a_k|$. Then

$$10 < r_i < |a_i - a_k| \le r_i + 1 \quad \text{and} \quad |a_i - a_j| \ge r_i$$

because $i \in I$, $a_k \notin B_j$, $B_j \cap B_k \neq \emptyset$, and $a_j \notin B_i$. Also

$$10 < r_i < d \le r_i + 1 < \frac{4}{3}r_i + 1$$

because $j \in I$, $a_k \notin B_j$, $B_j \cap B_k \neq \emptyset$, and (1.3.4) applies to r_i.

The law of cosines yields

$$\begin{aligned}\cos\alpha &= \frac{|a_i - a_k|^2 + d^2 - |a_i - a_j|^2}{2|a_i - a_k|d} \le \frac{(r_i+1)^2 + d^2 - r_i^2}{2r_i d} \\ &= \frac{2r_i + 1 + d^2}{2r_i d} = \frac{1}{d} + \frac{1}{2r_i d} + \frac{d}{2r_i} < \frac{1}{d} + \frac{1}{2r_i d} + \frac{4r_i}{6r_i} + \frac{1}{2r_i} \\ &\le \frac{1}{10} + \frac{1}{200} + \frac{4}{6} + \frac{1}{20} < .822;\end{aligned}$$

hence $|\alpha| > \arccos .822 > 0$. Consequently, the rays determined by $a_j - a_k$ and $a_i - a_k$ intersect the boundary of $B(a_k, 1)$ at points that are separated by a distance of at least $\sqrt{2(1 - \cos\alpha)}$. Since the boundary of $B(a_k, 1)$

has finite H^{n-1} measure, the number of points in I is no more than some constant depending only on n.

Step III. Choice of $\mathcal{B}_1, \dots, \mathcal{B}_M$ in case A is bounded.

With each positive integer j, we define an integer λ_j such that $\lambda_j = j$ whenever $1 \leq j \leq M$ and for $j > M$ we define λ_{j+1} inductively as follows. From Step II there is an integer $\lambda_{j+1} \in \{1, 2, \dots, M\}$ such that

$$B_{j+1} \cap \{\cup B_i : 1 \leq i \leq j, \lambda_i = \lambda_{j+1}\} = \emptyset.$$

Now deduce from (1.3.6) that the unions of the disjointed families

$$\mathcal{B}_1 = \{B_i : \lambda_i = 1\}, \dots, \mathcal{B}_M = \{B_i : \lambda_i = M\}$$

covers A.

Step IV. The case A is unbounded.

For each positive integer ℓ, apply Step III with A replaced by $E_\ell = A \cap \{x : 3(\ell - 1)R \leq |x| < 3\ell R\}$ and $\mathcal{B}$ replaced by the subfamily $\mathcal{C}_\ell$ of $\mathcal{B}$ of balls with centers in E_ℓ. We obtain disjointed subfamilies $\mathcal{B}_1^\ell, \dots, \mathcal{B}_M^\ell$ of $\mathcal{C}_\ell$ such that

$$E_\ell \subset \bigcup_{i=1}^{M} \{\cup B : B \in \mathcal{B}_i^\ell\}.$$

Since $P \cap Q = \emptyset$ whenever $P \in \mathcal{B}^\ell$, $Q \in \mathcal{B}^m$ and $m \geq \ell + 2$, the theorem follows with

$$\mathcal{B}_1 = \bigcup_{\ell=1}^{\infty} \mathcal{B}_1^{2\ell-1}, \dots, \mathcal{B}_M = \bigcup_{\ell=1}^{\infty} \mathcal{B}_M^{2\ell-1}$$

$$\mathcal{B}_{M+1} = \bigcup_{\ell=1}^{\infty} \mathcal{B}_1^{2\ell}, \dots, \mathcal{B}_{2M} = \bigcup_{\ell=1}^{\infty} \mathcal{B}_{2M}^{2\ell}$$

and $N = 2M$. □

We use this result to establish the following covering theorem which contains the classical result of Vitali involving Lebesgue measure. An interesting and novel aspect of the theorem is that the set A is not assumed to be μ-measurable. The thrust of the proof is that the previous theorem allows us to obtain a disjoint subfamily that provides a fixed percentage of the μ measure of the original set.

1.3.6. Theorem. *Let μ be a Radon measure on R^n and suppose $\mathcal{F}$ is a family of closed balls that covers a set $A \subset R^n$ finely, where $\mu(A) < \infty$. Then there exists a countable disjoint subfamily $\mathcal{G}$ of $\mathcal{F}$ such that*

$$\mu(A - \{\cup B : B \in \mathcal{G}\}) = 0.$$

Proof. Choose $\varepsilon > 0$ so that $\varepsilon < 1/N$, where N is the constant that appears in the previous theorem. Then $\mathcal{F}$ has disjointed subfamilies $\mathcal{B}_1, \ldots, \mathcal{B}_N$ such that

$$A \subset \bigcup_{i=1}^{N} \{\cup B : B \in \mathcal{B}_i\}$$

and therefore

$$\mu(A) \leq \sum_{i=1}^{N} \mu(\{\cup(A \cap B) : B \in \mathcal{B}_i\}).$$

Thus, there exists $1 \leq k \leq N$ such that

$$\mu(\{\cup(A \cap B) : B \in \mathcal{B}_k\} \geq 1/N \ \mu(A),$$

which implies

$$\mu(A - \{\cup B : B \in \mathcal{B}_k\}) \leq (1 - 1/N)\mu(A).$$

Hence, there is a finite subfamily $\mathcal{B}_{k_1}$ of $\mathcal{B}_k$ such that

$$\mu(A - \{\cup B : B \in \mathcal{B}_{k_1}\}) < (1 - 1/N + \varepsilon)\mu(A).$$

Now repeat this argument by replacing A with $A_1 = 1 - \{\cup B : B \in \mathcal{B}_{k_1}\}$ and $\mathcal{F}$ with $\mathcal{F}_1 = \mathcal{F} \cap \{B : B \cap \{\cup B : B \in \mathcal{B}_{k_1}\} = \emptyset\}$ to obtain a finite disjointed subfamily $\mathcal{B}_{k_2}$ of $\mathcal{F}_1$ such that

$$\mu(A_1 - \{\cup B : B \in \mathcal{B}_{k_2}\}) < (1 - 1/N + \varepsilon)\mu(A_1).$$

Thus,

$$\mu(A - \{\cup B : B \in \mathcal{B}_{k_1} \cup \mathcal{B}_{k_2}\}) \leq (1 - 1/N + \varepsilon)^2\mu(A).$$

Continue this process to obtain the conclusion of the theorem with

$$\mathcal{G} = \bigcup_{i=1}^{\infty} B_{k_i}. \qquad \square$$

1.3.7. Lemma. *Let μ and ν be Radon measures on R^n. For each positive number α let*

$$E_\alpha = \left\{x : \sup_{r>0} \frac{\mu[B(x,r)]}{\nu[B(x,r)]} > \alpha\right\}.$$

Then, $\mu(E_\alpha) \geq \alpha\nu(E_\alpha)$.

Proof. By restricting our attention to bounded subsets of E_α, we may assume that $\mu(E_\alpha)$, $\nu(E_\alpha) < \infty$. Let $U \supset E_\alpha$ be an open set. For $\varepsilon > 0$ and for each $x \in E_\alpha$, there exists a sequence of closed balls $B(x, r_i) \subset U$ with $r_i \to 0$ such that

$$\mu[B(x,r_i)] > (\alpha + \varepsilon)\nu[B(x,r_i)].$$

This produces a family $\mathcal{F}$ of closed balls that covers E_α finely. Hence, by Theorem 1.3.6, there exists a disjoint subfamily $\mathcal{G}$ that covers ν almost all of E_α. Consequently

$$(\alpha+\varepsilon)\nu(E_\alpha) \le (\alpha+\varepsilon)\sum_{B\in\mathcal{G}}\nu(B) \le \sum_{B\in\mathcal{G}}\mu(B) \le \mu(U).$$

Since ε and U are arbitrary, the conclusion follows. □

If f is a continuous function, then the integral average of f over a ball of small radius is nearly the same as the value of f at the center of the ball. A remarkable result of real analysis states that this is true at (Lebesgue) almost all points whenever f is integrable. The following result provides a proof relative to any Radon measure. The notation

$$⨍_{B(x,r)} f(y)\,d\mu(y)$$

denotes

$$\mu[B(x,r)]^{-1}\int_{B(x,r)} f(y)\,d\mu(y).$$

1.3.8. Theorem. *Let μ be a Radon measure on R^n and f a locally integrable function on R^n with respect to μ. Then*

$$\lim_{r\to 0} ⨍_{B(x,r)} f(y)\,d\mu(y) = f(x)$$

for μ almost all $x \in R^n$.

Proof. Note that

$$\left| ⨍_{B(x,r)} f(y)d\mu(y) - f(x)\right| \le ⨍_{B(x,r)} |f(y)-g(y)|d\mu(y)$$
$$+ ⨍_{B(x,r)} |g(y)-f(x)|d\mu(y)$$

and if g is continuous, the last term converges to $|g(x)-f(x)|$ as $r \to 0$. Letting $L(x)$ denote the upper limit of the term on the left, we obtain

$$L(x) \le \sup_{r>0} ⨍_{B(x,r)} |f(y)-g(y)|d\mu(y) + |g(x)-f(x)|.$$

Hence,

$$\{x : L(x) > \alpha\} \subset \left\{x : \sup_{r>0} ⨍_{B(x,r)} |f(y)-g(y)|d\mu(y) > \alpha/2\right\}$$

$$\cup \{x : |g(x) - f(x)| > \alpha/2\},$$

and therefore, by the previous lemma,

$$\mu(\{x : L(x) > \alpha\}) \leq 2/\alpha \int_{R^n} |f - g| d\mu + 2/\alpha \int_{R^n} |f - g| d\mu.$$

Since $\int_{R^n} |f - g| d\mu$ can be made arbitrarily small with appropriate choice of g, cf. Section 1.6, it follows that $\mu(\{x : L(x) > \alpha\}) = 0$ for each $\alpha > 0$.

□

1.3.9. Remark. If μ and ν are Radon measures with μ absolutely continuous with respect to ν, then the Radon–Nikodym theorem provides $f \in L^1(R^n, \nu)$ such that

$$\mu(E) = \int_E f(x)\, d\nu(x).$$

The results above show that the Radon–Nikodym derivative f can be taken as the derivative of μ with respect to ν; that is,

$$\lim_{r \to 0} \frac{\mu[B(x,r)]}{\nu[B(x,r)]} = f(x)$$

for ν almost all $x \in R^n$.

1.4 Hausdorff Measure

The purpose here is to define a measure on R^n that will assign a reasonable notion of "length," "area" etc. to sets of appropriate dimension. For example, if we would like to define the notion of length for an arbitrary set $E \subset R^n$, we might follow (1.2.1) and let

$$\lambda(E) = \inf \left\{ \sum_{i=1}^{\infty} \operatorname{diam} A_i : E \subset \bigcup_{i=1}^{\infty} A_i, \right\}.$$

However, if we take $n = 2$ and $E = \{(t, \sin(1/t)) : 0 \leq t \leq 1\}$, it is easily seen that $\lambda(E) < \infty$ whereas we should have $\lambda(E) = \infty$. The difficulty with this definition is that the approximating sets A_i are not forced to follow the geometry of the curve. This is changed in the following definition.

1.4.1. Definition. For each $\gamma \geq 0$, $\varepsilon > 0$, and $E \subset R^n$, let

$$H_\varepsilon^\gamma(E) = \inf \left\{ \sum_{i=1}^{\infty} \alpha(\gamma) 2^{-\gamma} \operatorname{diam}(A_i)^\gamma : E \subset \bigcup_{i=1}^{\infty} A_i, \operatorname{diam} A_i < \varepsilon \right\}.$$

Because $H^\gamma_\varepsilon(E)$ is non-decreasing in ε, we may define the γ *dimensional Hausdorff measure* of E as

$$H^\gamma(E) = \lim_{\varepsilon\to 0} H^\gamma_\varepsilon(E). \tag{1.4.1}$$

In case γ is a positive integer, $\alpha(\gamma)$ denotes the volume of the unit ball in R^γ. Otherwise, $\alpha(\gamma)$ can be taken as an arbitrary positive constant. The reason for requiring $\alpha(\gamma)$ to equal the volume of the unit ball in R^γ when γ is a positive integer is to ensure that $H^\gamma(E)$ agrees with intuitive notions of "γ-dimensional area" when E is a well-behaved set. For example, it can be shown that H^n agrees with the usual definition of n-dimensional area on an n-dimensional C^1 submanifold of R^{n+k}, $k \ge 0$. More generally, if $f\colon R^n \to R^{n+k}$ is a univalent, Lipschitz map and $E \subset R^n$ a Lebesgue measurable set, then

$$\int_E Jf = H^n[f(E)]$$

where Jf is the square root of the sum of the squares of the $n \times n$ determinants of the Jacobian matrix. The reader may consult [F4, Section 3.2] for a thorough treatment of this subject. Here, we will merely show that H^n defined on R^n is equal to Lebesgue measure.

1.4.2. Theorem. *If $E \subset R^n$, then $H^n(E) = |E|$.*

Proof. First we show that

$$H^n_\varepsilon(E) \le |E| \quad \text{for every} \quad \varepsilon > 0.$$

Consider the case where $|E| = 0$ and E is bounded. For each $\eta > 0$, let $U \supset E$ be an open set with $|U| < \eta$. Since U is open, U can be written as the union of closed balls, each of which has diameter less than ε. Theorem 1.3.1 states that there is a subfamily $\mathcal{F}$ of pairwise disjoint elements such that

$$U \subset \{\cup \hat{B} : B \in \mathcal{F}\}.$$

Therefore,

$$\begin{aligned} H^n_\varepsilon(E) \le H^n_\varepsilon(U) &\le \sum_{B_i\in\mathcal{F}} H^n_\varepsilon(\hat{B}_i) \le \sum_{B_i\in\mathcal{F}} 2^{-n}\alpha(n)(\operatorname{diam}\hat{B}_i)^n \\ &= \sum_{B_i\in\mathcal{F}} 2^{-n}\alpha(n)5^n(\operatorname{diam}B_i)^n \\ &= 5^n \sum_{B_i\in\mathcal{F}} |B_i| \\ &\le 5^n|U| < 5^n\eta, \end{aligned}$$

which proves that $H^n(E) = 0$ since ε and η are arbitrary. The case when E is unbounded is easily disposed of by considering $E \cap B(0,i)$, $i = 1, 2, \ldots$.

Each of these sets has zero n-dimensional Hausdorff measure, and thus so does E.

Now suppose E is an arbitrary set with $|E| < \infty$. Let $U \supset E$ be an open set such that

$$|U| < |E| + \eta. \tag{1.4.2}$$

Appealing to Theorem 1.3.6, it is possible to find a family $\mathcal{F}$ of disjoint closed balls $B_1, B_2, \ldots$, such that $\cup_{i=1}^{\infty} B_i \subset U$, diam $B_i < \varepsilon$, $i = 1, 2, \ldots$, and

$$\left| E - \bigcup_{i=1}^{\infty} B_i \right| = 0. \tag{1.4.3}$$

Let $E^* = \cup_{i=1}^{\infty}(E \cap B_i)$ and observe that $E = (E - E^*) \cup E^*$ with $|E - E^*| = 0$. Now apply (1.4.1) and (1.4.2) to conclude that

$$\begin{aligned} H_{\varepsilon}^{n}(E^*) &\leq \sum_{i=1}^{\infty} 2^{-n}\alpha(n)(\operatorname{diam} B_i)^n \\ &= \sum_{i=1}^{\infty} |B_i| \\ &= \left| \bigcup_{i=1}^{\infty} B_i \right| \\ &= |U| \leq |E| + \eta. \end{aligned}$$

Because ε and η are arbitrary, it follows that $H^n(E^*) \leq |E|$. However, $H^n(E) \leq H^n(E - E^*) + H^n(E^*)$ with $H^n(E - E^*) = 0$ because $|E - E^*| = 0$. Therefore, $H^n(E) \leq |E|$.

In order to establish the opposite inequality, we will employ the *isodiametric inequality* which states that among all sets $E \subset R^n$ with a given diameter, d, the ball with diameter d has the largest Lebesgue measure; that is,

$$|E| \leq 2^{-n}\alpha(n)(\operatorname{diam} E)^n \tag{1.4.4}$$

whenever $E \subset R^n$. For a proof of this fact, see [F4, p. 197]. From this the desired inequality follows immediately, for suppose

$$\sum_{i=1}^{\infty} 2^{-n}\alpha(n)(\operatorname{diam} E_i)^n < H_{\varepsilon}^{n}(E) + \eta$$

where $E \subset \cup_{i=1}^{\infty} E_i$. Applying (1.4.3) to each E_i yields

$$|E| \leq \sum_{i=1}^{\infty} |E_i| \leq \sum_{i=1}^{\infty} 2^{-n}\alpha(n)(\operatorname{diam} E_i)^n < H_{\varepsilon}^{n}(E) + \eta,$$

which implies, $|E| \leq H^n(E)$ since ε and η are arbitrary. □

1.4.3. Remark. The reader can easily verify that the outer measure, H^γ, has many properties in common with Lebesgue outer measure. For example, (1.2.4), (1.2.5), and (1.2.6) are also valid for H^γ as well as the analog of Corollary 1.2.2. However, a striking difference between the two is that $|E| < \infty$ whenever E is bounded whereas this may be false for $H^\gamma(E)$. One important ramification of this fact is the following. A Lebesgue measurable set, E, can be characterized by the fact that for every $\varepsilon > 0$, there exists an open set $U \supset E$ such that

$$|U - E| < \varepsilon. \tag{1.4.5}$$

This regularity property cannot hold in general for H^γ.

The fact that $H^\gamma(E)$ may be possibly infinite for bounded sets E can be put into better perspective by the following fact that the reader can easily verify. For every set E, there is a non-negative number, $d = d(E)$, such that

$$H^\gamma(E) = 0 \quad \text{if} \quad \gamma > d$$

$$H^\gamma(E) = \infty \quad \text{if} \quad \gamma < d.$$

The number $d(E)$ is called the *Hausdorff dimension* of E.

Finally, we make note of the following elementary but useful fact. Suppose $f\colon R^k \to R^{k+n}$ is a Lipschitz map with $\operatorname{Lip}(f) = M$. Then for any set $E \subset R^k$

$$H^k[f(E)] \leq MH^k(E). \tag{1.4.6}$$

In particular, sets of zero k-dimensional Hausdorff measure remain invariant under Lipschitz maps.

1.5 L^p Spaces

For $1 \leq p \leq \infty$, $L^p_{\text{loc}}(\Omega)$ will denote the space consisting of all measurable functions on Ω that are p^{th}-power integrable on each compact subset of Ω. $L^p(\Omega)$ is the subspace of functions that are p^{th}-power integrable on Ω. In case the underlying measure is μ rather than Lebesgue measure, we will employ the notation $L^p_{\text{loc}}(\Omega; \mu)$ and $L^p(\Omega, \mu)$ respectively. The norm on $L^p(\Omega)$ is given by

$$\|u\|_{p;\Omega} = \left(\int_\Omega |u|^p dx\right)^{1/p} \tag{1.5.1}$$

and in case $p = \infty$, it is defined as

$$\|u\|_{\infty,\Omega} = \text{ess}_\Omega \sup |u|. \tag{1.5.2}$$

Analogous definitions are used in the case of $L^p(\Omega;\mu)$ and then the norm is denoted by

$$\|u\|_{p,\mu;\Omega}.$$

The notation $\int u(x)\,dx$ or sometimes simply $\int u\,dx$ will denote integration with respect to Lebesgue measure and $\int u\,d\mu$ the integral with respect to the measure μ. Strictly speaking, the elements of $L^p(\Omega)$ are not functions but rather equivalence classes of functions, where two functions are said to be equivalent if they agree everywhere on Ω except possibly for a set of measure zero. The choice of a particular representative will be of special importance later in Chapters 3 and 5 when the pointwise behavior of functions in the spaces $W^{k,p}(\Omega)$ and $BV(\Omega)$ is discussed. Recall from Theorem 1.3.8 that if $u \in L^1(R^n)$, then for almost every $x_0 \in R^n$, there is a number z such that

$$\fint_{B(x_0,r)} u(y)dy \to z \quad \text{as} \quad r \to 0^+,$$

where $\fint$ denotes the integral average. We define $u(x_0) = z$, and in this way a canonical representative of u is determined. In those situations where no confusion can occur, the elements of $L^p(\Omega)$ will be regarded merely as functions defined on Ω.

The following lemma is very useful and will be used frequently throughout.

1.5.1. Lemma. *If $u \geq 0$ is measurable, $p > 0$, and $E_t = \{x : u(x) > t\}$, then*

$$\int_\Omega u(x)^p dx = \int_0^\infty |E_t| dt^p = p\int_0^\infty t^{p-1}|E_t|dt. \tag{1.5.3}$$

More generally, if μ is a measure defined on some σ-algebra of R^n, $u \geq 0$ is a μ-measurable function, and Ω is the countable union of sets of finite μ measure, then

$$\int_\Omega u^p d\mu = \int_0^\infty \mu(E_t)dt^p = p\int_0^\infty t^{p-1}\mu(E_t)dt. \tag{1.5.4}$$

The proof of this can be obtained in at least two ways. One method is to employ Fubini's Theorem on the product space $\Omega \times [0,\infty)$. Another is to observe that (1.5.3) is immediate when u is a simple function. The general case then follows by approximating u from below by simple functions.

The following algebraic and functional inequalities will be frequently used throughout the course of this book.

Cauchy's inequality: if $\varepsilon > 0$, $a, b \in R^1$, then

$$|ab| \leq \frac{\varepsilon}{2}|a|^2 + \frac{1}{2\varepsilon}|b|^2 \tag{1.5.5}$$

and more generally, *Young's inequality:*

$$|ab| \leq \frac{|\varepsilon a|^p}{p} + \frac{[b/\varepsilon]^{p'}}{p'} \tag{1.5.6}$$

where $p > 1$ and $1/p + 1/p' = 1$.

From Young's inequality follows *Hölder's inequality*

$$\int_\Omega uv\, dx \leq \|u\|_{p;\Omega}\|v\|_{p';\Omega}, \quad p \geq 1, \tag{1.5.7}$$

which holds for functions $u \in L^p(\Omega)$, $v \in L^{p'}(\Omega)$. In case $p = 1$, we take $p' = \infty$ and $\|v\|_{p';\Omega} = \text{ess}_\Omega \sup |v|$. Hölder's inequality can be extended to the case of k functions, $u_1, \ldots, u_k$ lying respectively in spaces $L^{p_1}(\Omega), \ldots, L^{p_k}(\Omega)$ where

$$\sum_{i=1}^{k} \frac{1}{p_i} = 1. \tag{1.5.8}$$

By an induction argument and (1.5.7) it follows that

$$\int_\Omega u_1 \ldots u_k dx \leq \|u_1\|_{p_1;\Omega} \ldots \|u_k\|_{p_k;\Omega}. \tag{1.5.9}$$

One important application of (1.5.7) is Minkowski's inequality, which states that (1.5.3) yields a norm on $L^p(\Omega)$. That is,

$$\|u + v\|_{p;\Omega} \leq \|u\|_{p;\Omega} + \|v\|_{p;\Omega} \tag{1.5.10}$$

for $p \geq 1$. Employing the notation

$$⨍_\Omega u\, dx = |\Omega|^{-1} \int_\Omega u\, dx,$$

another consequence of Hölder's inequality is

$$\left[⨍_\Omega u^p dx\right]^{1/p} \leq \left[⨍_\Omega u^q dx\right]^{1/q} \tag{1.5.11}$$

whenever $1 \leq p \leq q$ and $\Omega \subset R^n$ a measurable set with $|\Omega| < \infty$.

We also recall Jensen's inequality whose statement involves the notion of a convex function. A function $\Lambda\colon R^n \to R^1$ is said to be convex if

$$\Lambda[(1-t)x_1 + tx_2] \leq (1-t)\Lambda(x_1) + t\Lambda(x_2),$$

whenever $x_1, x_2 \in R^n$ and $0 \leq t \leq 1$. Jensen's inequality states that if Λ is a convex function on R^n and $E \subset R^n$ a bounded measurable set, then

$$\Lambda\left(⨍_E f(x)dx\right) \leq ⨍_E \Lambda[f(x)]dx \tag{1.5.12}$$

whenever $f \in L^1(E)$.

A further consequence of Hölder's inequality is

$$\|u\|_q \leq \|u\|_p^\lambda \|u\|_r^{1-\lambda}, \quad u \in L^r(\Omega), \tag{1.5.13}$$

where $p \leq q \leq r$, and $1/q = \lambda/p + (1-\lambda)/r$. In order to see this, let $\alpha = \lambda q$, $\beta = (1-\lambda)q$ and apply Hölder's inequality to obtain

$$\int_\Omega |u|^q dx = \int_\Omega |u|^\alpha |u|^\beta dx \leq \left(\int_\Omega |u|^{\alpha z} dx\right)^{1/z} \left(\int_\Omega |u|^{\beta y} dx\right)^{1/y}$$

where $z = p/\lambda q$ and $y = r/(1-\lambda)q$.

When endowed with the norm defined in (1.5.1), $L^p(\Omega)$, $1 \leq p \leq \infty$, is a Banach space; that is, a complete, linear space. If $1 \leq p < \infty$, it is also separable. The normed dual of $L^p(\Omega)$ consists of all bounded linear functionals on $L^p(\Omega)$ and is isometric to $L^{p'}(\Omega)$ provided $p < \infty$. Hence, $L^p(\Omega)$ is reflexive for $1 < p < \infty$. We recall the following fundamental result concerning reflexive Banach spaces, which is of considerable importance in the case of $L^p(\Omega)$.

1.5.2. Theorem. *A Banach space is reflexive if and only if its closed unit ball is weakly sequentially compact.*

1.6 Regularization

Let φ be a non-negative, real-valued function in $C_0^\infty(R^n)$ with the property that

$$\int_{R^n} \varphi(x)dx = 1, \quad \operatorname{spt}\varphi \subset \overline{B}(0,1). \tag{1.6.1}$$

An example of such a function is given by

$$\varphi(x) = \begin{cases} C\exp[-1/(1-|x|^2)] & \text{if } |x| < 1 \\ 0 & \text{if } |x| \geq 1 \end{cases} \tag{1.6.2}$$

where C is chosen so that $\int_{R^n} \varphi = 1$. For $\varepsilon > 0$, the function $\varphi_\varepsilon(x) \equiv \varepsilon^{-n}\varphi(x/\varepsilon)$ belongs to $C_0^\infty(R^n)$ and $\operatorname{spt}\varphi_\varepsilon \subset \overline{B}(0,\varepsilon)$. φ_ε is called a regularizer (or mollifier) and the convolution

$$u_\varepsilon(x) \equiv \varphi_\varepsilon * u(x) \equiv \int_{R^n} \varphi_\varepsilon(x-y)u(y)dy \tag{1.6.3}$$

defined for functions u for which the right side of (1.6.3) has meaning, is called the regularization (mollification) of u. Regularization has several important and useful properties that are summarized in the following theorem.

1.6.1. Theorem.

(i) *If* $u \in L^1_{\mathrm{loc}}(R^n)$*, then for every* $\varepsilon > 0$*,* $u_\varepsilon \in C^\infty(R^n)$ *and* $D^\alpha(\varphi_\varepsilon * u) = (D^\alpha \varphi_\varepsilon) * u$ *for each multi-index* α*.*

(ii) $u_\varepsilon(x) \to u(x)$ *whenever* x *is a Lebesgue point for* u*. In case* u *is continuous then* u_ε *converges uniformly to* u *on compact subsets of* R^n*.*

(iii) *If* $u \in L^p(R^n)$*,* $1 \leq p < \infty$*, then* $u_\varepsilon \in L^p(R^n)$*,* $\|u_\varepsilon\|_p \leq \|u\|_p$*, and* $\lim_{\varepsilon \to 0} \|u_\varepsilon - u\|_p = 0$*.*

Proof. For the proof of (i), it suffices to consider $|\alpha| = 1$, since the case of general α can be treated by induction. Let $e_1, \ldots, e_n$ be the standard basis of R^n and observe that

$$\begin{aligned} u_\varepsilon(x + he_i) - u_\varepsilon(x) &= \int_{R^n} \int_0^h D_i \varphi_\varepsilon(x - z + te_i) u(z) \, dt \, dz \\ &= \int_0^h \int_{R^n} D_i \varphi_\varepsilon(x - z + te_i) u(z) \, dz \, dt. \end{aligned}$$

As a function of t, the inner integral on the right is continuous, and thus (i) follows.

In case (ii) observe that

$$\begin{aligned} |u_\varepsilon(x) - u(x)| &\leq \int \varphi_\varepsilon(x - y) |u(y) - u(x)| \, dy \\ &\leq \sup \varphi \varepsilon^{-n} \int_{B(x,\varepsilon)} |u(x) - u(y)| \, dy \to 0 \end{aligned}$$

as $\varepsilon \to 0$ whenever x is a Lebesgue point for u. Clearly the convergence is locally uniform if u is continuous because u is uniformly continuous on compact sets.

For the proof of (iii), Hölder's inequality yields

$$\begin{aligned} |u_\varepsilon(x)| &= \left| \int \varphi_\varepsilon(x - y) u(y) \, dy \right| \\ &\leq \left(\int \varphi_\varepsilon(x - y) \, dy \right)^{1/p'} \left(\int \varphi_\varepsilon(x - y) |u(y)|^p \, dy \right)^{1/p} \end{aligned}$$

The first factor on the right is equal to 1 and hence, by Fubini's theorem,

$$\begin{aligned} \int_{R^n} |u_\varepsilon|^p \, dx &\leq \int_{R^n} \int_{R^n} \varphi_\varepsilon(x - y) |u(y)|^p \, dy \, dx \\ &\leq \int_{R^n} \int_{R^n} \varphi_\varepsilon(x - y) |u(y)|^p \, dx \, dy \\ &= \int_{R^n} |u(y)|^p \, dy. \end{aligned}$$

Consequently,

$$\|u_\varepsilon\|_p \le \|u\|_p. \tag{1.6.4}$$

To complete the proof, for each $\eta > 0$ let $v \in C_0(R^n)$ be such that

$$\|u - v\|_p < \eta. \tag{1.6.5}$$

Because v has compact support, it follows from (ii) that $\|v - v_\varepsilon\|_p < \eta$ for ε sufficiently small. Now apply (1.6.4) and (1.6.5) to the difference $v - u$ and obtain

$$\|u - u_\varepsilon\|_p \le \|u - v\|_p + \|v - v_\varepsilon\|_p + \|v_\varepsilon - u_\varepsilon\|_p \le 3\eta.$$

Hence $u_\varepsilon \to u$ in $L^p(R^n)$ as $\varepsilon \to 0$. □

1.6.2. Remark. If $u \in L^1(\Omega)$, then $u_\varepsilon(x) \equiv \varphi_\varepsilon * u(x)$ is defined provided $x \in \Omega$ and $\varepsilon < \operatorname{dist}(x, \partial\Omega)$. It is a simple matter to verify that Theorem 1.6.1 remains valid in this case with obvious modification. For example, if $u \in C(\Omega)$ and $\Omega' \subset\subset \Omega$, then u_ε converges uniformly to u on Ω' as $\varepsilon \to 0$.

Also note that (iii) of Theorem 1.6.1. implies that mollification does not increase the norm. This is intuitively clear since the norm must take into account the extremities of the function and mollification, which is an averaging operation, does not increase the extremities.

1.7 Distributions

In this section we present a very brief review of some of the elementary concepts and techniques of the Schwartz theory of distributions [SCH] that will be needed in subsequent chapters. The notion of weak or distributional derivative will be of special importance.

1.7.1. Definition. Let $\Omega \subset R^n$ be an open set. The space $\mathscr{D}(\Omega)$ is the set of all φ in $C_0^\infty(\Omega)$ endowed with a topology so that a sequence $\{\varphi_i\}$ converges to an element φ in $\mathscr{D}(\Omega)$ if and only if

(i) there exists a compact set $K \subset \Omega$ such that $\operatorname{spt}\varphi_i \subset K$ for every i, and

(ii) $\lim_{i\to\infty} D^\alpha\varphi_i = D^\alpha\varphi$ uniformly on K for each multi-index α.

The definition above does not attempt to actually define the topology on $\mathscr{D}(\Omega)$ but merely states a consequence of the rigorous definition which requires the concept of "generalized sequences" or "nets," a topic that we do not wish to pursue in this brief treatment. For our purposes, it will suffice to consider only ordinary sequences. It turns out that $\mathscr{D}(\Omega)$ is a topological vector space with a locally convex topology but is not a normable space. The dual space, $\mathscr{D}'(\Omega)$, of $\mathscr{D}(\Omega)$ is called the space of (Schwartz) distributions and is given the weak*-topology. Thus, $T_i \in \mathscr{D}(\Omega)$ converges to T if and only if $T_i(\varphi) \to T(\varphi)$ for every $\varphi \in \mathscr{D}(\Omega)$.

We consider some important examples of distributions. Let μ be a Radon measure on Ω and define the corresponding distribution by

$$T(\varphi) = \int \varphi(x) d\mu$$

for all $\varphi \in \mathscr{D}(\Omega)$. Clearly T is a linear functional on $\mathscr{D}(\Omega)$ and $|T(\varphi)| \leq |\mu|(\operatorname{spt} \varphi) \|\varphi\|_\infty$, from which it is easily seen that T is continuous, and thus a distribution. In this way we will make an identification of Radon measures and the associated distributions.

Similarly, let $f \in L^p_{\text{loc}}(\Omega)$, $p \geq 1$, and consider the corresponding signed measure μ defined for all Borel sets $E \subset R^n$ by

$$\mu(E) = \int_E f(x) dx$$

and pass to the associated distribution

$$f(\varphi) = \int_{R^n} \varphi(x) f(x) dx.$$

In the sequel we shall often identify locally integrable functions with their corresponding distributions without explicitly indicating the identification.

1.7.2. Remark. We recall two facts about distributions that will be of importance later. A distribution T on an open set Ω is said to be positive if $T(\varphi) \geq 0$ whenever $\varphi \geq 0$, $\varphi \in \mathscr{D}(\Omega)$. A fundamental result in distribution theory states that a positive distribution is a measure. Of course, not all distributions are measures. For example, the distribution defined on R^1 by

$$T(\varphi) = \int \varphi'(x) dx$$

is not a measure since it is not continuous on $\mathscr{D}(\Omega)$ when endowed with the topology of uniform convergence on compact sets.

Another important fact is that distributions are determined by their local behavior. By this we mean that if two distributions T and S on Ω have the property that for every $x \in \Omega$ there is a neighborhood U such that $T(\varphi) = S(\varphi)$ for all $\varphi \in \mathscr{D}(\Omega)$ supported by U, then $T = S$. For example, this implies that if $\{\Omega_\alpha\}$ is a family of open sets such that $\cup \Omega_\alpha = \Omega$ and T is a distribution on Ω such that T is a measure on each Ω_α, then T is a measure on Ω. This also implies that if a distribution T vanishes on each open set of some family $\mathcal{F}$, it then vanishes on the union of all elements of $\mathcal{F}$. The support of a distribution T is thus defined as the complement of the largest open set on which T vanishes.

We now proceed to define the convolution of a distribution with a test function $\varphi \in \mathscr{D}(\Omega)$. For this purpose, we introduce the notation $\tilde{\varphi}(x) = \varphi(-x)$ and $\tau_x\varphi(y) = \varphi(y-x)$. The convolution of a distribution T defined on R^n with $\varphi \in \mathscr{D}(\Omega)$ is a function of class C^∞ given by

$$T * \varphi(x) = T(\tau_x\tilde{\varphi}). \tag{1.7.1}$$

An obvious but important observation is

$$T * \varphi(0) = T(\tau_0\tilde{\varphi}) = T(\tilde{\varphi}).$$

If the distribution T is given by a locally integrable function f then we have

$$(T * \varphi)(x) = \int f(x-y)\varphi(y)dy$$

which is the usual definition for the convolution of two functions. It is easy to verify that

$$T * (\varphi * \psi) = (T * \varphi) * \psi$$

whenever $\varphi, \psi \in \mathscr{D}$.

Let T be a distribution on an open set Ω. The partial derivative of T is defined as

$$D_iT(\varphi) = -T(D_i\varphi)$$

for $\varphi \in \mathscr{D}(\Omega)$. Since $D_i\varphi \in \mathscr{D}(\Omega)$ it is clear that D_iT is again a distribution. Since the test functions φ are smooth, the mixed partial derivatives are independent of the order of differentiation:

$$D_iD_j\varphi = D_jD_i\varphi$$

and therefore the same equation holds for distributions:

$$D_iD_jT = D_jD_iT.$$

Consequently, for any multi-index α the corresponding derivative of T is given by the equation

$$D^\alpha T(\varphi) = (-1)^{|\alpha|}T(D^\alpha\varphi).$$

Finally, we note that a distribution on Ω can be multiplied by smooth functions. Thus, if $T \in \mathscr{D}'(\Omega)$ and $f \in C^\infty(\Omega)$, then the product fT is a distribution defined by

$$(fT)(\varphi) = T(f\varphi), \quad \varphi \in \mathscr{D}(\Omega).$$

The Leibniz formula is easily seen to hold in this context (see Exercise 1.5).

The reader is referred to [SCH] for a complete treatment of this topic.

1.8 Lorentz Spaces

We have seen in Lemma 1.5.1 that if $f \in L^1(R^n)$, $f \geq 0$, then its integral is completely determined by the measure of the sets $\{x : f(x) > t\}$, $t \in R^1$. The non-increasing rearrangement of f, (defined below) can be identified with a radial function $\overline{f}$ having the property that for all $t \in R^1$, $\{x : \overline{f}(x) > t\}$ is a ball centered at the origin with the same measure as $\{x : f(x) > t\}$. Consequently, f and $\overline{f}$ have the same integral. Because $\overline{f}$ can be thought of as a function of one variable, it is often easier to employ than f. We introduce a class of spaces called Lorentz spaces which are more general but closely related to L^p spaces. Their definition is based on the concept of non-increasing rearrangement. Later in Chapter 2, we will extend basic Sobolev inequalities in an L^p setting to that of Lorentz spaces.

1.8.1. Definition. If f is a measurable function defined on R^n, let

$$E_s^f = \{x : |f(x)| > s\}, \tag{1.8.1}$$

and let the distribution function of f be denoted by

$$\alpha_f(s) = |E_s^f|. \tag{1.8.2}$$

Note that the distribution function of f is non-negative, non-increasing, and continuous from the right. With the distribution function we associate the *non-increasing rearrangement of f* on $(0, \infty)$ defined by

$$f^*(t) = \inf\{s > 0 : \alpha_f(s) \leq t\}. \tag{1.8.3}$$

Clearly f^* is non-negative and non-increasing on $(0, \infty)$. Further, if α_f is continuous and strictly decreasing, then f^* is the inverse of α_f, that is, $f^* = \alpha_f^{-1}$. It follows immediately from the definition of $f^*(t)$ that

$$f^*(\alpha_f(s)) \leq s \tag{1.8.4}$$

and because α_f is continuous from the right, that

$$\alpha_f(f^*(t)) \leq t. \tag{1.8.5}$$

These two facts lead immediately to the following propositions.

1.8.2. Proposition. *f^* is continuous from the right.*

Proof. Clearly, $f^*(t) \geq f^*(t+h)$ for all $h > 0$. If f^* were not continuous at t, there would exist y such that $f^*(t) > y > f^*(t+h)$ for all $h > 0$. But then, (1.8.5) would imply that $\alpha_f(y) \leq \alpha_f(f^*(t+h)) \leq t+h$ for all $h > 0$.

Thus, $\alpha_f(y) \le t$ and therefore, $f^*(t) \le y$, a contradiction. $\square$

1.8.3. Proposition. *$\alpha_{f^*}(s) = \alpha_f(s)$ for all $s > 0$.*

Proof. Because f^* is non-increasing, it follows from the definition of $\alpha_{f^*}(t)$ that

$$\alpha_{f^*}(s) = \sup\{t > 0 : f^*(t) > s\}. \tag{1.8.6}$$

Hence, $f^*(\alpha_f(s)) \le s$ implies $\alpha_f(s) \ge \alpha_{f^*}(s)$. For the opposite inequality, note from (1.8.6) that if $t > \alpha_{f^*}(s)$, then $f^*(t) \le s$ and consequently, $\alpha_f(s) \le \alpha_f(f^*(t)) \le t$, by (1.8.5). Thus, $\alpha_f(s) \le \alpha_{f^*}(s)$ and the proposition is established. $\square$

1.8.4. Proposition. *Let $\{f_i\}$ be a sequence of measurable functions on R^n such that $\{|f_i|\}$ is a non-decreasing sequence. If $|f(x)| = \lim_{i\to\infty} |f_i(x)|$ for each $x \in R^n$, then α_{f_i} and f_i^* increase to α_f and f^* respectively.*

Proof. Clearly

$$E_s^{f_i} \subset E_s^f \quad \text{and} \quad \bigcup_{i=1}^{\infty} E_s^{f_i} = E_s^f$$

for each s and therefore $\alpha_{f_i}(s) \to \alpha_f(s)$ as $s \to \infty$. It follows from definition of non-increasing rearrangement, that $f_i^*(t) \le f_{i+1}^*(t) \le f^*(t)$ for each t and $i = 1, 2, \ldots$. Let $g(t) = \lim_{i\to\infty} f_i^*(t)$. Since $f_i^*(t) \le g(t)$ it follows from (1.8.5) that $\alpha_{f_i}[g(t)] \le \alpha_{f_i}[f_i^*(t)] \le t$. Therefore,

$$\alpha_f[g(t)] = \lim_{i\to\infty} \alpha_{f_i}[g(t)] \le t$$

which implies that $f^*(t) \le g(t)$. But $g(t) \le f^*(t)$ and therefore the proof is complete. $\square$

1.8.5. Theorem. *If $f \in L^p$, $1 \le p < \infty$, then*

$$\left[\int |f|^p\right]^{1/p} = \left[\int_0^\infty [f^*(t)]^p dt\right]^{1/p} \tag{1.8.7}$$

Proof. This follows immediately from Lemma 1.5.1 and the fact that f and f^* have the same distribution function (Proposition 1.8.3). $\square$

We now introduce Lorentz spaces and in order to motivate the following definition, we write (1.8.7) in a more suggestive form as

$$\|f\|_p = \left(\int_0^\infty [t^{1/p} f^*(t)]^p dt/t\right)^{1/p}$$

It is sometimes more convenient to work with the average of f^* than with f^* itself. Thus, we define

$$f^{**}(t) = \frac{1}{t}\int_0^t f^*(r)dr.$$

1.8.6. Definition. For $1 \leq p < \infty$ and $1 \leq q \leq \infty$, the Lorentz space $L(p,q)$ is defined as

$$L(p,q) = \{f : f \text{ measurable on } R^n, \|f\|_{(p,q)} < \infty\} \tag{1.8.8}$$

where $\|f\|_{(p,q)}$ is defined by

$$\|f\|_{(p,q)} = \begin{cases} \left(\int_0^\infty \left[[t^{1/p}f^{**}(t)]^q \frac{dt}{t}\right]\right)^{1/q}, & 1 \leq p < \infty, 1 \leq q < \infty \\ \sup\limits_{t>0} t^{1/p}f^{**}(t), & 1 \leq p \leq \infty, q = \infty. \end{cases}$$

It will be shown in Lemma 1.8.10 that

$$L(p,p) = L^p. \tag{1.8.9}$$

The norm above could be defined with f^{**} replaced by f^* in case $p > 1$ and $1 \leq q < \infty$. This alternate definition remains equivalent to the original one in view of Hardy's inequality (Lemma 1.8.11) and the fact that $f^{**} \geq f^*$ (since f^* is non-increasing). For $p > 1$, the space $L(p,\infty)$ is known as the Marcinkiewicz space and also as Weak L^p. In case $p = 1$, we clearly have $L(1,\infty) = L^1$. With the help of Lemma 1.5.1, observe that

$$\int_0^t f^*(r)dr = tf^*(t) + \int_{f^*(t)}^\infty \alpha_f(s)ds$$

and therefore

$$f^{**}(t) = f^*(t) + \frac{1}{t}\int_{f^*(t)}^\infty \alpha_f(s)ds. \tag{1.8.10}$$

For our applications it will be necessary to know how the non-increasing rearrangement behaves relative to the operation of convolution. The next two lemmas address this question. Because g^{**} is non-increasing, note that in the following lemma, the first and second conclusions are most interesting when $t \leq r$ and $t \geq r$, respectively.

1.8.7. Lemma. *Let f and g be measurable functions on R^n where* $\sup\{f(x)$: *$x \in R^n\} \leq \alpha$ and f vanishes outside of a measurable set E with $|E| = r$. Let $h = f * g$. Then, for $t > 0$,*

$$h^{**}(t) \leq \alpha r g^{**}(r)$$

and

$$h^{**}(t) \le \alpha r g^{**}(t).$$

Proof. For $a > 0$, define

$$g_a(x) = \begin{cases} g(x) & \text{if } |g(x)| \le a \\ a \operatorname{sgn} g(x) & \text{if } |g(x)| > a \end{cases}$$

and let

$$g^a(x) = g(x) - g_a(x).$$

Then, define functions h_1 and h_2 by

$$\begin{aligned} h &= f * g = f * g_a + f * g^a \\ &= h_1 + h_2. \end{aligned}$$

From elementary estimates involving the convolution and Lemma 1.5.1, we obtain

$$\sup\{h_2(x) : x \in R^n\} \le \sup\{f(x) : x \in E\} \|g^a\|_1 \le \alpha \int_a^\infty \alpha_g(s) ds \quad (1.8.11)$$

because $g^a(x) = 0$ whenever $|g(x)| \le a$. Also

$$\sup\{h_1(x) : x \in R^n\} \le \|f\|_1 \sup\{g_a(x) : x \in E\} \le \alpha r a, \quad (1.8.12)$$

and

$$\|h_2\|_1 \le \|f\|_1 \|g^a\|_1 \le \alpha r \int_a^\infty \alpha_g(s) ds. \quad (1.8.13)$$

Now set $a = g^*(r)$ in (1.8.11) and (1.8.12) and obtain

$$\begin{aligned} h^{**}(t) = \frac{1}{t} \int_0^t h^*(y) dy &\le \|h\|_\infty \\ &\le \|h_1\|_\infty + \|h_2\|_\infty \\ &\le \alpha r g^*(r) + \alpha \int_{g^*(r)}^\infty \alpha_g(s) ds \\ &\le \alpha \left[r g^*(r) + \int_{g^*(r)}^\infty \alpha_g(s) ds \right] \\ &= \alpha r g^{**}(r). \end{aligned}$$

The last equality follows from (1.8.10) and thus, the first inequality of the lemma is established.

To prove the second inequality, set $a = g^*(r)$ and use (1.8.12) and (1.8.13) to obtain

$$\begin{aligned} th^{**}(t) = \int_0^t h^*(y)dy &\leq \int_0^t h_1^*(y)dy + \int_0^t h_2^*(y)dy \\ &\leq t\|h_1\|_\infty + \int_0^\infty h_2^*(y)dy = t\|h_1\|_\infty + \|h_2\|_1 \\ &\leq t\alpha r g^*(t) + \alpha r \int_{g^*(r)}^\infty \alpha_g(s)ds \\ &\leq \alpha r \left[tg^*(t) + \int_{g^*(r)}^\infty \alpha_g(s)ds \right] \\ &\leq \alpha r t g^{**}(t) \end{aligned}$$

by (1.8.10). □

1.8.8. Lemma. *If h, f, and g are measurable functions such that $h = f * g$, then for any $t > 0$*

$$h^{**}(t) \leq tf^{**}(t)g^{**}(t) + \int_t^\infty f^*(u)g^*(u)du.$$

Proof. Fix $t > 0$.

Select a doubly infinite sequence $\{y_i\}$ whose indices ranges from $-\infty$ to $+\infty$ such that

$$\begin{aligned} y_0 &= f^*(t) \\ y_i &\leq y_{i+1} \\ \lim_{i\to\infty} y_i &= \infty \\ \lim_{i\to-\infty} y_i &= 0. \end{aligned}$$

Let

$$f(z) = \sum_{i=-\infty}^{\infty} f_i(z)$$

where

$$f_i(z) = \begin{cases} 0 & \text{if } |f(z)| \leq y_{i-1} \\ f(z) - y_{i-1} \operatorname{sgn} f(z) & \text{if } y_{i-1} < |f(z)| \leq y_i \\ y_i - y_{i-1} \operatorname{sgn} f(z) & \text{if } y_i < |f(z)|. \end{cases}$$

Clearly, the series converges absolutely and therefore,

$$\begin{aligned} h = f * g &= \left(\sum_{i=-\infty}^{\infty} f_i \right) * g \\ &= \left(\sum_{i=-\infty}^{0} f_i \right) * g + \left(\sum_{i=1}^{\infty} f_i \right) * g \\ &= h_1 + h_2 \end{aligned}$$

with

$$h^{**}(t) \le h_1^{**}(t) + h_2^{**}(t).$$

To evaluate $h_2^{**}(t)$ we use the second inequality of Lemma 1.8.7 with $E_i \equiv \{z : |f(z)| > y_{i-1}\} = E$ and $\alpha = y_i - y_{i-1}$ to obtain

$$\begin{aligned} h_2^{**}(t) &\le \sum_{i=1}^{\infty} (y_i - y_{i-1})\alpha_f(y_{i-1})g^{**}(t) \\ &= g^{**}(t) \sum_{i=1}^{\infty} \alpha_f(y_{i-1})(y_i - y_{i-1}). \end{aligned}$$

The series on the right is an infinite Riemann sum for the integral

$$\int_{f^*(t)}^{\infty} \alpha_f(y)dy,$$

and provides an arbitrarily close approximation with an appropriate choice of the sequence $\{y_i\}$. Therefore,

$$h_2^{**}(t) \le g^{**}(t) \int_{f^*(t)}^{\infty} \alpha_f(y)dy. \tag{1.8.14}$$

By the first inequality of Lemma 1.8.7,

$$h_1^{**}(t) \le \sum_{i=1}^{\infty} (y_i - y_{i-1})\alpha_f(y_{i-1})g^{**}(\alpha_f(y_{i-1})).$$

The sum on the right is an infinite Riemann sum tending (with proper choice of y_i) to the integral,

$$\int_0^{f^*(t)} \alpha_f(y)g^{**}(\alpha_f(y))dy.$$

We shall evaluate the integral by making the substitution $y = f^*(u)$ and then integrating by parts. In order to justify the change of variable in the integral, consider a Riemann sum

$$\sum_{i=1}^{\infty} \alpha_f(y_{i-1})g^{**}(\alpha_f(y_{i-1}))(y_i - y_{i-1})$$

that provides a close approximation to

$$\int_0^{f^*(t)} \alpha_f(y)g^{**}(\alpha_f(y))dy.$$

By adding more points to the Riemann sum if necessary, we may assume that the left-hand end point of each interval on which α_f is constant is included among the y_i. Then, the Riemann sum is not changed if each y_i that is contained in the interior of an interval on which α_f is constant, is deleted. It is now an easy matter to verify that for each of the remaining y_i there is precisely one element, u_i, such that $y_i = f^*(u_i)$ and that $\alpha_f(f^*(u_i)) = u_i$. Thus, we have

$$\sum_{i=1}^{\infty} \alpha_f(y_{i-1})g^{**}(\alpha_f(y_{i-1}))(y_i - y_{i-1})$$
$$= \sum_{i=1}^{\infty} u_{i-1}g^{**}(u_{i-1})(f^*(u_i) - f^*(u_{i-1}))$$

which, by adding more points if necessary, provides a close approximation to

$$-\int_t^{\infty} ug^{**}(u)df^*(u).$$

Therefore, we have

$$\begin{aligned} h_1^{**}(t) &\leq \int_0^{f^*(t)} \alpha_f(y)g^{**}(\alpha_f(y))dy \\ &= -\int_t^{\infty} ug^{**}(u)df^*(u) \\ &= -ug^{**}(u)f^*(u)|_t^{\infty} + \int_t^{\infty} f^*(u)g^*(u)du \\ &\leq tg^{**}(t)f^*(t) + \int_t^{\infty} f^*(u)g^*(u)du. \end{aligned} \tag{1.8.15}$$

To justify the integration by parts, let λ be an arbitrarily large number and choose u_j such that $t = u_1 \leq u_2 \leq \ldots \leq u_{j+1} = \lambda$. Observe that

$$\begin{aligned} \lambda g^{**}(\lambda)f^*(\lambda) - tg^{**}(t)f^*(t) &= \sum_{i=1}^{j} u_{i+1}g^{**}(u_{i+1})[f^*(u_{i+1}) - f^*(u_i)] \\ &\quad + \sum_{i=1}^{j} f^*(u_i)[g^{**}(u_{i+1})u_{i+1} - g^{**}(u_i)u_i] \\ &= \sum_{i=1}^{j} u_{i+1}g^{**}(u_{i+1})[f^*(u_{i+1}) - f^*(u_i)] \\ &\quad + \sum_{i=1}^{j} f^*(u_i)\left[\int_{u_i}^{u_{i+1}} g^*(r)dr\right] \\ &\leq \sum_{i=1}^{j} u_{i+1}g^{**}(u_{i+1})[f^*(u_{i+1}) - f^*(u_i)] \end{aligned}$$

$$+\sum_{i=1}^{j} f^*(u_i)g^*(u_i)[u_{i+1}-u_i].$$

This shows that

$$\lambda g^{**}(\lambda)f^*(\lambda) - tg^{**}(t)f^*(t) \le \int_t^\lambda ug^{**}(u)df^*(u) + \int_t^\lambda f^*(u)g^*(u)du.$$

To establish the opposite inequality, write

$$\begin{aligned}
\lambda g^{**}(\lambda)f^*(\lambda) - tg^{**}(t)f^*(t) &= \sum_{i=1}^{j} u_i g^{**}(u_i)[f^*(u_{i+1}) - f^*(u_i)] \\
&\quad + \sum_{i=1}^{j} f^*(u_{i+1})[g^{**}(u_{i+1})u_{i+1} - g^{**}(u_i)u_i] \\
&= \sum_{i=1}^{j} u_i g^{**}(u_i)[f^*(u_{i+1}) - f^*(u_i)] \\
&\quad + \sum_{i=1}^{j} f^*(u_{i+1})\left[\int_{u_i}^{u_{i+1}} g^*(r)dr\right] \\
&\ge \sum_{i=1}^{j} u_i g^{**}(u_i)[f^*(u_{i+1}) - f^*(u_i)] \\
&\quad + \sum_{i=1}^{j} f^*(u_{i+1})g^*(u_{i+1})[u_{i+1}-u_i].
\end{aligned}$$

Now let $\lambda \to \infty$ to obtain the desired equality. Thus, from (1.8.15), (1.8.14), and (1.8.10),

$$\begin{aligned}
h_1^{**}(t) + h_2^{**}(t) &\le g^{**}(t)\left[tf^*(t) + \int_{f^*(t)}^{\infty} \alpha_f(y)dy\right] + \int_t^\infty f^*(u)g^*(u)du \\
&\le tf^{**}(t)g^{**}(t) + \int_t^\infty f^*(u)g^*(u)du. \qquad \square
\end{aligned}$$

1.8.9. Lemma. *Under the hypotheses of Lemma* 1.8.8,

$$h^{**}(t) \le \int_t^\infty f^{**}(u)g^{**}(u)du.$$

Proof. We may as well assume the integral on the right is finite and then conclude

$$\lim_{u\to\infty} uf^{**}(u)g^{**}(u) = 0. \tag{1.8.16}$$

By Lemma 1.8.8 and the fact that $f^* \le f^{**}$, we have

$$h^{**}(t) \le tf^{**}(t)g^{**}(t) + \int_t^\infty f^*(u)g^*(u)du$$

$$\le tf^{**}(t)g^{**}(t) + \int_t^\infty f^{**}(u)g^*(u)du. \tag{1.8.17}$$

Note that since f^* and g^* are non-increasing,

$$\frac{d}{du}f^{**}(u) = \frac{1}{u}[f^*(u) - f^{**}(u)]$$

and

$$\frac{d}{du}ug^{**}(u) = g^*(u)$$

for almost all (in fact, all but countably many) u. Since f^{**} and g^{**} are absolutely continuous, we may perform integration by parts and employ (1.8.16) and (1.8.17) to obtain

$$\begin{aligned} h^{**}(t) &\le tf^{**}(t)g^{**}(t) + uf^{**}(u)g^{**}(u)|_t^\infty \\ &\quad + \int_t^\infty [f^{**}(u) - f^*(u)]g^{**}(u)du \\ &= \int_t^\infty [f^{**}(u) - f^*(u)]g^{**}(u)du \\ &\le \int_t^\infty f^{**}(u)g^{**}(u)du. \end{aligned}$$

□

We conclude this section by proving some lemmas that provide a comparison between various Lorentz spaces. We begin with the following that compares L^p and $L(p,p)$.

1.8.10. Lemma. *If* $1 < p < \infty$ *and* $1/p + 1/p' = 1$, *then*

$$\|f\|_p \le \|f\|_{(p,p)} \le p'\|f\|_p.$$

Proof. Since $f^* \le f^{**}$,

$$\|f\|_p^p = \int_0^\infty [f^*(t)]^p dt = \int_0^\infty [t^{1/p}f^*(t)]^p \frac{dt}{t} \le \int_0^\infty [t^{1/p}f^{**}(t)]^p \frac{dt}{t}$$
$$= (\|f\|_{(p,p)})^p.$$

The second inequality follows immediately from the definition of $f^{**}(t)$ and the inequality

$$\left[\int_0^\infty \left[\frac{x^{1/p}}{x}\int_0^x f(t)dt\right]^p \frac{dx}{x}\right]^{1/p} \le p'\left[\int_0^\infty [x^{1/p}f(x)]^p\frac{dx}{x}\right]^{1/p}$$

which is a consequence of the following lemma with $r = p - 1$. □

The next result is a classical estimate, known as Hardy's inequality, which gives information related to Jensen's inequality (1.5.12). If f is a non-negative measurable function defined on the positive real numbers, let

$$F(x) = \frac{1}{x}\int_0^x f(t)dt, \quad x > 0.$$

Jensen's inequality gives an estimate of the pth power of F; Hardy's inequality gives an estimate of the weighted integral of the pth power of F.

1.8.11. Lemma (Hardy). *If* $1 \le p < \infty$, $r \ge 0$ *and* f *is a non-negative measurable function on* $(0,\infty)$, *then with* F *defined as above,*

$$\int_0^\infty F(x)^p x^{p-r-1}dx \le \left(\frac{p}{r}\right)^p \int_0^\infty f(t)^p t^{p-r-1}dt.$$

Proof. By an application of Jensen's inequality (1.5.12) with the measure $t^{(r/p)-1}dt$, we obtain

$$\begin{aligned}\left(\int_0^x f(t)dt\right)^p &= \left(\int_0^x f(t)t^{1-(r/p)}t^{(r/p)-1}dt\right)^p \\ &\le \left(\frac{p}{r}\right)^{p-1} x^{r(1-1/p)}\int_0^x [f(t)]^p t^{p-r-1+r/p}dt.\end{aligned}$$

Then by Fubini's theorem,

$$\begin{aligned}&\int_0^\infty \left(\int_0^x f(t)dt\right)^p x^{-r-1}dx \\ &\le \left(\frac{p}{r}\right)^{p-1}\int_0^\infty x^{-1-(r/p)}\left(\int_0^x [f(t)]^p t^{p-r-1+(r/p)}dt\right)dx \\ &= \left(\frac{p}{r}\right)^{p-1}\int_0^\infty [f(t)]^p t^{p-r-1+(r/p)}\left(\int_t^\infty x^{-1-(r/p)}dx\right)dt \\ &= \left(\frac{p}{r}\right)^p\int_0^\infty [f(t)t]^p t^{-r-1}dt.\end{aligned}$$

□

The following two lemmas provide some comparison between the spaces $L(p,q)$ and $L(p,r)$.

1.8.12. Lemma.

$$f^{**}(x) \le \left(\frac{q}{p}\right)^{1/q} \frac{\|f\|_{(p,q)}}{x^{1/p}} \le e^{1/e}\frac{\|f\|_{(p,q)}}{x^{1/p}}.$$

Proof.

$$\begin{aligned}(\|f\|_{(p,q)})^q &= \int_0^\infty [t^{1/p} f^{**}(t)]^q \frac{dt}{t}\\ &\ge \int_0^x [f^{**}(t)]^q t^{(q/p)-1}\,dt\\ &\ge [f^{**}(x)]^q \int_0^x t^{(q/p)-1}\,dt\\ &= \frac{p}{q}[f^{**}(x)]^q x^{p/q}.\end{aligned}$$

The first inequality follows by solving for $f^{**}(x)$ and the second by observing that $\left(\frac{q}{p}\right)^{1/q} \le q^{1/q} \le e^{1/e}$. □

1.8.13. Lemma (Calderón). *If* $1 < p < \infty$ *and* $1 \le q < r < \infty$, *then*

$$\|f\|_{(p,r)} \le \left(\frac{q}{p}\right)^{(1/q)-(1/r)} \|f\|_{(p,q)} \le e^{1/e}\|f\|_{(p,q)}.$$

Proof.

$$\begin{aligned}(\|f\|_{(p,r)})^r &= \int_0^\infty [f^{**}(x)]^r x^{(r/p)-1}\,dx\\ &= \int_0^\infty [f^{**}(x)]^q [f^{**}(x)]^{r-q} x^{(r/p)-1}\,dx\\ &\le \int_0^\infty [f^{**}(x)]^q \left[\left(\frac{q}{p}\right)^{1/q} \frac{\|f\|_{(p,q)}}{x^{1/p}}\right]^{r-q} x^{(r/p)-1}\,dx\\ &= \left(\frac{q}{p}\right)^{(r/q)-1} (\|f\|_{(p,q)})^{r-q}(\|f\|_{(p,q)})^q,\end{aligned}$$

and the first inequality follows by taking the rth root of both sides. The second follows by the same reasoning as in the previous lemma. □

Exercises

1.1. Prove that if $E \subset R^n$ is an arbitrary set, then the distance function to E is Lipschitz with constant 1. That is, if $d(x) = d(x, E)$, then $|d(x) - d(y)| \leq |x - y|$ for all $x, y \in R^n$.

1.2. (a) Prove that if E is a set with $H^\alpha(E) < \infty$, then $H^\beta(E) = 0$ for every $\beta > \alpha$.

(b) Prove that any set $E \subset R^n$ has a unique Hausdorff dimension. See Remark 1.4.3.

1.3. Give a proof of Lemma 1.5.1. More generally, prove the following: Let $\varphi\colon [0, \infty] \to [0, \infty]$ be a monotonic function which is absolutely continuous on every closed interval of finite length. Then, under the conditions of Lemma 1.5.1, prove that

$$\int_{R^n} \varphi \circ u \, d\mu = \int_0^\infty \mu(E_t)\varphi'(t)dt.$$

1.4. Prove that $C^{k,\alpha}(\overline{\Omega})$ is a Banach space with the norm defined in Section 1.1.

1.5. Let $f \in C_0^\infty(R^n)$ and T a distribution. Verify the Leibniz formula

$$D^\alpha(fT) = \sum_{\beta \leq \alpha} \frac{\alpha!}{\beta!(\alpha - \beta)!} D^\beta f D^{\alpha-\beta} T$$

where we say $\beta \leq \alpha$ provided $\beta_i \leq \alpha_i$ for $1 \leq i \leq n$.

1.6. Prove that if T is a distribution and $\varphi \in C_0^\infty(R^n)$, then $T * \varphi \in C_0^\infty(R^n)$ and $D(T * \varphi) = (DT) * \varphi$ where D denotes any partial derivative of the first order. This may be accomplished by analyzing difference quotients and using the fact that $\tau_h(DT) = D(\tau_h T)$.

1.7. Lemma 1.8.13 shows that if $1 < p < \infty$ and $1 < q < r < \infty$, then

$$L(p, q) \subset L(p, q) \subset L(p, r) \subset L(p, \infty).$$

Give examples that show the above inclusions are strict.

1.8. As we have noted in Remark 1.4.3, the measure H^γ does not satisfy the regularity property analogous to (1.4.5). However, it does have other approximation properties. Prove that if $A \subset R^n$ is an arbitrary set, there exists a G_δ-set $G \supset A$ such that

$$H^\gamma(A) = H^\gamma(G).$$

It can also be shown (although the proof is not easy) that if A is a Suslin set, then

$$H^\gamma(A) = \sup\{H^\gamma(K) : K \subset A, K \text{ compact}, H^\gamma(K) < \infty\}.$$

See [F4, 2.10.48].

1.9. Prove the statement that leads to (1.8.10), namely, if $f \in L^1(R^n)$, then

$$\int_0^t f^*(r)dr = tf^*(t) + \int_{f^*(t)}^\infty \alpha_f(s)ds.$$

Hint: Consider the graph of f and employ Lemma 1.5.1.

1.10. Another Hausdorff-type measure often used in the literature is *Hausdorff spherical measure,* H_S^γ. It is defined in the same manner as H^γ (see Definition 1.4.1) except that the sets A_i are taken as n-balls. Clearly, $H^\gamma(E) \leq H_S^\gamma(E)$ for any set E. Prove that $H_S^\gamma(E) = 0$ whenever $H^\gamma(E) = 0$.

1.11. Suppose u is a function defined on an open set $\Omega \subset R^n$ with the property that it is continuous almost everywhere. Prove that u is measurable.

1.12. Using only basic information, prove that the class of simple functions is dense in the Lorentz space $L(p, q)$.

1.13. Let μ be a Radon measure on R^n. As an application of Theorem 1.3.6 prove that any open set $U \subset R^n$ is essentially (with respect to μ) the disjoint union of n-balls. That is, prove that there is a sequence of disjoint n-balls $B_i \subset U$ such that

$$\mu\left[U - \bigcup_{i=1}^\infty B_i\right] = 0.$$

1.14. Let μ be a Radon measure on R^n. Let I be an arbitrary index set and suppose for each $\alpha \in I$, that E_α is an μ-measurable set with the property that

$$\lim_{r \to 0} \frac{\mu[E_\alpha \cap B(x,r)]}{\mu[B(x,r)]} = 1$$

for every $x \in E_\alpha$. Prove that $\cup_{\alpha \in I} E_\alpha$ is μ-measurable.

1.15. From Exercise 1.1 we know that the distance function, d, to an arbitrary set E is Lipschitz with constant 1. Looking ahead to Theorem 2.2.1, we then can conclude that d is differentiable almost everywhere.

Prove that if E is a closed set and d is differentiable at a point $x \notin E$, then there exists a unique point $\xi(x) \in E$ nearest x. Also prove that

$$Dd(x) = \frac{x - \xi(x)}{d(x)}.$$

1.16. (a) If f is a continuous function defined on R^n, prove that its non-increasing rearrangement f^* is also continuous. Thus, continuous functions remain invariant under the operation of rearrangement.

(b) Now prove that Lipschitz functions also remain invariant under rearrangement. For this it will be necessary to use the Brunn–Minkowski inequality. It states that if E and F are nonempty subsets of R^n, then

$$|E + F|^{1/n} \geq |E|^{1/n} + |F|^{1/n}$$

where $E + F = \{x + y : x \in E, y \in F\}$.

(c) Looking ahead to Chapter 2, prove that if $f \in W^{1,p}(R^n)$, then $f^* \in W^{1,p}(R^n)$. Use part (b) and Theorem 2.5.1.

(d) Show by an example that $C^1(R^n)$ does not remain invariant under the operation of rearrangement.

1.17. Let $u \in C^0(R^1)$. For each $h \neq 0$, let u_h be the function defined by

$$u_h(x) = \frac{u(x+h) - u(x)}{h}.$$

Prove that $u_h \to u'$ in the sense of distributions.

1.18. Let $\{u_i\}$ be a sequence in $L^p(R^n)$ that converges weakly to u in $L^p(R^n)$, $p > 1$. That is,

$$\lim_{i \to \infty} \int_{R^n} u_i v \, dx \to \int_{R^n} uv \, dx$$

for every $v \in L^{p'}(R^n)$. Prove that $D^\alpha u_i \to D^\alpha u$ in the sense of distributions for each multi-index α.

Historical Notes

1.2. The notion of measures has two fundamental applications: one can be used for estimating the size of sets while the other can be used to define integrals. In his 1894 thesis, E. Borel (cf. [BO]) essentially introduced what is now known as Lebesgue outer measure to estimate the size of sets to assist his investigation of certain pathological functions. Lebesgue [LE1] used measures as a device to construct his integral. Later, when more general measures were studied, Radon (1913) for example, emphasized measure as a countably additive set function defined on a σ-ring of sets whereas Carathéodory (1914) pursued the notion of outer measures defined on all sets.

1.3. The material in this section represents only a very small portion of the literature devoted to differentiation theory and the related subject of covering theorems. Central to this theory is the celebrated theorem of Lebesgue [LE2] which states that a locally integrable function can be represented by the limit of its integral averages over concentric balls whose radii tend to zero. Theorem 1.3.8 generalizes this result to the situation in which Lebesgue measure is replaced by a Radon measure. This result and the covering theorems (Theorems 1.3.5 and 1.3.6) which lead to it are due to Besicovitch, [BE1], [BE2]. The proof of Theorem 1.3.5 was communicated to the author by Robert Hardt. The original version of Theorem 1.3.6 is due to Vitali [VI] who employed closed cubes and Lebesgue measure. Lebesgue [LE2] observed that the result is still valid if cubes are replaced by general sets that are "regular" when compared to cubes. A sequence of sets $\{E_k\}$ is called regular at a point x_0 if $x_0 \in \cap_{k=1}^{\infty} E_k$, $\operatorname{diam}(E_k) \to 0$ and $\liminf_{k\to 0} \rho(E_k) > 0$ where $\rho(E_k)$ is defined as the infimum of the numbers $|C|/|E_k|$ with C ranging over all cubes containing E_k. In particular, one is allowed to consider coverings by nested cubes or balls that are not necessarily concentric. However, in the case when Lebesgue measure is replaced by a Radon measure, Theorem 1.3.6 no longer remains valid if the balls in the covering are allowed to become too non-concentric. At about the time that Besicovitch made his contributions, A.P. Morse developed a theory which allowed coverings by a general class of sets rather than by concentric closed balls. The following typifies the results obtained by Morse [MSE2]: Let $A \subset R^n$ be a bounded set. Suppose for each $x \in A$ there is a set $H(x)$ satisfying the following two properties: (i) there exist $M > 0$ independent of x and $r(x) > 0$ such that

$$\overline{B}(x, r(x)) \subset H(x) \subset \overline{B}(x, Mr(x));$$

(ii) $H(x)$ contains the convex hull of the set $\{y\} \cup \overline{B}(x, r(x))$ whenever $y \in H(x)$. Then a conclusion similar to that in Theorem 1.3.5 holds.

Another useful covering theorem due to Whitney [WH] states than an open set in R^n can be covered by non-overlapping cubes that become smaller as they approach the boundary. Theorem 1.3.5 is a similar result where balls are used instead of cubes and where the requirement of disjointness is replaced by an estimate of the amount of overlap. This treatment is found in [F4, Section 3.1].

Among the many results concerning differentiation with respect to irregular families is the following interesting theorem proved in [JMZ]: Suppose

u is a measurable function defined on R^n such that

$$\int |u|(1 + \log^+ |u|)^{n-1} dx < \infty.$$

Then, for almost every $x \in R^n$,

$$\lim |I|^{-1} \int_I |u(y) - u(x)| dy = 0$$

where the limit is taken over all bounded open intervals I containing the point x. This result is false if u is assumed only to be integrable. Such irregular intervals are useful in applications concerning parabolic differential equations, where it is natural to consider intervals of the form $C \times [0, r^2]$, where C is an $(n-1)$-cube of side-length r.

For further information pertaining to differentiation and coverings, the reader may consult [DG], [F4, Section 2.8].

1.4. Carathéodory [CAY] was the first to introduce "Hausdorff" measure in his work on the general theory of outer measure. He only developed linear measure in R^n although he indicated how k-dimensional measure could be defined for integer values of k. k-dimensional measure for general positive values of k was introduced by Hausdorff [HAU] who illustrated the use of these measures by showing that the Cantor ternary set has fractional dimension $\log 2/\log 3$.

1.7. There are various ways of presenting the theory of distributions, but the method employed in this section is the one that reflects the original theory of Schwartz [SCH] which is based on the duality of topological vector spaces. The reader may wish to consult the monumental work of Gelfand and his collaborators which contains a wealth of material on "generalized functions" [GE1], [GE2], [GE3], [GE4], [GE5].

1.8. Fundamental to the notion of Lorentz spaces is the classical concept of the non-increasing rearrangement of a function which, in turn, is based upon a notion of symmetrization which transforms a given solid in R^3 into a ball with the same volume. There are a variety of symmetrization procedures including the one introduced by J. Steiner [ST] in 1836 which changes a solid into one with the same volume and at least one plane of symmetry. The reader may consult the works by Pólya and Szegö [PS] or Burago and Zalgaller [BUZ] for excellent accounts of isoperimetric inequalities and their connection with symmetrization techniques. In 1950 G.G. Lorentz [LO1], [LO2], first discussed the spaces that are now denoted by $L(p, 1)$ and $L(p, \infty)$. Papers by Hunt [HU] and O'Neil [O] present interesting developments of Lorentz spaces. Much of this section is based on the work of O'Neil and the main results of this section, Lemmas 1.8.7–1.8.9, were first proved in [O]. The reader may consult [CA2], [CA3], [LP], [PE] for further developments in this area.

2

Sobolev Spaces and Their Basic Properties

This chapter is concerned with the fundamental properties of Sobolev spaces including the Sobolev inequality and its associated imbedding theorems. The basic Sobolev inequality is proved in two ways, one of which employs the co-area formula (Section 2.7) to obtain the best constant in the inequality. This method relates the Sobolev inequality to the isoperimetric inequality.

The point-wise behavior of Sobolev functions will be discussed in Chapters 3 and 4 and this will entail a method of defining Sobolev functions on large sets, sets larger than the complement of sets of Lebesgue measure zero. It turns out that the appropriate null sets for this purpose are described in terms of sets of Bessel capacity zero. This capacity is introduced and developed in Section 2.6 but only to the extent needed for the analysis in Chapters 3 and 4. The theory of capacity is extensive and there is a vast literature that relates Bessel capacity to non-linear potential theory. It is beyond the scope of this book to give a thorough treatment of this topic.

One of the interesting aspects of Sobolev theory is the behavior of the Sobolev inequality in the case of critical indices. In order to gain a better appreciation of this phenomena, we will include a treatment in the context of Lorentz spaces.

2.1 Weak Derivatives

Let $u \in L^1_{\text{loc}}(\Omega)$. For a given multi-index α, a function $v \in L^1_{\text{loc}}(\Omega)$ is called the αth *weak derivative* of u if

$$\int_\Omega \varphi v \, dx = (-1)^{|\alpha|} \int_\Omega u D^\alpha \varphi \, dx \tag{2.1.1}$$

for all $\varphi \in C_0^\infty(\Omega)$. v is also referred to as the *generalized derivative* of u and we write $v = D^\alpha u$. Clearly, $D^\alpha u$ is uniquely determined up to sets of Lebesgue measure zero. We say that the αth *weak derivative* of u *is a measure* if there exists a regular Borel (signed) measure μ on Ω such that

$$\int_\Omega \varphi \, d\mu = (-1)^{|\alpha|} \int_\Omega u D^\alpha \varphi \, dx \tag{2.1.2}$$

for all $\varphi \in C_0^\infty(\Omega)$. In most applications, $|\alpha| = 1$ and then we speak of u whose partial derivatives are measures.

2.1.1. Definition. For $p \geq 1$ and k a non-negative integer, we define the Sobolev space

$$W^{k,p}(\Omega) = L^p(\Omega) \cap \{u : D^\alpha u \in L^p(\Omega), |\alpha| \leq k\}. \tag{2.1.3}$$

The space $W^{k,p}(\Omega)$ is equipped with a norm

$$\|u\|_{k,p;\Omega} = \left(\int_\Omega \sum_{|\alpha| \leq k} |D^\alpha u|^p dx \right)^{1/p} \tag{2.1.4}$$

which is clearly equivalent to

$$\sum_{|\alpha| \leq k} \|D^\alpha u\|_{p;\Omega}. \tag{2.1.5}$$

It is an easy matter to verify that $W^{k,p}(\Omega)$ is a Banach space. The space $W_0^{k,p}(\Omega)$ is defined as the closure of $C_0^\infty(\Omega)$ relative to the norm (2.1.4). We also introduce the space $BV(\Omega)$ of integrable functions whose partial derivatives are (signed measures) with finite variation; thus,

$$BV(\Omega) = L^1(\Omega) \cap \{u : D^\alpha u \text{ is a measure, } |D^\alpha u|(\Omega) < \infty, |\alpha| = 1\}. \tag{2.1.6}$$

A norm on $BV(\Omega)$ is defined by

$$\|u\|_{BV(\Omega)} = \|u\|_{1;\Omega} + \sum_{|\alpha|=1} |D^\alpha u|(\Omega). \tag{2.1.7}$$

2.1.2. Remark. Observe that if $u \in W^{k,p}(\Omega) \cup BV(\Omega)$, then u is determined only up to a set of Lebesgue measure zero. We agree to call these functions u continuous, bounded, etc. if there is a function $\overline{u}$ such that $\overline{u} = u$ a.e. and $\overline{u}$ has these properties.

We will show that elements in $W^{k,p}(\Omega)$ have representatives that permit us to regard them as generalizations of absolutely continuous functions on R^1. First, we prove an important result concerning the convergence of regularizers of Sobolev functions.

2.1.3. Lemma. *Suppose $u \in W^{k,p}(\Omega)$, $p \geq 1$. Then the regularizers of u (see Section 1.6), u_ε, have the property that*

$$\lim_{\varepsilon \to 0} \|u_\varepsilon - u\|_{k,p;\Omega'} = 0$$

whenever $\Omega' \subset\subset \Omega$. *In case* $\Omega = R^n$, *then* $\lim_{\varepsilon\to 0} \|u_\varepsilon - u\|_{k,p} = 0$.

Proof. Since Ω' is a bounded domain, there exists $\varepsilon_0 > 0$ such that $\varepsilon_0 < \operatorname{dist}(\Omega', \partial\Omega)$. For $\varepsilon < \varepsilon_0$, differentiate under the integral sign and refer to (2.1.1) to obtain for $x \in \Omega'$ and $|\alpha| \le k$,

$$\begin{aligned} D^\alpha u_\varepsilon(x) &= \varepsilon^{-n} \int_\Omega D_x^\alpha \varphi\left(\frac{x-y}{\varepsilon}\right) u(y)dy \\ &= (-1)^{|\alpha|}\varepsilon^{-n} \int_\Omega D_y^\alpha \varphi\left(\frac{x-y}{\varepsilon}\right) u(y)dy \\ &= \varepsilon^{-n} \int_\Omega \varphi\left(\frac{x-y}{\varepsilon}\right) D^\alpha u(y)dy \\ &= (D^\alpha u)_\varepsilon(x) \end{aligned}$$

for each $x \in \Omega'$. The result now follows from Theorem 1.6.1(iii). □

Since the definition of a Sobolev function requires that its distributional derivatives belong to L^p, it is natural to inquire whether the function possesses any classical differentiability properties. To this end, we begin by showing that its partial derivatives exist almost everywhere. That is, in keeping with Remark 2.1.2, we will show that there is a function $\overline{u}$ such that $\overline{u} = u$ a.e. and that the partial derivatives of $\overline{u}$ exist almost everywhere. However, the result does not give any information concerning the most useful concept of total differential, the linear approximation of the difference quotient. This topic will be pursued in Chapter 3.

2.1.4. Theorem. *Suppose* $u \in L^p(\Omega)$. *Then* $u \in W^{1,p}(\Omega)$, $p \ge 1$, *if and only if* u *has a representative* $\overline{u}$ *that is absolutely continuous on almost all line segments in* Ω *parallel to the coordinate axes and whose (classical) partial derivatives belong to* $L^p(\Omega)$.

Proof. First, suppose $u \in W^{1,p}(\Omega)$. Consider a rectangular cell in Ω

$$R \equiv [a_1, b_1] \times \ldots \times [a_n, b_n]$$

all of whose side lengths are rational. We know from Lemma 2.1.3 that the regularizers of u converge to u in the $W^{1,p}_{\text{loc}}(\Omega)$ norm. Thus, writing $x \in R$ as $x = (\tilde{x}, x_i)$ where $\tilde{x} \in R^{n-1}$ and $x_i \in [a_i, b_i]$, $1 \le i \le n$, it follows from Fubini's Theorem that there is a sequence $\{\varepsilon_k\} \to 0$ such that

$$\lim_{k\to\infty} \int_{a_i}^{b_i} |u_k(\tilde{x}, x_i) - u(\tilde{x}, x_i)|^p + |Du_k(\tilde{x}, x_i) - Du(\tilde{x}, x_i)|^p dx_i = 0$$

for almost all $\tilde{x}$. Here, we denote $u_{\varepsilon_k} = u_k$. Since u_k is smooth, for each such $\tilde{x}$ and for every $\eta > 0$, there is $M > 0$ such that for $b \in [a_i, b_i]$,

$$|u_k(\tilde{x}, b) - u_k(\tilde{x}, a_i)| \le \int_{a_i}^{b_i} |Du_k(\tilde{x}, x_i)| dx_i$$

$$\leq \int_{a_i}^{b_i} |Du(\tilde{x}, x_i)| dx_i + \eta$$

for $k > M$. If $\{u_k(\tilde{x}, a_i)\}$ converges as $k \to \infty$, (which may be assumed without loss of generality), this shows that the sequence $\{u_k\}$ is uniformly bounded on $[a_i, b_i]$. Moreover, as a function of x_i, the u_k are absolutely continuous, uniformly with respect to k, because the L^1 convergence of Du_k to Du implies that for each $\varepsilon > 0$, there is a $\delta > 0$ such that $\int_E |Du_k(\tilde{x}, x_i)| dx_i < \varepsilon$ whenever $H^1(E) < \delta$ for all positive integers k. Thus, by the Arzelà–Ascoli theorem, $\{u_k\}$ converges uniformly on $[a_i, b_i]$ to an absolutely continuous function that agrees almost everywhere with u. This shows that u has the desired representative on R. The general case follows from the familiar diagonalization process.

Now suppose that u has such a representative $\overline{u}$. Then $\overline{u}\varphi$ also possesses the absolute continuity properties of $\overline{u}$, whenever $\varphi \in C_0^\infty(\Omega)$. Thus, for $1 \leq i \leq n$, it follows that

$$\int \overline{u} D_i \varphi \, dx = -\int D_i \overline{u} \varphi \, dx$$

on almost every line segment in Ω whose end-points belong to $R^n - \operatorname{spt} \varphi$ and is parallel to the ith coordinate axis. Fubini's Theorem thus implies that the weak derivative $D_i u$ has $D_i \overline{u}$ as a representative. □

2.1.5. Remark. Theorem 2.1.4 can be stated in the following way. If $u \in L^p(\Omega)$, then $u \in W^{1,p}(\Omega)$ if and only if u has a representative $\overline{u}$ such that $\overline{u} \in W^{1,p}(\Lambda)$ for almost all line segments Λ in Ω parallel to the coordinate axes and $|D\overline{u}| \in L^p(\Omega)$. For an equivalent statement, an application of Fubini's Theorem allows us to replace almost all line segments Λ by almost all k-dimensional planes Λ_k in Ω that are parallel to the coordinate k-planes.

It is interesting to note that the proof of Theorem 2.1.4 reveals that the regularizers of u converge *everywhere* on almost all lines parallel to the coordinate axes. If u were not an element of $W^{1,p}(\Omega)$, but merely an element of $L^1(\Omega)$, Fubini's theorem would imply that the convergence occurs only H^1–a.e. on almost all lines. Thus, the assumption $u \in W^{1,p}(\Omega)$ implies that the regularizers converge on a relatively large set of points. This is an interesting facet of Sobolev functions that will be pursued later in Chapter 3.

Recall that if $u \in L^p(R^n)$, then $\|u(x+h) - u(x)\|_p \to 0$ as $h \to 0$. A similar result provides a very useful characterization of $W^{1,p}(R^n)$.

2.1.6. Theorem. *Let* $1 < p < \infty$. *Then* $u \in W^{1,p}(R^n)$ *if and only if* $u \in L^p(R^n)$ *and*

$$\left(\int \left| \frac{u(x+h) - u(x)}{|h|} \right|^p dx \right)^{1/p} = |h|^{-1} \|u(x+h) - u(x)\|_p$$

remains bounded for all $h \in R^n$.

Proof. First assume $u \in C_0^\infty(R^n)$. Then

$$\frac{u(x+h)-u(x)}{|h|} = \frac{1}{|h|}\int_0^{|h|} Du\left(x+t\frac{h}{|h|}\right)\cdot\frac{h}{|h|}dt,$$

so by Jensen's inequality (1.5.12),

$$\left|\frac{u(x+h)-u(x)}{h}\right|^p \le \frac{1}{|h|}\int_0^{|h|}\left|Du\left(x+t\frac{h}{|h|}\right)\right|^p dt.$$

Therefore,

$$\|u(x+h)-u(x)\|_p^p \le |h|^p\frac{1}{|h|}\int_0^{|h|}\int_{R^n}\left|Du\left(x+t\frac{h}{|h|}\right)\right|^p dxdt,$$

or

$$\|u(x+h)-u(x)\|_p \le |h|\,\|Du\|_p.$$

By Lemma 2.1.3, this holds whenever $u \in W^{1,p}(R^n)$.

Conversely, if e_i is the ith unit basis vector, then the sequence

$$\left\{\frac{u(x+e_i/k)-u(x)}{1/k}\right\}$$

is bounded in $L^p(R^n)$. Hence, by Theorem 1.5.2, there exists a subsequence (which will be denoted by the full sequence) and $u_i \in L^p(R^n)$ such that

$$\frac{u(x+e_i/k)-u(x)}{1/k} \to u_i$$

weakly in $L^p(R^n)$. Thus, for $\varphi \in \mathscr{D}$,

$$\begin{aligned}\int_{R^n} u_i\varphi\,dx &= \lim_{k\to\infty}\int_{R^n}\left[\frac{u(x+e_i/k)-u(x)}{1/k}\right]\varphi(x)dx\\ &= \lim_{k\to\infty}\int_{R^n} u(x)\left[\frac{\varphi(x-e_i/k)-\varphi(x)}{1/k}\right]dx\\ &= -\int_{R^n} uD_i\varphi\,dx.\end{aligned}$$

This shows that

$$D_i u = u_i$$

in the sense of distributions. Hence, $u \in W^{1,p}(R^n)$. □

2.1.7. Definition. For a measurable function $u\colon \Omega \to R^1$, let

$$u^+ = \max\{u, 0\}, \quad u^- = \min\{u, 0\}.$$

2.1.8. Corollary. *Let $u \in W^{1,p}(\Omega)$, $p \geq 1$. Then $u^+, u^- \in W^{1,p}(\Omega)$ and*

$$Du^+ = \begin{cases} Du \text{ if } & u > 0 \\ 0 \quad \text{if } & u \leq 0 \end{cases}$$

$$Du^- = \begin{cases} 0 \quad \text{if } & u \geq 0 \\ Du \text{ if } & u < 0. \end{cases}$$

Proof. Because u has a representative that has the absolute continuity properties stated in Theorem 2.1.4, it follows immediately that $u^+, u^- \in W^{1,p}(\Omega)$. The second part of the theorem is reduced to the observation that if f is a function of one variable such that f' exists a.e., then $(f^+)' = f' \cdot \chi_{\{f>0\}}$. □

2.1.9. Corollary. *If Ω is connected, $u \in W^{1,p}(\Omega)$, $p \geq 1$, and $Du = 0$ a.e. on Ω, then u is constant on Ω.*

Proof. Appealing to Theorem 2.1.4, we see that u has a representative that assumes a constant value on almost all line segments in Ω parallel to the coordinate axes. □

2.1.10. Remark. The corollary states that elements of $W^{1,p}(\Omega)$ remain invariant under the operation of truncation. One of the interesting aspects of the theory is that this, in general, is no longer true for the space $W^{k,p}(\Omega)$. Motivated by the observation that $u^+ = H \circ u$ where H is defined by

$$H(t) = \begin{cases} t \; t \geq 0 \\ 0 \; t < 0 \end{cases}$$

we consider the composition $H \circ u$ where H is a smooth function. It was shown in [MA2] and [MA3] that it is possible to smoothly truncate non-negative functions in $W^{2,p}$. That is, if $H \in C^\infty(R^1)$ and

$$\sup |t^{j-1} H^{(j)}(t)| \leq M < \infty$$

for $j = 1, 2$, then there exists $C = C(p, M)$ such that for any non-negative $v \in C_0^\infty(R^n)$

$$\|D^\alpha H(v)\|_p \leq C \|D^2 v\|_p$$

for $1 < p < n/2$ and any multi-index α with $|\alpha| = 2$. Here $D^2 v$ denotes the vector whose components consist of all second derivatives of v. However, it is surprising to find that this is not true for all spaces $W^{k,p}$. Indeed, it was established in [DA1] that if $1 \leq p < n/k$, $2 < k < n$, or $1 < p < n/k$, $k = 2$, and $H \in C^\infty(R^1)$ with $H^{(k)}(t) \geq 1$ for $|t| \leq 1$, then there exists a function $u \in W^{k,p}(R^n) \cap C^\infty(R^n)$ such that $H(u) \notin W^{k,p}(R^n)$. The most general result available in the positive direction is stated in terms of Riesz potentials, $I_\alpha * f$ (see Section 2.6), where f is a non-negative function in

L^p. The following result is due to Dahlberg [DA2]. Let $0 < \alpha < n$ and $1 < p < n/\alpha$. Let $H \in C^\infty(R^1)$ have the property that

$$\sup_{t>0} |t^{j-1} H^{(j)}(t)| \le M < \infty$$

for $j = 0, 1, \ldots, \alpha^*$, where α^* is the smallest integer $\ge \alpha$. If $f \in L^p(R^n)$ and $f \ge 0$, then there exists $g \in L^p(R^n)$ such that

$$H(I_\alpha * g) = I_\alpha * g \quad \text{a.e.}$$

and $\|g\|_p \le C\|f\|_p$ where $C = C(\alpha, p, n, M)$. The case of integral α was treated in [AD4] and in this situation the result can be formulated as

$$\|D^\gamma[H(I_\alpha * f)]\|_p \le C\|f\|_p^p$$

for any multi-index γ with $|\gamma| = k$.

To continue our investigation of the calculus of Sobolev functions, we consider the problem of composition of a suitable function with $u \in W^{1,p}(\Omega)$. Before doing so, we remind the reader of the analogous problem in Real Variable theory. In general, if f and g are both absolutely continuous functions, then the composition, $f \circ g$, need not be absolutely continuous. Recall that a function, f, is absolutely continuous if and only if it is continuous, of bounded variation, and has the property that $|f(E)| = 0$ whenever $|E| = 0$. Thus, the consideration that prevents $f \circ g$ from being absolutely continuous is that $f \circ g$ need not be of bounded variation. A result of Vallée Poussin [PO] states that $f \circ g$ is absolutely continuous if and only if $f' \circ g \cdot g'$ is integrable. An analogous result is valid in the context of Sobolev theory, cf. [MM1], [MM2], but we will consider only the case when the outer function is Lipschitz.

2.1.11. Theorem. *Let $f : R^1 \to R^1$ be a Lipschitz function and $u \in W^{1,p}(\Omega)$, $p \ge 1$. If $f \circ u \in L^p(\Omega)$, then $f \circ u \in W^{1,p}(\Omega)$ and for almost all $x \in \Omega$,*

$$D(f \circ u)(x) = f'[u(x)] \cdot Du(x).$$

Proof. By Theorem 2.1.4, we may assume that u is absolutely continuous on almost all line segments in Ω. Select a coordinate direction, say the ith, and consider the partial derivative operator, D_i. On almost all line segments, λ, in Ω parallel to the ith coordinate axis, $f \circ u$ is clearly absolutely continuous because f is Lipschitz. Moreover,

$$D_i(f \circ u)(x) = f'[u(x)] \cdot D_i u(x) \tag{2.1.8}$$

holds at all $x \in \lambda$ such that $D_i u(x)$ and $f'[u(x)]$ both exist. Note that if $D_i u(x) = 0$, then $D_i(f \circ u)(x) = 0$ because

$$\frac{|f[u(x + he_i) - f[u(x)]|}{|h|} \le M \frac{|u(x + he_i) - u(x)|}{|h|}$$

where M is the Lipschitz constant of f and e_i is the ith coordinate vector. Thus, letting $N = \lambda \cap \{x : D_i u(x) = 0\}$, we have that (2.1.8) holds on N. Now let

$$P = (\lambda - N) \cap \{x : D_i u(x) \text{ exists and } D_i u(x) \neq 0\}$$

and note that $P \cup N$ occupies H^1-almost all of λ. From classical considerations, we have that if $S \subset P$ and $H^1[u(S)] = 0$, then $H^1(S) = 0$. In particular, if we let $E = \{y : f'(y) \text{ fails to exist}\}$, then $H^1[u^{-1}(E) \cap P] = 0$. Since (2.1.8) holds if $x \in \lambda - u^{-1}(E) \cap P$ and $D_i u(x)$ exists, it follows therefore that (2.1.8) holds at H^1-almost all points of λ. At all such x, we may conclude that

$$|D_i(f \circ u)(x)|^p \leq M^p |D_i u(x)|^p. \tag{2.1.9}$$

Once it is known that the set of $x \in \Omega$ for which (2.1.8) holds is a measurable set, we may apply Fubini's Theorem to conclude that $f \circ u$ satisfies the hypotheses of Theorem 2.1.4. This is a consequence of the fact that the functions on both sides of (2.1.8) are measurable. In particular, $f' \circ u$ is measurable because f' agrees with one of its Borel measurable Dini derivates almost everywhere. □

2.2 Change of Variables for Sobolev Functions

In addition to the basic facts considered in the previous section, it is also useful to know what effect a change of variables has on a Sobolev function. For this purpose, we consider a bi-Lipschitzian map

$$T: \Omega \to \Omega'.$$

That is, for some constant M, we assume that both T and T^{-1} satisfy,

$$|T(x) - T(y)| \leq M|x - y|, \quad \text{for all } x, y \in \Omega,$$

$$|T^{-1}(x') - T^{-1}(y')| \leq M|x' - y'|, \quad \text{for all } x', y' \in \Omega'. \tag{2.2.1}$$

In order to proceed, we will need an important result of Rademacher which states that a Lipschitz map $T: R^n \to R^m$ is differentiable at almost all points in R^n. That is, there is a set $E \subset R^n$ with $|E| = 0$ such that for each $x \in R^n - E$, there is a linear map $dT(x): R^n \to R^m$ (the differential of T at x) with the property that

$$\lim_{y \to 0} \frac{|T(x + y) - T(x) - dT(x, y)|}{|y|} = 0. \tag{2.2.2}$$

In order to establish (2.2.2) it will be sufficient to prove the following result.

2.2.1. Theorem. *If $f: R^n \to R^1$ is Lipschitz, then for almost all $x \in R^n$,*

$$\lim_{y \to 0} \frac{f(x+y) - f(x) - Df(x) \cdot y}{|y|} = 0.$$

Proof. For $v \in R^n$ with $|v| = 1$, and $x \in R^n$, let $\gamma(t) = f(x + tv)$. Since f is Lipschitz, γ is differentiable for almost all t.

Let $df(x, v)$ denote the directional derivative of f at x. Thus, $df(x, v) = \gamma'(0)$ whenever $\gamma'(0)$ exists. Let

$$N_v = R^n \cap \{x : df(x, v) \text{ fails to exist}\}.$$

Note that

$$N_v = \left\{ x : \limsup_{t \to 0} \frac{f(x+tv) - f(x)}{t} > \liminf_{t \to 0} \frac{f(x+tv) - f(x)}{t} \right\},$$

and is therefore a Borel measurable set. However, for each line λ whose direction is v, we have $H^1(N_v \cap \lambda) = 0$, because f is Lipschitz on λ. Therefore, by Fubini's theorem, $|N_v| = 0$. Note that on each line λ parallel to v,

$$\int_\lambda df(x, v)\varphi(x)dx = -\int_\lambda f(x)d\varphi(x, v)dx$$

for $\varphi \in C_0^\infty(R^n)$. Because Lebesgue measure remains invariant under orthogonal transformations, it follows by Fubini's Theorem that

$$\begin{aligned}
\int_{R^n} df(x, v)\varphi(x)dx &= -\int_{R^n} f(x)d\varphi(x, v)dx \\
&= -\int_{R^n} f(x)D\varphi(x) \cdot v\, dx \\
&= -\sum_{j=1}^n \int_{R^n} f(x)D_j\varphi(x) \cdot v_j\, dx \\
&= \sum_{j=1}^n \int_{R^n} D_j f(x)\varphi(x) \cdot v_j\, dx \\
&= \int_{R^n} \varphi(x)Df(x) \cdot v\, dx.
\end{aligned}$$

Because this is valid for all $\varphi \in C_0^\infty(R^n)$, we have that

$$df(x, v) = Df(x) \cdot v, \quad \text{a.e. } x \in R^n. \tag{2.2.3}$$

Now let $v_1, v_2, \ldots$ be a countable dense subset of S^{n-1} and observe that there is a set E with $|E| = 0$ such that

$$df(x, v_k) = Df(x) \cdot v_k \tag{2.2.4}$$

for all $x \in R^n - E$, $k = 1, 2, \ldots$.

We will now show that our result holds at all points of $R^n - E$. For this purpose, let $x \in R^n - E$, $|v| = 1$, $t > 0$ and consider the difference quotient

$$Q(x, v, t) \equiv \frac{f(x + tv) - f(x)}{t} - Df(x) \cdot v.$$

For $v, v' \in S^{n-1}$ and $t > 0$ note that

$$|Q(x, v, t) - Q(x, v', t)| = \frac{|f(x + tv) - f(x + tv') + (v - v') \cdot Df(x)|}{t}$$

$$\leq M|v - v'| + |v - v'| \cdot |Df(x)| \leq M(n + 1)|v - v'| \tag{2.2.5}$$

where M is the Lipschitz constant of f. Since the sequence $\{v_i\}$ is dense in S^{n-1}, there exists an integer K such that

$$|v - v_k| < \frac{\varepsilon}{2(n + 1)M} \quad \text{for some} \quad k \in \{1, 2, \ldots, K\} \tag{2.2.6}$$

whenever $v \in S^{n-1}$. For $x_0 \in R^n - E$, we have from (2.2.4) the existence of $\delta > 0$ such that

$$|Q(x_0, v_k, t)| < \frac{\varepsilon}{2} \quad \text{for} \quad 0 < t < \delta, \; k \in \{1, 2, \ldots, K\}. \tag{2.2.7}$$

Since

$$|Q(x_0, v, t)| \leq |Q(x_0, v_k, t)| + |Q(x_0, v, t) - Q(x_0, v_k, t)|$$

for $k \in \{1, 2, \ldots, K\}$, it follows from (2.2.7), (2.2.5), and (2.2.6) that

$$|Q(x_0, v, t)| < \frac{\varepsilon}{2} + \frac{\varepsilon}{2} = \varepsilon$$

whenever $|v| = 1$ and $0 < t < \delta$. □

Recall that if $L: R^n \to R^n$ is a linear mapping and $E \subset R^n$ a measurable set, then

$$|L(E)| = |\det L|\,|E|.$$

It is not difficult to extend this result to more general transformations. Indeed, if $T: R^n \to R^n$ is Lipschitz, we now know from Theorem 2.2.1 that T has a total differential almost everywhere. Moreover, if T is also univalent, one can show that

$$H^n[T(E)] = \int_E JT\,dx \text{ for every measurable set } E, \tag{2.2.8}$$

where JT is the Jacobian of T. From this follows the general transformation formula

$$\int_E f \circ T \, JT \, dx = \int_{T(E)} f \, dx \tag{2.2.9}$$

whenever f is a measurable function. We refer the reader to [F4; 3.2.3] for a proof.

We are now in a position to discuss a bi-Lipschitzian change of coordinates for Sobolev functions.

2.2.2. Theorem. *Let $T: R^n \to R^n$ be a bi-Lipschitzian mapping as in (2.2.1). If $u \in W^{1,p}(\Omega)$, $p \geq 1$, then $v = u \circ T \in W^{1,p}(V)$, $V \equiv T^{-1}(\Omega)$, and*

$$Du[T(x)] \cdot dT(x, \xi) = Dv(x) \cdot \xi \tag{2.2.10}$$

for a.e $x \in \Omega$ and for all $\xi \in R^n$.

Proof. Let u_ε be a sequence of regularizers for u, defined on $\Omega' \subset\subset \Omega$, (see Section 1.6). Then $v_\varepsilon \equiv u_\varepsilon \circ T$ is Lipschitz on $V' = T^{-1}(\Omega')$ and because v_ε is differentiable almost everywhere (Theorem 2.2.1), it follows that

$$D_i v_\varepsilon(x) = \sum_{j=1}^{n} D_j u_\varepsilon[T(x)] D_i T^j(x) \tag{2.2.11}$$

for a.e. $x \in V'$. Here we have used the notation $T = (T^1, T^2, \ldots, T^n)$ where the T^j are the coordinate functions of T. They too are Lipschitz. (2.2.11) holds at all points x at which the right side is meaningful, i.e., at all points at which T is differentiable. If M denotes the Lipschitz constant of T, we have from (2.2.11) that

$$|Dv_\varepsilon(x)| \leq n^2 M |Du_\varepsilon[T(x)]| \quad \text{for a.e. } x \in V'. \tag{2.2.12}$$

In view of the fact that

$$M^{-n} \leq JT(x) \leq M^n \quad \text{for a.e. } x \in R^n,$$

(2.2.12) implies that there exists a constant $C = C(n, M)$ such that

$$|Dv_\varepsilon(x)|^p \leq C |Du_\varepsilon[T(x)]|^p \cdot JT(x), \quad \text{a.e. } x,$$

and therefore

$$\int_{V'} |Dv_\varepsilon|^p dx \leq C \int_\Omega |Du_\varepsilon|^p dx$$

from (2.2.9). A quick review of the above analysis shows that in fact, we have

$$\int_{V'} |Dv_\varepsilon - Dv_{\varepsilon'}|^p dx \leq C \int_\Omega |Du_\varepsilon - Du_{\varepsilon'}|^p dx. \tag{2.2.13}$$

Also,

$$\int_{V'} |v_\varepsilon - v_{\varepsilon'}|^p dx \le C \int_\Omega |u_\varepsilon - u_{\varepsilon'}|^p dx. \tag{2.2.14}$$

From 2.2.11 we see that the regularizers u_ε converge to u in the norm of $W^{1,p}(\Omega')$ whenever $\Omega' \subset\subset \Omega$. Thus, (2.2.13) and (2.2.14) imply that $\{v_\varepsilon\}$ is a Cauchy sequence in $W^{1,p}(\Omega')$, and thus converges to some element $v \in W^{1,p}(V')$ with

$$\|v\|_{1,p;V'} \le C\|u\|_{1,p;\Omega'} \le C\|u\|_{1,p;\Omega}. \tag{2.2.15}$$

Since $u_\varepsilon(x) \to u(x)$ for a.e. $x \in \Omega$, it is clear that v is defined on V with $v = u \circ T$. Moreover, $v \in W^{1,p}(V')$ whenever $V' \subset\subset V$ and (2.2.15) shows that, in fact, $v \in W^{1,p}(V)$. Finally, observe that (2.2.10) holds by letting $\varepsilon \to 0$ in (2.2.11). □

2.3 Approximation of Sobolev Functions by Smooth Functions

From Theorem 1.6.1, we see that for each $u \in W^{k,p}(\Omega)$, there is a sequence of $C_0^\infty(\Omega)$ functions, $\{u_\varepsilon\}$, such that $u_\varepsilon \to u$ in $W^{k,p}(\Omega')$ for $\Omega' \subset\subset \Omega$. The purpose of the next important result is to show that a similar approximation exists on all of Ω and not merely on compact subsets of Ω.

We first require a standard result which concerns the existence of a C^∞ partition of unity subordinate to an open cover.

2.3.1. Lemma. *Let $E \subset R^n$ and let $\mathcal{G}$ be a collection of open sets U such that $E \subset \{\cup U : U \in G\}$. Then, there exists a family $\mathcal{F}$ of non-negative functions $f \in C_0^\infty(R^n)$ such that $0 \le f \le 1$ and*

(i) *for each $f \in \mathcal{F}$, there exists $U \in \mathcal{G}$ such that* spt $f \subset U$,

(ii) *if $K \subset E$ is compact, then* spt $f \cap K \neq 0$ *for only finitely many $f \in \mathcal{F}$,*

(iii) $\sum_{f\in\mathcal{F}} f(x) = 1$ *for each $x \in E$.*

Proof. Suppose first that E is compact, so that there exists a positive integer N such that $E \subset \cup_{i=1}^N U_i$, $U_i \in \mathcal{G}$. Clearly, there exist compact sets $E_i \subset U_i$ such that $E \subset \cup_{i=1}^N E_i$. By regularizing χ_{E_i}, the characteristic function of E_i, there exists $g_i \in C_0^\infty(U_i)$ such that $g_i > 0$ on E_i. Let $g = \sum_{i=1}^N g_i$ and note that $g \in C^\infty(R^n)$ and that $g > 0$ on some neighborhood of E. Consequently, it is not difficult to construct a function $h \in C^\infty(R^n)$ such that $h > 0$ everywhere and that $h = g$ on E. Now let $\mathcal{F} = \{f_i : f_i = g_i/h, 1 \le i \le N\}$ to obtain the desired result in case E is compact.

If E is open, let

$$E_i = E \cap \overline{B}(0,i) \cap \left\{ x : \operatorname{dist}(x, \partial E) \geq \frac{1}{i} \right\}.$$

Thus, E_i is compact and $E = \cup_{i=1}^{\infty} E_i$. Let $\mathcal{G}_i$ be the collection of all open sets of the form

$$U \cap \{\operatorname{int} E_{i+1} - E_{i-2}\}$$

where $U \in \mathcal{G}$. (We take $E_0 = E_{-1} = \emptyset$). The elements of $\mathcal{G}_i$ provide an open cover for $E_i - \operatorname{int} E_{i-1}$ and therefore possess a partition of unity $\mathcal{F}_i$ with finitely many elements. Let

$$s(x) = \sum_{i=1}^{\infty} \sum_{g \in \mathcal{F}_i} g(x)$$

and observe that only finitely many positive terms are represented and that $s(x) > 0$ for $x \in E$. A partition of unity for the open set E is obtained by defining

$$\mathcal{F} = \left\{ f : \begin{array}{ll} f(x) = \frac{g(x)}{s(x)} & \text{for some } g \in \mathcal{F}_i \text{ if } x \in E, \\ f(x) = 0 & \text{if } x \notin E. \end{array} \right\}$$

If $E \subset R^n$ is arbitrary, then any partition of unity for the open set $\{\cup U : U \in \mathcal{G}\}$ provides one for E. □

Clearly, the set

$$S = C^k(\Omega) \cap \{u : \|u\|_{k,p;\Omega} < \infty\}$$

is contained in $W^{k,p}(\Omega)$ and therefore, since $W^{k,p}(\Omega)$ is complete, $\overline{S} \subset W^{k,p}(\Omega)$. The next result shows that $\overline{S} = W^{k,p}(\Omega)$.

2.3.2. Theorem. *The space*

$$C^{\infty}(\Omega) \cap \{u : \|u\|_{k,p;\Omega} < \infty\}$$

is dense in $W^{k,p}(\Omega)$.

Proof. Let Ω_i be subdomains of Ω such that $\Omega_i \subset\subset \Omega_{i+1}$ and $\cup_{i=1}^{\infty} \Omega_i = \Omega$. Let $\mathcal{F}$ be a partition of unity of Ω subordinate to the covering $\{\Omega_{i+1} - \overline{\Omega}_{i-1}\}$, $i = 0, 1, \ldots$, where Ω_0 and Ω_{-1} are taken as the null set. Thus, if we let f_i denote the sum of the finitely many $f \in \mathcal{F}$ with $\operatorname{spt} f \subset \Omega_{i+1} - \overline{\Omega}_{i-1}$, then $f_i \in C_0^{\infty}(\Omega_{i+1} - \overline{\Omega}_{i-1})$ and

$$\sum_{i=1}^{\infty} f_i \equiv 1 \quad \text{on} \quad \Omega. \tag{2.3.1}$$

Choose $\varepsilon > 0$. For $u \in W^{k,p}(\Omega)$, there exists $\varepsilon_i > 0$ such that

$$\operatorname{spt}\left((f_i u)_{\varepsilon_i}\right) \subset \Omega_{i+1} - \overline{\Omega}_{i-1}, \tag{2.3.2}$$

$$\|(f_i u)_{\varepsilon_i} - f_i u\|_{k,p;\Omega} < \varepsilon 2^{-i}.$$

With $v_i \equiv (f_i u)_{\varepsilon_i}$, (2.3.2) implies that only a finite number of the v_i can fail to vanish on any given $\Omega' \subset\subset \Omega$, and therefore $v \equiv \sum_{i=1}^{\infty} v_i$ is defined and belongs to $C^\infty(\Omega)$. For $x \in \Omega_i$, we have

$$u(x) = \sum_{j=1}^{i} f_j(x)u(x),$$

$$v(x) = \sum_{j=1}^{i} (f_j u)_{\varepsilon_j}(x) \quad \text{by (2.3.2)}$$

and consequently,

$$\|u - v\|_{k,p;\Omega_i} \le \sum_{j=1}^{i} \|(f_j u)_{\varepsilon_j} - f_j u\|_{k,p;\Omega} < \varepsilon.$$

The conclusion follows from the Monotone Convergence theorem. □

The approximating space $C^\infty(\Omega) \cap \{u : \|u\|_{k,p;\Omega} < \infty\}$ admits functions that are not smooth across the boundary of Ω and therefore it is natural to ask whether it is possible to approximate functions in $W^{k,p}(\Omega)$ by a nicer space, say

$$C^\infty(\overline{\Omega}) \cap \{u : \|u\|_{k,p;\Omega} < \infty\}. \tag{2.3.3}$$

In general, this is easily seen to be false by considering the domain Ω defined as an n-ball with its equatorial $(n-1)$-plane deleted. The function u defined by $u \equiv 1$ on the top half-ball and $u \equiv -1$ on the bottom half-ball is clearly an element of $W^{k,p}(\Omega)$ that cannot be closely approximated by an element in (2.3.3). The difficulty here is that the domain lies on both sides of part of its boundary. If the domain Ω possesses the *segment property,* it has been shown in [AR2, Theorem 3.18] that the space (2.3.3) is then dense in $W^{k,p}(\Omega)$. A domain Ω has the segment property if for each $x \in \partial\Omega$, there is an $r > 0$ and a vector $v_x \in R^n$ such that if $y \in \overline{\Omega} \cap B(x,r)$, then $y + tv_x \in \Omega$ for all $0 < t < 1$.

2.4 Sobolev Inequalities

One of the main objectives of this monograph is to investigate the many inequalities that allow the L^p-norm of a function to be estimated by the norm of its partial derivatives. In this section the Sobolev inequality, which is of fundamental importance, will be established for functions in the space $W_0^{1,p}(\Omega)$. We will return to the topic of Sobolev-type inequalities in Chapter 4.

2.4.1. Theorem. *Let $\Omega \subset R^n$, $n > 1$, be an open domain. There is a constant $C = C(n,p)$ such that if $n > p$, $p \geq 1$, and $u \in W_0^{1,p}(\Omega)$, then*

$$\|u\|_{np/(n-p);\Omega} \leq C\|Du\|_{p;\Omega}.$$

If $p > n$ and Ω bounded, then $u \in C(\overline{\Omega})$ and

$$\sup_{\Omega} |u| \leq C|\Omega|^{1/n-1/p}\|Du\|_{p;\Omega}.$$

Proof. First assume that $u \in C_0^\infty(\Omega)$ and that $p = 1$. Clearly, for each i, $1 \leq i \leq n$,

$$|u(x)| \leq \int_{-\infty}^{x_i} |D_i u(x_1, \ldots, t, \ldots, x_n)| dt$$

where t occupies the ith component of the vector in the integrand. Therefore

$$|u(x)|^{n/n-1} \leq \left(\prod_{i=1}^{n} \int_{-\infty}^{\infty} |D_i u| dx_i \right)^{1/n-1}. \tag{2.4.1}$$

If this inequality is integrated with respect to the first variable, x_1, and then Hölder's inequality is applied, we obtain

$$\begin{aligned}
&\int_{-\infty}^{+\infty} |u(x)|^{n/n-1} dx_1 \\
&\quad \leq \left(\int_{-\infty}^{\infty} |D_1 u(t, x_2, \ldots, x_n)| dt \right)^{1/(n-1)} \\
&\qquad \cdot \int_{-\infty}^{\infty} \prod_{i=2}^{n} \left(\int_{-\infty}^{\infty} |D_i u| dx_i \right)^{1/(n-1)} dx_1 \\
&\quad \leq \left(\int_{-\infty}^{\infty} |D_1 u(t, x_2, \ldots, x_n)| dt \right)^{1/(n-1)} \\
&\qquad \cdot \prod_{i=2}^{n} \left(\int_{-\infty}^{\infty} \int_{-\infty}^{\infty} |D_i u| dx_i dx_1 \right)^{1/(n-1)}.
\end{aligned} \tag{2.4.2}$$

Continuing this procedure and thus integrating (2.4.1) successively with respect to each variable, we obtain

$$\int_{R^n} |u(x)|^{n/(n-1)} dx \leq \prod_{i=1}^{n} \left(\int_{R^n} |D_i u| dx \right)^{1/(n-1)}$$

and therefore, using the fact that the geometric mean is dominated by the arithmetic mean,

$$(\Pi a_j)^{1/n} \leq \frac{1}{n}\sum_{j=1}^{n} a_j, \quad a_j \geq 0,$$

we have

$$\|u\|_{n/(n-1)} \leq \prod_{i=1}^{n}\left(\int_{R^n} |D_i u| dx\right)^{1/n} \leq \frac{1}{n}\int_{R^n} \sum_{i=1}^{n} |D_i u| dx$$

$$\leq \frac{\sqrt{n}}{n}\|Du\|_1. \tag{2.4.3}$$

This establishes the result in case $p = 1$. The result in full generality can be obtained from (2.4.3) by replacing $|u|$ by powers of $|u|$. Thus, if $q > 1$,

$$\begin{aligned} \|\,|u^q|\,\|_{n/(n-1)} &\leq \frac{\sqrt{n}}{n}\int_{R^n} \|D(|u|^q)\| dx \\ &\leq q\frac{\sqrt{n}}{n}\int_{R^n} |u|^{q-1}|Du| dx \\ &\leq \frac{q\sqrt{n}}{n}\|\,|u|^{q-1}\|_{p'}\|Du\|_p, \end{aligned}$$

by Hölder's inequality. Now let $q = (n-1)p/(n-p)$ to obtain the desired result for the case $1 \leq p < n$ and $u \in C_0^\infty(\Omega)$. Now assume $u \in W_0^{1,p}(\Omega)$ and let $\{u_i\}$ be a sequence of functions in $C_0^\infty(\Omega)$ converging to u strongly in $W_0^{1,p}(\Omega)$. Then, with $p^* = np/(n-p)$, an application of the inequality to $u_i - u_j$ yields

$$\|u_i - u_j\|_{p^*} \leq C\|u_i - u_j\|_{1,p}.$$

Thus, $u_i \to u$ in $L^{p^*}(\Omega)$ and the desired result follows. This completes the proof in case $1 \leq p < n$.

In case $p > n$ and Ω bounded, let $\{u_i\}$ be a sequence such that $u_i \in C_0^\infty(\Omega)$ and $u_i \to u$ $W^{1,p}(\Omega)$. The proof is thus reduced to the case when $u \in C_0^\infty(\Omega)$. Now select $x \in R^n$ and because u has compact support, note that

$$|u(x)| \leq \int_{\lambda_x} |Du(r)| dH^1(r) \tag{2.4.4}$$

where λ_x is any ray whose end-point is x. Let $S^{n-1}(x)$ denote the $(n-1)$-sphere of radius 1 centered at x and denote by $\lambda_x(\theta)$ the ray with end-point x that passes through θ, where $\theta \in S^{n-1}(x)$. By integrating (2.4.4) over $S^{n-1}(x)$ we obtain

$$\int_{S^{n-1}(x)} |u(x)| dH^{n-1}(\theta) \leq \int_{S^{n-1}(x)} \cdot \int_{\lambda_x(\theta)} |Du(r)| dH^1(r) dH^{n-1}(\theta)$$

$$= \int_{S^{n-1}(x)} \int_{\lambda_x(\theta)} \frac{|Du(r)|}{r^{n-1}} r^{n-1} dH^1(r) dH^{n-1}(\theta)$$

$$= \int_{R^n} \frac{|Du(y)|}{|x-y|^{n-1}} dy \tag{2.4.5}$$

where $r = |x-y|$. Thus, for any $x \in R^n$,

$$\omega(n-1)|u(x)| \leq \|Du\|_p \left(\int_{\operatorname{spt} u} |x-y|^{(1-n)p'} dy \right)^{1/p'}, \tag{2.4.6}$$

where $\omega(n-1) = H^{n-1}[S^{n-1}]$. We estimate the potential on the right side of (2.4.6) in the following way. Let $B(x,R)$ be the ball such that $|B(x,R)| = |\operatorname{spt} u|$. Observe that for each $y \in \operatorname{spt} u - B(x,R)$ and $z \in B(x,R) - \operatorname{spt} u$, we have

$$|x-y|^{(1-n)p'} \leq |x-z|^{(1-n)p'}$$

and because $|\operatorname{spt} u - B(x,R)| = |B(x,R) - \operatorname{spt} u|$, it therefore follows that

$$\int_{\operatorname{spt} u - B(x,R)} |x-y|^{(1-n)p'} dy \leq \int_{B(x,R) - \operatorname{spt} u} |x-y|^{(1-n)p'} dy.$$

Consequently,

$$\int_{\operatorname{spt} u} |x-y|^{(1-n)p'} dy \leq \int_{B(x,R)} |x-y|^{(1-n)p'} dy. \tag{2.4.7}$$

However,

$$\left(\int_{B(x,R)} |x-y|^{(1-n)p'} dy \right)^{1/p'} = (\gamma^{-1} \alpha(n) R^{\gamma})^{1/p'} \tag{2.4.8}$$

where $\gamma = (1-n)p' + n$ and $\alpha(n)$ is the volume of the unit n-ball. But $\alpha(n)R^n = |\operatorname{spt} u|$ and therefore

$$(\alpha(n) R^{\gamma})^{1/p'} = \alpha^{(n-1)/n} |\operatorname{spt} u|^{1/n - 1/p}. \tag{2.4.9}$$

The second inequality of the theorem follows from (2.4.9), (2.4.8), and (2.4.6). To show that $u \in C(\overline{\Omega})$ when $p > n$, let $\{u_i\} \in C_0^\infty(\Omega)$ be a sequence converging to u in $W_0^{1,p}(\Omega)$. Apply the second inequality of the theorem to the difference $u_i - u_j$ and obtain that $\{u_i\}$ is fundamental in the sup norm on $\overline{\Omega}$. □

The first part of Theorem 2.4.1 states that the L^{p^*} norm of u can be bounded by $\|u\|_{1,p}$, the Sobolev norm of u, where $p^* = np/(n-p)$. It is possible to bound a higher L^p norm of u by utilizing higher order derivatives of u as shown in the next theorem. Observe that the proof is slightly different from that of Theorem 2.4.1 in case $k = 1$, $p > n$.

2.4.2. Theorem. *Let $\Omega \subset R^n$ be an open set. There is a constant $C = C(n,k,p)$ such that if $kp < n$, $p \geq 1$, and $u \in W_0^{k,p}(\Omega)$, then*

$$\|u\|_{p^*;\Omega} \leq C\|u\|_{k,p;\Omega}, \text{ where } p^* = np/(n-kp). \tag{2.4.10}$$

If $kp > n$, then $u \in C(\overline{\Omega})$ and

$$\sup_{\Omega}|u| \leq C|K|^{1/p'}\left[\sum_{|\alpha|=0}^{k-1}(\operatorname{diam}K))^{|\alpha|}\frac{1}{\alpha!}\|D^\alpha u\|_{p;K} + (\operatorname{diam}(K))^k\frac{1}{(k-1)!}\left(k-\frac{n}{p}\right)^{-1}\|D^k u\|_{p;K}\right] \tag{2.4.11}$$

where $K = \operatorname{spt} u$ and $C = C(k,p,n)$.

Proof. When $kp < n$, the proof proceeds by induction on k. Observe that Theorem 2.4.1 establishes the case $k = 1$.

Now assume for every $v \in W_0^{k-1,p}(\Omega)$ that

$$\|v\|_{q_{k-1}} \leq C\|v\|_{k-1,p} \tag{2.4.12}$$

where

$$q_{k-1} = np/(n-kp+p).$$

An application of (2.4.12) to $v = D_j u$, $1 \leq j \leq n$, yields

$$\|D_j u\|_{q_{k-1}} \leq C\|D_j u\|_{k-1,p} \leq C\|u\|_{k,p}. \tag{2.4.13}$$

However, (2.4.12) holds with v replaced by u and this, combined with (2.4.13), implies

$$\|u\|_{1,q_{k-1}} \leq C\|u\|_{k,p}. \tag{2.4.14}$$

Since $kp < n$, we have $q_{k-1} < n$ and therefore, Theorem 2.4.1 implies

$$\|u\|_q \leq C\|u\|_{1,q_{k-1}} \tag{2.4.15}$$

where $q = nq_{k-1}/(n-q_{k-1}) = np/(n-kp)$. (2.4.14) and (2.4.15) give the desired conclusion.

In order to treat the case $kp > n$, first assume $u \in C_0^\infty(\Omega)$ and for each $y \in \Omega$ use the Taylor expansion of u to obtain, with the notation of Section 1.1,

$$u(y) = P_x(y) + R_x(y)$$

where

$$P_x(y) = \sum_{|\alpha|=0}^{k-1} \frac{1}{\alpha!} D^\alpha u(x)(y-x)^\alpha$$

and

$$R_x(y) = k \sum_{|\alpha|=k} \frac{1}{\alpha!} \left[\int_0^1 (1-t)^{k-1} D^\alpha u((1-t)x + ty) dt \right] (y-x)^\alpha.$$

To estimate $|u(y)|$, note that

$$|K|\,|u(y)| \leq \int_K [|P_x(y)| + |R_x(y)|]\,dx \tag{2.4.16}$$

and employ Hölder's inequality to obtain

$$\begin{aligned}\int_K |P_x(y)| dx &\leq \int_K \left| \sum_{|\alpha|=0}^{k-1} \frac{1}{\alpha!} D^\alpha u(x)(y-x)^\alpha \right| dx \\ &\leq |K|^{1/p'} \sum_{|\alpha|=0}^{k-1} (\operatorname{diam} K)^{|\alpha|} \frac{1}{\alpha!} \|D^\alpha u\|_{p;K}. \end{aligned} \tag{2.4.17}$$

Similarly, to estimate the remainder term, we have

$$\begin{aligned}\int_K |R_x(y)| dx &\leq (\operatorname{diam}(K))^k k \sum_{|\alpha|=k} \frac{1}{\alpha!} \int_0^1 \int_K (1-t)^{k-1} \\ &\quad \cdot |D^\alpha u((1-t)x + ty)| dx dt \\ &\leq (\operatorname{diam}(K))^k k \sum_{|\alpha|=k} \frac{1}{\alpha!} \int_0^1 \int_{K_t} (1-t)^{k-1} (1-t)^{-n} \\ &\quad \cdot |D^\alpha u(z)| dz dt,\end{aligned}$$

where $K_t = T_t(K)$ and $T_t(x) = (1-t)x + ty$. Note that $|K_t| = (1-t)^n |K|$. Consequently, by Hölder's inequality and $kp > n$, we obtain

$$\begin{aligned}\int_K |R_x(y)| dx &\leq |K|^{1/p'} (\operatorname{diam}(K))^k k \sum_{|\alpha|=k} \frac{1}{\alpha!} \\ &\quad \int_0^1 (1-t)^{k-1} (1-t)^{-n} \|D^\alpha u\|_{p;K} (1-t)^{n/p'} dt \\ &\leq |K|^{1/p'} (\operatorname{diam}(K))^k k \sum_{|\alpha|=k} \frac{1}{\alpha!} \left(k - \frac{n}{p}\right)^{-1} \|D^\alpha u\|_{p;K},\end{aligned}$$

which, along with (2.4.16) and (2.4.17), establishes the desired inequality. If $u \in W_0^{k,p}(\Omega)$, let $\{u_i\}$ be a sequence of smooth functions converging to

u in $W_0^{k,p}(\Omega)$. The application of (2.4.1) to each u_i thus establishes the inequality for $u \in W_0^{k,p}(\Omega)$. To conclude that $u \in C^0(\overline{\Omega})$, apply (2.4.11) to the difference $u_i - u_j$ and obtain that $\{u_i\}$ is fundamental in the sup norm on $\overline{\Omega}$. □

2.4.3. Remark. An important case to consider in the previous two theorems is $\Omega = R^n$. In this situation, $W^{k,p}(R^n) = W_0^{k,p}(R^n)$ (see Exercise 2.1) and therefore the results apply to $W^{k,p}(R^n)$.

Observe that for $p > n$, the proof of Theorem 2.4.1 as well as that of Theorem 2.4.2 yields more than the fact that u is bounded. Indeed, u is Hölder continuous, which we state as a separate result.

2.4.4. Theorem. *If $u \in W_0^{1,p}(\Omega)$, $p > n$, then $u \in C^{0,\alpha}(\overline{\Omega})$, where $\alpha = 1 - n/p$.*

Proof. Assume $u \in C_0^1(\Omega)$ and select $x \in \Omega$. Let $B = B(x,r)$ be an arbitrary ball and choose $z \in B \cap \Omega$. Then,

$$|u(x) - u(z)| \leq \int_{\lambda_x(\theta)} |Du(r)|\chi_B(r)dH^{n-1}(r)$$

where $\lambda_x(\theta)$ is the ray whose end-point is x and passes through the point θ, $\theta \in S^{n-1}(x)$. Proceeding as in (2.4.5) and (2.4.6), we obtain

$$\omega(n-1)|u(x) - u(z)| \leq \|Du\|_p \left(\int_B |x - y|^{(1-n)p'} dy\right)^{1/p'} \tag{2.4.18}$$

But,

$$\left(\int_B |x - y|^{(1-n)p'} dy\right)^{1/p'} = (\gamma^{-1}\alpha(n))^{1/p'} r^{1-n/p}$$

where γ and $\alpha(n)$ are as in (2.4.8). Since the smooth functions are dense in $W_0^{1,p}(\Omega)$, we find that (2.4.18) holds for $u \in W_0^{1,p}(\Omega)$ and for almost all x, z. □

An interesting aspect of the Sobolev inequality is the limiting case $kp = n$. This will be considered separately in Chapter 2, Section 2.4.

2.5 The Rellich–Kondrachov Compactness Theorem

As a result of the inequalities proved in the previous section, it follows that the Sobolev spaces $W_0^{k,p}(\Omega)$ are continuously imbedded in $L^{p^*}(\Omega)$ where $p^* = np/(n - kp)$, if $kp < n$. In case $kp > n$, the imbedding is into the space $C^0(\overline{\Omega})$, and if $kp > n + mp$, it can easily be shown that the imbedding is into $C^m(\overline{\Omega})$. In this section it will be shown that the imbedding possesses a compactness property if we allow a slightly larger target space. Specifically, we will show that the injection map from $W_0^{k,p}(\Omega)$ into either $L^q(\Omega)$, $q < p^*$, or $C^m(\overline{\Omega})$ has the property that the closure of an arbitrary closed set in $W_0^{k,p}(\Omega)$ is compact in the range space. That is, the image sets are precompact. We recall here that a set S in a metric space is said to be totally bounded if for each $\varepsilon > 0$, there are a finite number of points in S such that the union of balls of radius ε with centers at these points contains S.

2.5.1. Theorem. *Let $\Omega \subset R^n$ be a bounded domain. Then, if $kp < n$ and $p \geq 1$, $W_0^{k,p}(\Omega)$ is compactly imbedded in $L^q(\Omega)$ where $q < np/(n-kp)$. If $kp > n + mp$, $W_0^{k,p}(\Omega)$ is compactly imbedded in $C^m(\overline{\Omega})$.*

Proof. Consider the first part of the theorem and let $B \subset W_0^{k,p}(\Omega)$ be a bounded set. We will show that $\overline{B}$ is a compact set in $L^q(\Omega)$. Since $C_0^\infty(\Omega)$ is dense in $W_0^{k,p}(\Omega)$, we may assume without loss of generality that $B \subset C_0^\infty(\Omega)$. For convenience, we will also assume that $\|u\|_{k,p;\Omega} \leq 1$ for all $u \in B$.

For $\varepsilon > 0$, let u_ε be the regularization of u. That is, $u_\varepsilon = u * \varphi_\varepsilon$ where φ_ε is the regularizer (see Section 1.6). If $u \in B$, then

$$\begin{aligned}|u_\varepsilon(x)| &\leq \int_{B(0,\varepsilon)} |u(x-y)|\varphi_\varepsilon(y)dy \\ &\leq \varepsilon^{-n} \sup \varphi \|u\|_1 \\ &\leq \varepsilon^{-n} \sup\{\varphi(y) : y \in R^n\},\end{aligned}$$

and

$$\begin{aligned}|Du_\varepsilon(x)| &\leq \int_{B(0,\varepsilon)} |u(x-y)|\,|D\varphi_\varepsilon(y)|dy \\ &\leq \varepsilon^{-n-1} \sup\{|D\varphi(y)| : y \in R^n\}\|u\|_1 \\ &\leq \varepsilon^{-n-1} \sup\{|D\varphi(y)| : y \in R^n\}.\end{aligned}$$

Therefore, if we let $B_\varepsilon = \{u_\varepsilon : u \in B\}$, it follows that B_ε is a bounded, equicontinuous subset of $C^0(\overline{\Omega})$. With the help of Arzela's theorem, it follows that B_ε is precompact in $L^1(\Omega)$. Next, observe that

$$\begin{aligned}|u(x) - u_\varepsilon(x)| &\leq \int_{B(0,\varepsilon)} |u(x) - u(x-y)|\varphi_\varepsilon(y)dy \\ &\leq \int_{B(0,\varepsilon)} \int_0^1 |Du \circ \gamma(t) \cdot \gamma'(t)|\varphi_\varepsilon(y)dtdy \\ &\leq \int_{B(0,\varepsilon)} \int_0^1 |Du(x-ty)|\,|y|\varphi_\varepsilon(y)dtdy\end{aligned}$$

where $\gamma(t) = t(x-y) + (1-t)x = x - ty$. Consequently, Fubini's theorem leads to

$$\int_{R^n} |u(x) - u_\varepsilon(x)| dx \leq \int_{B(0,\varepsilon)} \int_0^1 \int_{R^n} |Du(x-ty)|\, |y| \varphi_\varepsilon(y) dx dt dy$$
$$\leq \varepsilon \int_\Omega |Du| dx \leq \varepsilon.$$

Thus, B is contained within an ε-neighborhood of B_ε in $L^1(\Omega)$. Since B_ε is precompact in $L^1(\Omega)$ it is totally bounded. That is, for every $r > 0$, there exist a finite number of balls in $L^1(\Omega)$ of radius r whose union contains B_ε. Hence, B is totally bounded and therefore precompact in $L^1(\Omega)$. This establishes the theorem in case $q = 1$.

If $1 \leq q < np/(n-kp)$, refer to (1.5.13) to obtain

$$\|u\|_q \leq \|u\|_1^\lambda \|u\|_{np/(n-kp)}^{1-\lambda}$$

where

$$\lambda = \frac{1/q - (n-kp)/np}{1 - (n-kp)/np}.$$

Then, by Theorem (2.4.2)

$$\|u\|_q \leq C \|u\|_1^\lambda \|u\|_{k,p}^{1-\lambda}$$

which implies that bounded sets in $W_0^{k,p}(\Omega)$ are totally bounded in $L^q(\Omega)$ and therefore precompact.

The second part of the theorem follows immediately from Theorem 2.4.4 and Arzela's theorem in case $k = 1$. The general case follows from repeated applications of this and Theorem 2.4.1. □

2.5.2. Remark. The results of Sections 2.4 and 2.5 are stated in terms of functions in $W_0^{k,p}(\Omega)$. A natural and important question is to identify those domains Ω for which the results are valid for functions in $W^{k,p}(\Omega)$. One answer can be formulated in terms of those domains of Ω having the property that there exists a bounded linear operator

$$L : W^{k,p}(\Omega) \to W^{k,p}(R^n) \tag{2.5.1}$$

such that $L(u)|_\Omega = u$ for all $u \in W^{k,p}(\Omega)$. We say that Ω is an (k,p)-*extension domain* for $W^{k,p}(\Omega)$ if there exists an extension operator for $W^{k,p}(\Omega)$ with $1 \leq p \leq \infty$, k a non-negative integer. We will refer to this definition extensively in Chapter 4, and if the context makes it clear what indices k and p are under consideration, for brevity we will use the term extension domain rather than (k,p)-extension domain. Clearly, the results of the previous two sections are valid for $u \in W^{k,p}(\Omega)$ when Ω is a bounded extension domain. Indeed, by Lemma 2.3.1 there exists a function $f \in C_0^\infty(R^n)$ such that $f \equiv 1$ on Ω. Thus, if $u \in W^{k,p}(\Omega)$, then $f \cdot L(u) \in W_0^{k,p}(\Omega')$ where Ω' is some bounded domain containing spt f. It is now an easy matter to check that the results of the previous two sections are valid for the space $W^{k,p}(\Omega)$ by employing $W_0^{k,p}(\Omega')$.

A fundamental result of Calderón–Stein states that every Lipschitz domain is an extension domain. An open set Ω is a Lipschitz domain if its boundary can be locally represented as the graph of a Lipschitz function defined on some open ball of R^{n-1}. This result was proved by Calderón [CA1] when $1 < p < n$ and Stein [ST] extended Calderón's result to $p = 1, \infty$. Later, Jones [JO] introduced a class of domains that includes Lipschitz domains, called (ε, δ) domains, which he proved are extension domains for Sobolev functions. A domain Ω is called an (ε, δ) domain if whenever $x, y \in R^n$ and $|x-y| < \delta$, there is a rectifiable arc $\gamma \subset \Omega$ joining x to y and satisfying

$$\operatorname{length} \gamma \le \varepsilon^{-1}|x-y|$$

and

$$d(z, R^n - \Omega) \ge \frac{\varepsilon|x-z|\,|y-z|}{|x-y|} \quad \text{for all } z \text{ on } \gamma.$$

Among the interesting results he obtained is the following: If $\Omega \subset R^2$ is finitely connected, then Ω is an extension domain if and only if it is an (ε, δ) domain for some values of $\varepsilon, \delta > 0$.

2.6 Bessel Potentials and Capacity

In this section we introduce the notion of capacity which is critical in describing the appropriate class of null sets for the treatment of pointwise behavior of Sobolev functions which will be discussed in the following chapter. We will not attempt a complete development of capacity and non-linear potential theory which is closely related to the theory of Sobolev spaces, for these topics deserve a treatment that lies beyond the scope of this exposition. Instead, we will develop the basic properties of Bessel capacity and refer the reader to other sources for further information, cf. [HM], [ME1], [AD6].

The Riesz kernel, I_α, $0 < \alpha < n$, is defined by

$$I_\alpha(x) = \gamma(\alpha)^{-1}|x|^{\alpha-n}$$

where

$$\gamma(\alpha) = \frac{\pi^{n/2}2^\alpha\Gamma(\alpha/2)}{\Gamma(n/2-\alpha/2)}.$$

The Riesz potential of a function f is defined as the convolution

$$I_\alpha * f(x) = \frac{1}{\gamma(\alpha)}\int_{R^n}\frac{f(y)dy}{|x-y|^{n-\alpha}}.$$

The precise value of $\gamma(\alpha)$ is not important for our purposes except for the role it plays in the Riesz composition formula:

$$I_\alpha * I_\beta = I_{\alpha+\beta}, \quad \alpha > 0,\ \beta > 0,\ \alpha + \beta < n$$

cf. [ST, p. 118].

Observe that $I_\alpha * f$ is lower semicontinuous whenever $f \geq 0$. Indeed, if $x_i \to x$, then $|x_i - y|^{\alpha-n} f(y) \to |x-y|^{\alpha-n} f(y)$ for all $y \in R^n$, and lower semicontinuity thus follows from Fatou's lemma.

The Riesz potential leads to many important applications, but for the purpose of investigating Sobolev functions, the Bessel potential is more suitable. For an analysis of the Bessel kernel, we refer the reader to [ST, Chapter 5] or [DO, Part III] and quote here without proof the facts relevant to our development.

The Bessel kernel, g_α, $\alpha > 0$, is defined as that function whose Fourier transform is

$$\hat{g}_\alpha(x) = (2\pi)^{-n/2}(1 + |x|^2)^{-\alpha/2}$$

where the Fourier transform is

$$\hat{f}(x) = (2\pi)^{-n/2} \int e^{-ix\cdot y} f(y)dy. \tag{2.6.1}$$

It is known that g_α is a positive, integrable function which is analytic except at $x = 0$. Similar to the Riesz kernel, we have

$$g_\alpha * g_\beta = g_{\alpha+\beta}, \quad \alpha, \beta \geq 0. \tag{2.6.2}$$

There is an intimate connection between Bessel and Riesz potentials which is exhibited by g_α near the origin and infinity. Indeed, an analysis shows that for some $C > 0$,

$$g_\alpha(x) \sim C|x|^{(1/2)(\alpha-n-1)} e^{-|x|} \quad \text{as} \quad |x| \to \infty.$$

Here, $a(x) \sim b(x)$ means that $a(x)/b(x)$ is bounded above and below for all large $|x|$. Moreover, it can be shown that

$$g_\alpha(x) = \frac{|x|^{\alpha-n}}{\gamma(\alpha)} + o(|x|^{\alpha-n}) \quad \text{as} \quad |x| \to 0$$

if $0 < \alpha < n$. Thus, it follows for some constants C_1 and C_2, that

$$g_\alpha(x) \leq \frac{C_1}{|x|^{n-\alpha}} e^{-C_2|x|} \tag{2.6.3}$$

for all $x \in R^n$. Moreover, it also can be shown that

$$|Dg_\alpha(x)| \leq \frac{C_1}{|x|^{n-\alpha+1}} e^{-C_2|x|}. \tag{2.6.4}$$

From our point of view, one of the most interesting facts concerning Bessel potentials is that they can be employed to characterize the Sobolev spaces $W^{k,p}(R^n)$. This is expressed in the following theorem where we employ the notation

$$L^{\alpha,p}(R^n), \quad \alpha > 0,\ 1 \le p \le \infty$$

to denote all functions u such that

$$u = g_\alpha * f$$

for some $f \in L^p(R^n)$.

2.6.1. Theorem. *If k is a positive integer and $1 < p < \infty$, then*

$$L^{k,p}(R^n) = W^{k,p}(R^n).$$

Moreover, if $u \in L^{k,p}(R^n)$ with $u = g_\alpha * f$, then

$$C^{-1}\|f\|_p \le \|u\|_{k,p} \le C\|f\|_p$$

where $C = C(\alpha, p, n)$.

Remark. The equivalence of the spaces $L^{k,p}$ and $W^{k,p}$ fails when $p = 1$ or $p = \infty$.

It is also interesting to observe the following dissimilarity between Bessel and Riesz potentials. In view of the fact that $\|g_\alpha\|_1 \le C$, Young's inequality for convolutions implies

$$\|g_\alpha * f\|_p \le C\|f\|_p, \quad 1 \le p \le \infty. \tag{2.6.5}$$

On the other hand, we will see in Theorem 2.8.4 that the Riesz potential satisfies

$$\|I_\alpha * f\|_q \le C\|f\|_p, \quad p > 1 \tag{2.6.6}$$

where $q = np/(n - \alpha p)$. However, an inequality of type (2.6.6) is possible for only such q, cf. (Exercise 2.19), thus disallowing an inequality of type (2.6.5) for I_α and for every $f \in L^p$.

We now introduce the notion of capacity, which we develop in terms of the Bessel and Riesz potentials.

2.6.2. Definition. For $\alpha > 0$ and $p > 1$, the Bessel capacity is defined as

$$B_{\alpha,p}(E) = \inf\{\|f\|_p^p : g_\alpha * f \ge 1 \text{ on } E, f \ge 0\},$$

whenever $E \subset R^n$. In case $\alpha = 0$, we take $B_{\alpha,p}$ as Lebesgue measure. The Riesz capacity, $R_{\alpha,p}$, is defined in a similar way, with g_α replaced by I_α.

Since $g_\alpha(x) \leq I_\alpha(x)$, $x \in R^n$, it follows immediately from definitions that for $0 < \alpha < n$, $1 < p < n$, there exists a constant $C = C(\alpha, p, n)$ such that

$$R_{\alpha,p}(E) \leq C B_{\alpha,p}(E), \quad \text{whenever} \quad E \subset R^n. \tag{2.6.7}$$

Moreover, it can easily be shown that

$$R_{\alpha,p}(E) = 0 \quad \text{if and only if} \quad B_{\alpha,p}(E) = 0, \tag{2.6.8}$$

(Exercise 2.5).

We now give some elementary properties of capacity.

2.6.3. Lemma. *For $0 \leq \alpha < n$ and $1 < p < \infty$, the following hold:*

(i) $B_{\alpha,p}(\emptyset) = 0$,

(ii) *If $E_1 \subset E_2$, then $B_{\alpha,p}(E_1) \leq B_{\alpha,p}(E_2)$,*

(iii) *If $E_i \subset R^n$, $i = 1, 2, \ldots$, then*

$$B_{\alpha,p}\left(\bigcup_{i=1}^{\infty} E_i\right) \leq \sum_{i=1}^{\infty} B_{\alpha,p}(E_i).$$

Proof. (i) and (ii) are trivial to verify. For the proof of (iii), we may assume that $\sum_{i=1}^{\infty} B_{\alpha,p}(E_i) < \infty$. Since each term in the series is finite, for each $\varepsilon > 0$ there is a non-negative function $f_i \in L^p(R^n)$ such that

$$g_\alpha * f_i > 1 \quad \text{on} \quad E_i, \quad \|f_i\|_p < B_{\alpha,p}(E_i) + 2^{-i}\varepsilon.$$

Let $f(x) = \sup\{f_i(x) : i = 1, 2, \ldots\}$. Clearly, $g_\alpha * f \geq 1$ on $\cup_{i=1}^{\infty} E_i$ and $f(x)^p \leq \sum_{i=1}^{\infty} f_i(x)^p$. Therefore,

$$B_{\alpha,p}\left(\bigcup_{i=1}^{\infty} E_i\right) \leq \|f\|_p \leq \sum_{i=1}^{\infty} \|f_i\|_p \leq \sum_{i=1}^{\infty} B_{\alpha,p}(E_i) + \varepsilon. \qquad \square$$

Another useful characterization of capacity is the following:

$$B_{\alpha,p}(E) = \inf_f \{\inf_{x \in E} g_\alpha * f(x)\}^{-p} = \{\sup_f \inf_{x \in E} g_\alpha * f(x)\}^{-p} \tag{2.6.9}$$

where $f \in L^p(R^n)$, $f \geq 0$ and $\|f\|_p \leq 1$ (Exercise 2.4).

Although Lemma 2.6.3 states that $B_{\alpha,p}$ is an outer measure, it is fruitless to attempt a development in the context of measure theory because it can be shown that there is no adequate supply of measurable sets. Rather, we will establish other properties that show that the appropriate context for

$B_{\alpha,p}$ is the theory of capacity, as developed by Brelot, Choquet, [BRT], [CH].

2.6.4. Lemma. *If $\{f_i\}$ is a sequence in $L^p(R^n)$ such that $\|f_i - f\|_p \to 0$ as $i \to \infty$, $p > 1$, then there is a subsequence $\{f_{i_j}\}$ such that*

$$g_\alpha * f_{i_j}(x) \to g_\alpha * f(x)$$

for $B_{\alpha,p}$-q.e. $x \in R^n$.

(We employ the time-honored convention of stating that a condition holds $B_{\alpha,p}$-q.e., an abbreviation for $B_{\alpha,p}$-quasi everywhere, if it holds at all points except possibly for a set of $B_{\alpha,p}$-capacity zero.)

Proof. It follows easily from the definition of $B_{\alpha,p}$ capacity that if $f \in L^p(R^n)$, then $|g_\alpha * f(x)| < \infty$ for $B_{\alpha,p}$-q.e. $x \in R^n$. Thus, for $\varepsilon > 0$,

$$\begin{aligned} B_{\alpha,p}(\{x : |g_\alpha * f_i(x) - g_\alpha * f(x)| \geq \varepsilon\} &= B_{\alpha,p}(\{x : |g_\alpha * (f_i - f)(x)| \geq \varepsilon\}) \\ &\leq \varepsilon^{-p}\|f_i - f\|_p^p. \end{aligned}$$

Consequently, there exists a subsequence $\{f_{i_j}\}$ and a sequence of sets E_j such that

$$|g_\alpha * f_{i_j}(x) - g_\alpha * f(x)| \leq j^{-1}, \quad x \in R^n - E_j,$$

with

$$B_{\alpha,p}(E_j) \leq \varepsilon 2^{-j}.$$

Hence, $g_\alpha * f_{i_j} \to g_\alpha * f$ uniformly on $R^n - \cup_{j=1}^\infty E_j$, where $B_{\alpha,p}\left(\cup_{j=1}^\infty E_j\right) \leq \varepsilon$. Now a standard diagonalization process yields the conclusion. □

2.6.5. Lemma. *If $\{f_i\}$ is a sequence in $L^p(R^n)$, $p > 1$, such that $f_i \to f$ weakly in $L^p(R^n)$, then*

$$\liminf_{i\to\infty} g_\alpha * f_i(x) \leq g_\alpha * f(x) \leq \limsup_{i\to\infty} g_\alpha * f_i(x) \tag{2.6.10}$$

for $B_{\alpha,p}$-q.e. $x \in R^n$. If in addition, it is assumed that each $f_i \geq 0$, then

$$g_\alpha * f(x) \leq \liminf_{i\to\infty} g_\alpha * f_i(x) \quad \textit{for} \quad x \in R^n \tag{2.6.11}$$

and

$$g_\alpha * f(x) = \liminf_{i\to\infty} g_\alpha * f_i(x) \tag{2.6.12}$$

for $B_{\alpha,p}$-q.e. $x \in R^n$.

Proof. Under the assumption that $f_i \to f$ weakly in $L^p(R^n)$, by the Banach–Saks theorem there exists a subsequence of $\{f_i\}$ (which will be denoted by the full sequence) such that

$$g_i = i^{-1} \sum_{j=1}^{i} f_j$$

converges strongly in $L^p(R^n)$ to f. Lemma 2.6.4 thus yields a subsequence of $\{g_i\}$ (denoted by the full sequence) such that

$$g_\alpha * f(x) = \lim_{i\to\infty} g_\alpha * g_i(x)$$

for $B_{\alpha,p}$-q.e. $x \in R^n$. However, for each $x \in R^n$,

$$\liminf_{i\to\infty} g_\alpha * f_i(x) \le \lim_{i\to\infty} g_\alpha * g_i(x),$$

which establishes the first inequality in (2.6.10). The second part of (2.6.10) follows from the first by replacing f_i and f by $-f_i$ and $-f$ respectively.

In the complement of any ball, B, containing the origin, $\|g_\alpha\|_{p';R^n-B} \le \infty$, by (2.6.3). Thus, (2.6.11) follows from the weak convergence of f_i to f. (2.6.12) follows from (2.6.11) and (2.6.10). □

2.6.6. Lemma. *For every set $E \subset R^n$*

$$B_{\alpha,p}(E) = \inf\{B_{\alpha,p}(U) : U \supset E, U \text{ open}\}.$$

Proof. Since g_α is continuous away from the origin, the proof of the lower semicontinuity of $g_\alpha * f$ when $f \ge 0$ is similar to that for the Riesz potential given at the beginning of this section. The lemma follows immediately from this observation. □

The lemma states that $B_{\alpha,p}$ is outer regular. To obtain inner regularity on a large class of sets, we will require the following continuity properties of $B_{\alpha,p}$.

2.6.7. Theorem. *If $\{E_i\}$ is a sequence of subsets of R^n, then*

$$B_{\alpha,p}\left(\liminf_{i\to\infty} E_i\right) \le \liminf_{i\to\infty} B_{\alpha,p}(E_i). \tag{2.6.13}$$

If $E_1 \subset E_2 \subset \ldots$, then

$$B_{\alpha,p}\left(\bigcup_{i=1}^{\infty} E_i\right) = \lim_{i\to\infty} B_{\alpha,p}(E_i). \tag{2.6.14}$$

If $K_1 \supset K_2 \supset \ldots$ are compact sets, then

$$B_{\alpha,p}\left(\bigcap_{i=1}^{\infty} K_i\right) = \lim_{i\to\infty} B_{\alpha,p}(K_i). \tag{2.6.15}$$

Proof. For the proof of (2.6.14) assume that the limit is finite and let f_i be a non-negative function in $L^p(R^n)$ such that $g_\alpha * f_i \geq 1$ on E_i with

$$\|f_i\|_p^p < B_{\alpha,p}(E_i) + 1/i. \tag{2.6.16}$$

Since $\|f_i\|_p^p$ is a bounded sequence of real numbers, Theorem 1.5.2 asserts the existence of $f \in L^p(R^n)$ and a subsequence of $\{f_i\}$ that converges weakly to f. Hence, (2.6.12) implies that there exists a set $B \subset E = \cup_{i=1}^\infty E_i$ with $B_{\alpha,p}(E - B) = 0$ such that $g_\alpha * f \geq 1$ on B. Therefore,

$$\begin{aligned} B_{\alpha,p}(E) = B_{\alpha,p}(B) &\leq \|f\|_p^p \\ &\leq \liminf_{i\to\infty} \|f_i\|_p^p \\ &\leq \lim_{i\to\infty} B_{\alpha,p}(E_i), \end{aligned}$$

from (2.6.16). If

$$A_i = \bigcup_{j=1}^{i} \bigcap_{k=j}^{\infty} E_k,$$

then $\{A_i\}$ is an ascending sequence of sets whose union equals lim inf E_i. Therefore, since $A_i \subset E_i$ for $i \geq 1$, (2.6.14) implies (2.6.13) because

$$\begin{aligned} B_{\alpha,p}\left(\liminf_{i\to\infty} E_i\right) &= B_{\alpha,p}\left(\bigcup_{i=1}^{\infty} A_i\right) \\ &= \lim_{i\to\infty} B_{\alpha,p}(A_i) \\ &\leq \liminf_{i\to\infty} B_{\alpha,p}(E_i). \end{aligned}$$

Finally, it $\{K_i\}$ is a descending sequence of compact sets, Lemma 2.6.6 provides an open set $U \supset \cap_{i=1}^\infty K_i$ such that

$$B_{\alpha,p}(U) < B_{\alpha,p}\left(\bigcap_{i=1}^{\infty} K_i\right) + \varepsilon$$

for an arbitrarily chosen $\varepsilon > 0$. However, $K_i \subset U$ for all sufficiently large i and consequently $B_{\alpha,p}(K_i) \leq B_{\alpha,p}(U)$. (2.6.15) is now immediate and the proof is complete. □

(2.6.14) states that $B_{\alpha,p}$ is left-continuous on arbitrary sets whereas (2.6.15) states that $B_{\alpha,p}$ is right continuous on compact sets. The importance of these two facts is seen in a fundamental result of Choquet [CH, Theorem 1] which we state without proof.

2.6.8. Theorem. *Let C be a non-negative set function defined on the Borel sets in R^n with the following properties:*

(i) $C(\emptyset) = 0$,

(ii) *If* $B_1 \subset B_2$ *are Borel sets, then* $C(B_1) \leq C(B_2)$,

(iii) *If* $\{B_i\}$ *is a sequence of Borel sets, then* $C\left(\cup_{i=1}^{\infty} B_i\right) \leq \sum_{i=1}^{\infty} C(B_i)$,

(iv) C *is left continuous on arbitrary sets and right continuous on compact sets.*

Then, for any Suslin set $A \subset R^n$,

$$\sup\{C(K) : K \subset A, K \textit{ compact}\} = \inf\{C(U) : U \supset A, U \textit{ open}\}.$$

Any set A for which the conclusion of the theorem applies is called C-capacitable. In view of Lemma 2.6.3 and Theorem 2.6.8, the following is immediate.

2.6.9. Corollary. *All Suslin sets are* $B_{\alpha,p}$*-capacitable.*

The usefulness of Theorem 2.6.8 and its attending corollary is quite clear, for it reduces many questions concerning capacity to the analysis of its behavior on compact sets.

We now introduce what will eventually result in an equivalent formulation of Bessel capacity.

2.6.10. Definition. For $1 < p < \infty$, and $E \subset R^n$ a Suslin set, let $\mathcal{M}(E)$ denote the class of Radon measures μ on R^n such that $\mu(R^n - E) = 0$. We define

$$b_{\alpha,p}(E) = \sup\{\mu(R^n)\} \tag{2.6.17}$$

where the supremum is taken over all $\mu \in \mathcal{M}(E)$ such that

$$\|g_\alpha * \mu\|_{p'} \leq 1. \tag{2.6.18}$$

Clearly,

$$b_{\alpha,p}(E) = (\inf\{\|g_\alpha * \nu\|_{p'}\})^{-1} \tag{2.6.19}$$

where the infimum is taken over all $\nu \in \mathcal{M}(E)$ with $\nu(R^n) = 1$. We have that

$$\begin{aligned} \|g_\alpha * \nu\|_{p'} &= \sup\left\{\int_{R^n} g_\alpha * \nu \cdot f\,dx : f \geq 0, \|f\|_p \leq 1\right\} \\ &= \sup\left\{\int_{R^n} g_\alpha * f\,d\nu : f \geq 0, \|f\|_p \leq 1\right\}, \end{aligned}$$

and thus obtain

$$b_{\alpha,p}(E)^{-1} = \left(\inf_\nu \sup_f \int g_\alpha * f\,d\nu\right) \tag{2.6.20}$$

where $\nu \in \mathcal{M}(E)$, $\nu(R^n) = 1$, and $f \geq 0$ with $\|f\|_p \leq 1$.

Recall from (2.6.9) that if $E \subset R^n$, then

$$B_{\alpha,p}(E) = \{\sup_f \inf_{x\in E} g_\alpha * f(x)\}^{-p}$$

where $f \in L^p(R^n)$, $f \geq 0$ and $\|f\|_p \leq 1$. By considering measures concentrated at points, this is easily seen to be

$$B_{\alpha,p}(E)^{-1/p} = \sup_f \inf_\nu \int g_\alpha * f \, d\nu \tag{2.6.21}$$

where f and ν are the same as in (2.6.20).

We would like to conclude that there is equality between (2.6.20) and (2.6.21). For this purpose, assume $E \subset R^n$ is a compact set and let

$$F(f,\nu) = \int g_\alpha * f \, d\nu \tag{2.6.22}$$

where $f \in L^p(R^n)$, $f \geq 0$, $\|f\|_p \leq 1$ and $\nu \in \mathcal{M}(E)$, $\nu(R^n) = 1$. Clearly F is linear in each variable and is lower semicontinuous in ν relative to weak convergence. Since the spaces in which f and ν vary are compact we may apply the following minimax theorem, which we state without proof, to obtain our conclusion, [FA].

Minimax Theorem. *Let X be a compact Hausdorff space and Y an arbitrary set. Let F be a real-valued function on $X \times Y$ such that, for every $y \in Y$, $F(x,y)$ is lower semicontinuous on X. If F is convex on X and concave on Y, then*

$$\inf_{x\in X} \sup_{y\in Y} F(x,y) = \sup_{y\in Y} \inf_{x\in X} F(x,y).$$

We thus obtain the following result.

2.6.11. Lemma. *If $K \subset R^n$ is compact, then*

$$[b_{\alpha,p}(K)]^p = B_{\alpha,p}(K). \tag{2.6.23}$$

Our next task is to extend (2.6.23) to a more general class of sets. For this purpose, observe that if $E \subset R^n$ is a Suslin set, then

$$b_{\alpha,p}(E) = \sup\{b_{\alpha,p}(K) : K \subset E, K \text{ compact}\}. \tag{2.6.24}$$

To see this, for each Suslin set E, let $\mu \in \mathcal{M}(E)$ with $\|g_\alpha * \mu\|_{p'} \leq 1$. If $K \subset E$ is compact, then $\nu = \mu \lfloor K$ has the property that $\nu \in \mathcal{M}(E)$ with $\|g_\alpha * \nu\|_{p'} \leq 1$. Since μ is a regular measure, we have

$$\mu(E) = \sup\{\mu(K) : K \subset E, K \text{ compact}\},$$

and therefore

$$b_{k,p}(E) = \sup\{b_{k,p}(K) : K \subset E, K \text{ compact}\}. \tag{2.6.25}$$

From (2.6.25), (2.6.23), and Corollary 2.6.9 we conclude the following.

2.6.12. Theorem. *If $E \subset R^n$ is a Suslin set, then*

$$[b_{\alpha,p}(E)]^p = B_{\alpha,p}(E).$$

Thus far, we have developed the set-theoretic properties of $B_{\alpha,p}$. We now will investigate its metric properties.

2.6.13. Theorem. *For $p > 1$, $\alpha p < n$, there exists a constant $C = C(\alpha, p, n)$ such that*

$$C^{-1} r^{n-\alpha p} \leq B_{\alpha,p}[B(x,r)] \leq C r^{n-\alpha p}$$

whenever $x \in R^n$ and $0 < r \leq 1/2$.

Proof. Without loss of generality, we will prove the theorem only for $B(0,r)$ and write $B(r) = B(0,r)$. Let $f \in L^p(R^n)$, $f \geq 0$, have the property that

$$g_\alpha * f \geq 1 \quad \text{on} \quad B(2). \tag{2.6.26}$$

By a change of variable, this implies

$$\int_{R^n} g_\alpha\left(\frac{x-y}{r}\right) f\left(\frac{y}{r}\right) r^{-n} dy \geq 1 \tag{2.6.27}$$

for $x \in B(2r)$. From (2.6.3) and (2.6.4), there exists $C = C(\alpha, p, n)$ such that

$$C^{-1}|x-y|^{\alpha-n} e^{-2|x-y|} \leq g_\alpha(x-y) \leq C|x-y|^{\alpha-n} e^{-|x-y|},$$

and therefore

$$\begin{aligned} g_\alpha\left(\frac{x-y}{r}\right) &\leq C|x-y|^{\alpha-n} r^{n-\alpha} e^{-|x-y|r^{-1}} \\ &\leq C|x-y|^{\alpha-n} r^{n-\alpha} e^{-2|x-y|} \qquad (r \leq 1/2) \\ &\leq C^2 r^{n-\alpha} g_\alpha(x-y) \qquad (r \leq 1/2). \end{aligned}$$

Consequently, from (2.6.27),

$$C^2 \int_{R^n} g_\alpha(x-y) f\left(\frac{y}{r}\right) r^{-\alpha} dy \geq 1 \quad \text{for} \quad x \in B(2r), \qquad (r \leq 1/2).$$

However,

$$\int_{R^n} \left[C^2 r^{-\alpha} f\left(\frac{y}{r}\right)\right]^p dy = C^{2p} r^{n-\alpha p} \|f\|_p^p.$$

Hence,

$$B_{\alpha,p}[B(2r)] \leq C^{2p} r^{n-\alpha p} \|f\|_p^p, \qquad r \leq 1/2,$$

for every $f \in L^p(R^n)$ satisfying (2.6.26). Thus,

$$B_{\alpha,p}[B(2r)] \leq C^{2p} r^{n-\alpha p} B_{\alpha,p}[B(2)], \qquad r \leq 1/2,$$

from which the conclusion follows.

For the proof of the first inequality of the theorem, let $f \in L^p(R^n)$, $f \geq 0$, be such that $g_\alpha * f \geq 1$ on $B(r)$. Then

$$|B(r)| \leq \int_{B(r)} g_\alpha * f \, dx \leq |B(r)|^{1/q'} \|g_\alpha * f\|_q,$$

where $q = p^* = np/(n - \alpha p)$. It follows from (2.6.3) that $g_\alpha \leq C I_\alpha$. Because there is no danger of a circular argument, we employ the Sobolev inequality for Riesz potentials (Theorem 2.8.4) to obtain

$$r^{n-\alpha p} \leq C \|f\|.$$

Taking the infimum over all such f establishes the desired inequality. □

The case $\alpha p \geq n$ requires special treatment.

2.6.14. Theorem. *If $p > 1$, $\alpha p = n$ and $0 < \overline{r} < 1$, there exists $C = C(n, \overline{r})$ such that*

$$C^{-1}(\log r^{-1})^{1-p} \leq B_{\alpha,p}[B(x,r)] \leq C(\log r^{-1})^{1-p}$$

whenever $0 < r \leq \overline{r} < 1$ and $x \in R^n$.

Proof. As in the proof of the previous theorem, it suffices to consider only the case $x = 0$. Let μ be a Radon measure such that $\mu[R^n - B(r)] = 0$ and $\|g_\alpha * \mu\|_{p'} \leq 1$, where we write $B(r) = B(x,r)$. Because of the similarity between the Riesz and Bessel kernels discussed at the beginning of this section, there exists a constant C independent of r such that

$$\int_{B(1)} (I_\alpha * \mu)^{p'} dx \leq C \int_{R^n} (g_\alpha * \mu)^{p'} dx \leq C.$$

If $|y| \leq r$ and $|x| \geq r$, then $|x - y| \leq |x| + |y| \leq |x| + r \leq 2|x|$ and therefore

$$\begin{aligned} C &\geq \int_{r \leq |x| \leq 1} (I_\alpha * \mu)^{p'} dx = \int_{r \leq |x| \leq 1} \left(\int_{R^n} |x - y|^{\alpha - n} d\mu(y)\right)^{p'} dx \\ &\geq C_1 [\mu(R^n)]^{p'} \int_{r \leq |x| \leq 1} |x|^{-n} dx \\ &= C_1 [\mu(R^n)]^{p'} [\log r^{-1}]. \end{aligned}$$

Thus, by Theorem 2.6.12, it follows that

$$B_{\alpha,p}[B(r)] \le C(\log r^{-1})^{1-p}.$$

To establish the opposite inequality, let λ_r denote the restriction of Lebesgue measure to $B(r)$. Since $g_\alpha \le I_\alpha$, we have

$$g_\alpha * \lambda_r(x) \le C \int_{B(r)} |x-y|^{\alpha-n}\,dy. \tag{2.6.28}$$

If $|x| \le r/\overline{r}$, $|y| \le r$, then $|x-y| \le cr$ where $c(\overline{r}) = 1 + 1/\overline{r}$. That is, $B(r) \subset B(x, cr)$. Therefore,

$$\begin{aligned}\int_{B(r)} |x-y|^{\alpha-n}\,dy &\le \int_{B(x,cr)} |x-y|^{\alpha-n}\,dy \\ &\le C(\overline{r})r^\alpha\end{aligned}$$

which, by (2.6.28), implies

$$g_\alpha * \lambda_r(x) \le C(\overline{r})r^\alpha \quad \text{if} \quad |x| \le r/\overline{r}. \tag{2.6.29}$$

If $|y| \le r$ and $r/\overline{r} < |x| \le 1$, then $|x-y| \ge |x| - |y| \ge |x| - r > c(\overline{r})|x|$, where now $c(\overline{r}) = 1 - \overline{r}$. Hence,

$$g_\alpha * \lambda_r(x) \le C \int_{B(r)} |x-y|^{\alpha-n}\,dy \le C_1 r^n |x|^{\alpha-n} \text{ if } r/\overline{r} < |x| \le 1. \tag{2.6.30}$$

If $|x| > 1$, then (2.6.3) yields

$$g_\alpha * \lambda_r(x) \le C r^n e^{-|x|}. \tag{2.6.31}$$

Thus, (2.6.29), (2.6.30), and (2.6.31) yield

$$\|g_\alpha * \lambda_r\|_{p'} \le C r^n (\log r^{-1})^{1/p'}.$$

Appealing again to Theorem 2.6.12, we establish the desired result. □

2.6.15. Remark. In case $\alpha p > n$, it is not difficult to show that there is a constant $C = C(\alpha, p, n)$ such that

$$B_{\alpha,p}(E) \ge C$$

whenever $E \ne \emptyset$. See Exercise 2.6.

Because $B_{\alpha,p}[B(x,r)] \approx r^{n-\alpha p}$ one would expect that Bessel capacity and Hausdorff measure are related. This is indeed the case as seen by the following theorem that we state without proof, [ME1], [HM]. See Exercises 2.15 and 2.16.

2.6.16. Theorem. *If $p > 1$ and $\alpha p \le n$, then $B_{\alpha,p}(E) = 0$ if $H^{n-\alpha p}(E) < \infty$. Conversely, if $B_{\alpha,p}(E) = 0$, then $H^{n-\alpha p+\varepsilon}(E) = 0$ for every $\varepsilon > 0$.*

2.7 The Best Constant in the Sobolev Inequality

There is a fundamental relationship between the classical isoperimetric inequality for subsets of Euclidean space and the Sobolev inequality in the case $p = 1$. Indeed, it was shown in [FF] that the former implies the latter and, as we shall see in Remark 2.7.5 below, the converse is easily seen to hold.

We will give a method that gives the best constant in the Sobolev Inequality (Theorem 2.4.1), by employing an argument that depends critically on a suitable interpretation of the total variation for functions of several variables. This is presented in Theorem 2.7.1 and equality (2.7.1) if referred to as the co-area formula. This is a very useful tool in analysis that has seen many applications. We will give a proof for only smooth functions but this will be sufficient for our purposes.

2.7.1. Theorem. *Let* $u \in C_0^n(R^n)$. *Then*

$$\int_\Omega |Du|dx = \int_{-\infty}^{+\infty} H^{n-1}[u^{-1}(t) \cap \Omega]dt. \tag{2.7.1}$$

Before giving the proof of this theorem, let us first consider some of its interpretations. In case $n = 1$, the integrand on the right-hand side involves Hausdorff 0-dimensional measure, H^0. $H^0(E)$ is merely the number of points (including ∞) in E and thus, the integrand on the right side of (2.7.1) gives the number of points in the set $u^{-1}(t) \cap \Omega$. This is equivalent to the number of times the graph of u, when considered as a subset of $R^2 = \{(x, y)\}$, intersects the line $y = t$. In this case (2.7.1) becomes

$$\int_\Omega |u'|dx = \int N(y)dy \tag{2.7.2}$$

where $N(y)$ denotes the number of points in $u^{-1}(y) \cap \Omega$. (2.7.2) is known as the Banach Indicatrix formula, [SK, p. 280].

The Morse–Sard Theorem [MSE1], [SA], states that a real-valued function u of class C^n defined on R^n has the property that $H^1[u(N)] = 0$ where $N = \{x : Du(x) = 0\}$. For example, if we consider a function $u \in C_0^2(R^2)$, an application of the Implicit Function theorem implies that $u^{-1}(t) \cap \Omega$ is a 1-dimensional class C^2 manifold for a.e. t. In this case, $H^1[u^{-1}(t) \cap \Omega]$ is the length of the curve obtained by intersecting the graph of u in R^3 by the hyperplane $z = t$. Thus, the variation of u, $\int_\Omega |Du|dx$, is obtained by integrating the length of the curves, $\Omega \cap u^{-1}(t)$, with respect to t.

The co-area formula is known to be valid for Lipschitzian functions. (We will see in Chapter 5, that another version is valid for BV functions.) The proof in its complete generality requires a delicate argument from geometric measure theory that will not be given here. The main obstacle in the proof is to show that if u is Lipschitz, then

$$\int_{R^1} H^{n-1}[u^{-1}(t) \cap N]dt = 0$$

where $N = \{x : D(x) = 0\}$. Once this has been established, the remainder of the proof follows from standard arguments. Because our result assumes that $u \in C^n$, we avoid this difficulty by appealing to the Morse–Sard theorem referred to above. In preparation for the proof, we first require the following lemma.

2.7.2. Lemma. *If $U \subset R^n$ is a bounded, open set with C^2 boundary, then*

$$\sup\left\{\int_\Omega \operatorname{div}\varphi\, dx : \varphi \in C_0^1(R^n; R^n), \sup|\varphi| \leq 1\right\} = H^{n-1}[\partial U].$$

Proof. By the Gauss–Green theorem,

$$\int_U \operatorname{div}\varphi\, dx = \int_{\partial U} \varphi(x)\cdot\nu(x) dH^{n-1}(x)$$

where ν is the unit exterior normal. Hence

$$\sup\left\{\int_U \operatorname{div}\varphi\, dx : \varphi \in C_0^1(R^n; R^n), \sup|\varphi| \leq 1\right\} \leq H^{n-1}(\partial U).$$

To prove the opposite inequality, note that ν is a C^1 vector field of unit length defined on ∂U and so may be extended to a C^1 vector field V defined on R^n such that $|V(x)| \leq 1$ for all $x \in R^n$, cf. Theorem 3.6.2. If $\psi \in C_0^\infty(R^n)$ and $|\psi| \leq 1$, then with $\varphi = \psi V$, we have

$$\int_U \operatorname{div}\varphi\, dx = \int_{\partial U} \psi(y) dH^{n-1}(y)$$

so that

$$\sup\left\{\int_U \operatorname{div}\varphi : \varphi \in C_0^1(R^n; R^n), \sup|\varphi| \leq 1\right\}$$

$$\geq \sup\left\{\int_{\partial U} \psi\, dH^{n-1} : \psi \in C_0^\infty(R^n), \sup|\psi| \leq 1\right\} = H^{n-1}(\partial U). \quad \square$$

Proof of Theorem 2.7.1. We first consider linear maps $L: R^n \to R^1$. Then there exists an orthogonal transformation $f: R^n \to R^n$ and a nonsingular transformation g such that $f(N^\perp) = R^1$, $f(N) = R^{n-1}$, ($N = \ker L$) and

$$L = g \circ p \circ f$$

where $p: R^n \to R^1$ is the projection. For each $y \in R^1$, $p^{-1}(y)$ is a hyperplane that is a translate of the subspace $p^{-1}(0)$. The inverse images $p^{-1}(y)$ decompose R^n into parallel $(n-1)$-dimensional slices and an easy application of Fubini's theorem yields

$$|E| = \int_{R^1} H^{n-1}[E \cap p^{-1}(y)]dy \tag{2.7.3}$$

whenever E is a measurable subset of R^n. Therefore

$$\begin{aligned} |f(E)| = |E| &= \int_{R^1} H^{n-1}[E \cap p^{-1}(y)]dy \\ &= \int_{R^1} H^{n-1}[f(E) \cap p^{-1}(y)]dy \\ &= \int_{R^1} H^{n-1}[E \cap f^{-1}(p^{-1}(y))]dy. \end{aligned}$$

Now use the change of variables $z = g(y)$ and observe that the last integral above becomes

$$\begin{aligned} |g'|\,|E| &= \int_{R^1} H^{n-1}[E \cap f^{-1}(p^{-1}(g^{-1}(z)))]dz \\ &= \int_{R^1} H^{n-1}[E \cap L^{-1}(z)]dz. \end{aligned} \tag{2.7.4}$$

But $|g'| = |DL|$ and thus (2.7.4) establishes Theorem 2.7.1 for linear maps.

We now proceed to prove the result for general u as stated in the theorem. Let $N = \{x : Du(x) = 0\}$ and for each $t \in R^1$, let

$$E_t = R^n \cap \{x : u(x) > t\}$$

and define a function $f_t : R^n \to R^1$ by

$$f_t = \begin{cases} \chi_{E_t} & \text{if } t \geq 0 \\ -\chi_{R^n - E_t} & \text{if } t < 0. \end{cases}$$

Thus,

$$u(x) = \int_{R^1} f_t(x)dt, \quad x \in R^n.$$

Now consider a test function $\varphi \in C_0^\infty(R^n - N)$, such that $\sup|\varphi| \leq 1$. Then, by Fubini's theorem,

$$\begin{aligned} \int_{R^n} u(x)\varphi(x)dx &= \int_{R^n}\int_{R^1} f_t(x)\varphi(x)dtdx \\ &= \int_{R^1}\int_{R^n} f_t(x)\varphi(x)dxdt. \end{aligned} \tag{2.7.5}$$

Now (2.7.5) remains valid if φ is replaced by any one of its first partial derivatives. Since $Du \neq 0$ in the open set $R^n - N$, the Implicit Function theorem implies that $u^{-1}(t) \cap (R^n - N)$ is an $(n-1)$-manifold of class C^n. In addition, since $\operatorname{spt}\varphi \subset R^n - N$, it follows from the Divergence theorem that

$$\int_{E_t} \operatorname{div}\varphi\, dx = \int_{(\partial E_t)\cap(R^n - N)} \varphi(x)\cdot\nu(x)dH^{n-1}(x).$$

Therefore, if φ is now taken as $\varphi \in C_0^\infty(R^n - N; R^n)$ with $\sup|\varphi| \leq 1$, we have

$$\begin{aligned}
-\int_{R^n} Du\cdot\varphi\, dx &= \int_{R^n} u\cdot \operatorname{div}\varphi\, dx = \int_{R^1}\int_{E_t} \operatorname{div}\varphi\, dxdt \\
&= \int_{R^1}\int_{(R^n-N)\cap\partial E_t} \varphi(x)\cdot\nu(x)dH^{n-1}(x)dt \\
&\leq \int_{R^1} H^{n-1}[(R^n - N)\cap u^{-1}(t)]dt \\
&\leq \int_{R^1} H^{n-1}[u^{-1}(t)]dt. \qquad (2.7.6)
\end{aligned}$$

However, the sup of (2.7.6) over all such φ equals

$$\int_{R^n - N} |Du|dx = \int_{R^n} |Du|dx.$$

In order to prove the opposite inequality, let $L_k\colon R^n \to R^1$ be piecewise linear maps such that

$$\lim_{k\to\infty}\int_{R^n} |L_k - u|dx = 0 \qquad (2.7.7)$$

and

$$\lim_{k\to\infty}\int_{R^n} |DL_k|dx = \int_{R^n} |Du|dx. \qquad (2.7.8)$$

Let

$$E_t^k = R^n \cap \{x : L_k(x) > t\},$$
$$\chi_t^k = \chi_{E_t^k}.$$

From (2.7.7) it follows that there is a countable set $S \subset R^1$ such that

$$\lim_{k\to\infty}\int_{R^n} |\chi_t - \chi_t^k|dx = 0 \qquad (2.7.9)$$

whenever $t \notin S$. By the Morse–Sard theorem and the Implicit Function theorem, we have that $u^{-1}(t)$ is a closed manifold of class C^n for all $t \in R^1 - T$ where $H^1(T) = 0$. Redefine the set S to also include T. Thus, for

$t \notin S$, and $\varepsilon > 0$, refer to Lemma 2.7.2 to find $\varphi \in C_0^\infty(R^n; R^n)$ such that $|\varphi| \le 1$ and

$$H^{n-1}[u^{-1}(t)] - \int_{E_t} \operatorname{div} \varphi \, dx < \frac{\varepsilon}{2}. \tag{2.7.10}$$

Let $M = \int_{R^n} |\operatorname{div} \varphi| dx$ and choose k_0 such that for $k \ge k_0$,

$$\int_{R^n} |\chi_t - \chi_t^k| dx < \frac{\varepsilon}{2M}.$$

For $k \ge k_0$,

$$\left| \int_{E_t} \operatorname{div} \varphi \, dx - \int_{E_t^k} \operatorname{div} \varphi \, dx \right| \le M \int_{R^n} |\chi_t - \chi_t^k| dx < \frac{\varepsilon}{2}. \tag{2.7.11}$$

Therefore, from (2.7.10) and (2.7.11)

$$\begin{aligned} H^{n-1}[u^{-1}(t)] &\le \int_{E_t} \operatorname{div} \varphi \, dx + \frac{\varepsilon}{2} \\ &\le \int_{E_t^k} \operatorname{div} \varphi \, dx + \varepsilon \\ &= \int_{\partial E_t^k} \varphi \cdot \nu \, dH^{n-1} + \varepsilon \\ &\le H^{n-1}[L_k^{-1}(t)] + \varepsilon. \end{aligned}$$

Thus, for $t \notin S$,

$$H^{n-1}[u^{-1}(t)] \le \liminf_{k \to \infty} H^{n-1}[L_k^{-1}(t)].$$

Fatou's lemma, (2.7.8), and (2.7.4) imply

$$\begin{aligned} \int_{R^1} H^{n-1}[u^{-1}(t)] dt &\le \liminf_{k \to \infty} \int_{R^1} H^{n-1}[L_k^{-1}(t)] dt \\ &\le \liminf_{k \to \infty} \int_{R^n} |DL_k| dx \\ &= \int_{R^n} |Du| dx. \end{aligned}$$

□

Theorem 2.7.1 is a special case of a more general version developed by Federer [F1] which we state without proof.

2.7.3. Theorem. *If X and Y are separable Riemannian manifolds of class 1 with*

$$\dim X = m \ge k = \dim Y$$

and $f : X \to Y$ is a Lipschitzian map, then

$$\int_A Jf(x)dH^m(x) = \int_Y H^{m-k}[A \cap f^{-1}(y)]dH^k(y)$$

whenever $A \subset X$ is an H^m-measurable set. Moreover, if g is an H^m integrable function on X, then

$$\int_X g(x)Jf(x)dH^m(x) = \int_Y \int_{f^{-1}(y)} g(x)dH^{m-k}(x)dH^k(y).$$

Here, $Jf(x)$ denotes the square root of the sum of squares of the determinant of the $k \times k$ minors of Jacobian matrix of f at x.

The proof of Theorem 2.7.1 above is patterned after the one by Fleming and Rishel [FR] which establishes a similar result for BV functions. Their result will be presented in Chapter 5.

We now give another proof of Theorem 2.4.1 that yields the best constant in the case $p = 1$.

2.7.4. Theorem. *If $u \in C_0^\infty(R^n)$, then*

$$\|u\|_{n/(n-1)} \leq n^{-1}\alpha(n)^{-1/n}\|Du\|.$$

Proof. For $t \geq 0$, let

$$A_t = \{x : |u(x)| > t\}, \quad B_t = \{x : |u(x)| = t\}$$

and let u_t be the function obtained from u by truncation at heights t and $-t$. If

$$f(t) = \|u_t\|_{n/(n-1)},$$

then clearly

$$|u_{t+h}| \leq |u_t| + h\chi_{A_t}$$

$$f(t+h) \leq f(t) + h|A_t|^{(n-1)/n} \tag{2.7.12}$$

for $h > 0$. It follows from the Morse–Sard theorem that for a.e. $t > 0$, B_t is an $(n-1)$-dimensional manifold of class ∞ and therefore, an application of the classical isoperimetric inequality yields

$$|A_t|^{(n-1)/n} \leq n^{-1}\alpha(n)^{-1/n}H^{n-1}(B_t). \tag{2.7.13}$$

It follows from (2.7.12) that f is an absolutely continuous function with

$$f'(t) \leq |A_t|^{(n-1)/n}$$

for a.e. t. Therefore, with the aid of (2.7.13), it follows that

$$\begin{aligned}\left(\int_{R^n} |u|^{n/(n-1)}dx\right)^{(n-1)/n} &= f(\infty) - f(0)\\ &= \int_0^\infty f'(t)dt\\ &\leq n^{-1}\alpha(n)^{-1/n}\int_0^\infty H^{n-1}[B_t]dt.\end{aligned}$$

The co-area formula, Theorem 2.7.1, shows that the last integral equals

$$\int_{R^n} |Du|dx,$$

thus establishing the theorem. □

From the inequality

$$\|u\|_{n/(n-1)} \leq n^{-1}\alpha(n)^{-1/n}\|Du\| \tag{2.7.14}$$

one can deduce the inequality

$$\|u\|_{p^*} \leq np(n-1)/(n-p)\,\|Du\|_p \tag{2.7.15}$$

by replacing u in (2.7.14) by u^q where $q = p(n-1)/(n-p)$. Then

$$\begin{aligned}\left(\int u^{np/(n-p)}dx\right)^{(n-1)/n} &\leq n^{-1}\alpha(n)^{-1/n}q\int |u|^{q-1}|Du|dx\\ &\leq n^{-1}\alpha(n)^{-1/n}q\left(\int u^{np/(n-p)}dx\right)^{(p-1)/p}\|Du\|_p\end{aligned}$$

by Hölder's inequality.

Of course, one cannot expect the constant in (2.7.15) to be optimal. Indeed, Talenti [TA] has shown that the best constant $C(n,p)$ is

$$C(n,p) = \pi^{-1/2}n^{-1/2}\left(\frac{p-1}{n-p}\right)^{1-(1/p)}\left[\frac{\Gamma(1+(n/2))\Gamma(n)}{\Gamma(n/p)\Gamma(1+n-(n/p))}\right]^{1/n}$$

where $1 < p < n$.

2.7.5. Remark. The proof of Theorem 2.7.4 reveals that the classical isoperimetric inequality implies the validity of the Sobolev inequality when $p = 1$. It is not difficult to see that the converse is also true.

To that end let $K \subset R^n$ be a compact set with smooth boundary. Let $d_K(x)$ denote the distance from x to K,

$$d_K(x) = \inf\{|x-y| : y \in K\}.$$

It is well-known and easy to verify that $d_K(x)$ is a Lipschitz function with Lipschitz constant 1. (See Exercise 1.1.) Moreover, Rademacher's theorem (Theorem 2.2.1) implies that d_K is totally differentiable at almost every point x with $|Dd_k(x)| = 1$ for a.e. $x \in R^n$. For each $h > 0$, let

$$F_h(x) = 1 - \min[d_K(x), h] \cdot h^{-1}$$

and observe that F_h is a Lipschitz function such that

(i) $F_h(x) = 1$ if $x \in K$

(ii) $F_h(x) = 0$ if $d_K(x) \geq h$

(iii) $|DF_h(x)| \leq h^{-1}$ for a.e. $x \in R^n$.

By standard smoothing techniques, Theorem 2.7.4 is valid for F_h because F_h is Lipschitz. Therefore

$$(|K|)^{(n-1)/n)} \leq n^{-1}\alpha(n)^{-1/n}\frac{|\{x : 0 < d_K(x) < h\}|}{h}.$$

Since $|Dd_K(x)| = 1$ for a.e. $x \in R^n$, the co-area formula for Lipschitz maps, Theorem 2.7.3, implies that

$$\begin{aligned}\frac{|\{x : 0 < d_K(x) < h\}|}{h} &= \frac{1}{h}\int_{\{0<d_K<h\}} |Dd_K|dx \\ &= \frac{1}{h}\int_0^h H^{n-1}[d_K^{-1}(t)]dt \\ &= H^{n-1}[d_K^{-1}(t_h)]\end{aligned}$$

where $0 < t_h < h$. Because K is smoothly bounded, it follows that

$$H^{n-1}[d_K^{-1}(t_h)] \to H^{n-1}(\partial K) \quad \text{as} \quad h \to 0$$

and thus, the isoperimetric inequality is established.

Of course, by appealing to some of the more powerful methods in geometric measure theory, the argument above could be employed to cover the case where the compact set K is a Lipschitz domain. By appealing to the properties of Minkowski content, cf. [F4, Section 3.2.39], it can be shown that the above proof still remains valid.

2.8 Alternate Proofs of the Fundamental Inequalities

In this section another proof of the Sobolev inequality (2.4.10) is given which is based on the Hardy–Littlewood–Wiener maximal theorem. This approach will be used in Section 2.9, where the inequality will be treated in the case of critical indices, $kp = n$.

We begin by proving the Hardy–Littlewood–Wiener maximal theorem.

2.8.1. Definition. Let f be a locally integrable function defined on R^n. The maximal function of f, $M(f)$, is defined by

$$M(f)(x) = \sup\left\{ ⨍_{B(x,r)} |f(y)| dy : r > 0 \right\}.$$

2.8.2. Theorem. *If $f \in L^p(R^n)$, $1 < p \leq \infty$, then $M(f) \in L^p(R^n)$ and there exists a constant $C = C(p, n)$ such that*

$$\|M(f)\|_p \leq C\|f\|_p.$$

Proof. For each $t \in R^1$, let $A_t = \{x : M(f)(x) > t\}$. From Definition 2.8.1 it follows that for each $x \in A_t$, there exists a ball with center $x \in A_t$, such that

$$⨍_{B_x} |f|\, dy > t. \tag{2.8.1}$$

If we let $\mathcal{F}$ be the family of n-balls defined by $\mathcal{F} = \{B_x : x \in A_t\}$, then Theorem 1.3.1 provides the existence of a disjoint subfamily $\{B_1, B_2, \dots, B_k, \dots\}$ such that

$$\sum_{k=1}^{\infty} |B_k| \geq 5^{-n}|A_t|$$

and therefore, from (2.8.1),

$$\|f\|_1 \geq \int_{\cup_{k=1}^{\infty} B_k} |f| dy > t \sum_{k=1}^{\infty} |B_k| \geq t5^{-n}|A_t|,$$

or

$$|A_t| \leq \frac{5^n}{t}\|f\|_1 \quad \text{whenever} \quad t \in R^1. \tag{2.8.2}$$

We now assume that $1 < p < \infty$, for the conclusion of the theorem obviously holds in case $p = \infty$. For each $t \in R^1$, define

$$f_t(x) = \begin{cases} f(x) & \text{if } |f(x)| \geq t/2 \\ 0 & \text{if } |f(x)| < t/2. \end{cases}$$

Then, for all x,

$$|f(x)| \leq |f_t(x)| + t/2,$$
$$M(f)(x) \leq M(f_t)(x) + t/2$$

and thus,

$$\{x : M(f)(x) > t\} \subset \{M(f_t)(x) > t/2\}.$$

Applying (2.8.2) with f replaced by f_t yields

$$|A_t| \le |\{M(f_t)(x) > t/2\}| \le \frac{2\cdot 5^n}{t}\int_{R^n}|f_t|dy = \frac{2\cdot 5^n}{t}\int_{\{|f|\ge t/2\}}|f|dy. \tag{2.8.3}$$

Now, from Lemma 1.5.1, and (2.8.3),

$$\begin{aligned}\int_{R^n}(Mf)^p dy &= \int_0^\infty |A_t|dt^p \\ &= p\int_0^\infty t^{p-1}|A_t|dt \\ &\le p2\cdot 5^n\int_0^\infty t^{p-2}\left(\int_{\{|f|>t/2\}}|f|dx\right)dt \\ &= p2^p\cdot 5^n\int_0^\infty t^{p-2}\left(\int_{\{|f|>t\}}|f|dx\right)dt \\ &= p2^p\cdot 5^n\int_0^\infty t^{p-2}\mu(\{|f|>t\})dt\end{aligned}$$

where μ is a measure defined by $\mu(E) = \int_E |f|dx$ for every Borel set E. Thus, appealing again to Lemma 1.5.1, we have

$$\begin{aligned}\int_{R^n}(Mf)^p dx &= \frac{p2^p\cdot 5^n}{p-1}\int_0^\infty \mu(\{|f|>t\})dt^{p-1} \\ &= \frac{p2^p\cdot 5^n}{p-1}\int_{R^n}|f|^{p-1}d\mu \\ &= \frac{p2^p\cdot 5^n}{p-1}\int_{R^n}|f|^p dx < \infty.\end{aligned}$$

Since $p > 1$. This establishes the theorem. □

For $0 < \alpha < n$, we recall from Section 2.6 the definition of the Riesz potential of f of order α:

$$I_\alpha * f(x) = I_\alpha f(x) = \frac{1}{\gamma(\alpha)}\int_{R^n}\frac{f(y)dy}{|x-y|^{n-\alpha}}.$$

The following lemma is the final ingredient necessary to establish the Sobolev inequality for Riesz potentials.

2.8.3. Lemma. *If $0 < \alpha < n$, $\beta > 0$, and $\delta > 0$, then there is a constant $C = C(n)$ such that for each $x \in R^n$,*

(i) $$\int_{B(x,\delta)}\frac{|f(y)|dy}{|x-y|^{n-\alpha}} \le C\delta^\alpha M(f)(x)$$

(ii) $\displaystyle\int_{R^n - B(x,\delta)} \frac{|f(y)|dy}{|x-y|^{\beta+n}} \le C\delta^{-\beta} M(f)(x).$

Proof. Only (i) will be proved since the proof of (ii) is similar. For $x \in R^n$ and $\delta > 0$, let the annulus be denoted by

$$\mathcal{A}\left(x, \frac{\delta}{2^k}, \frac{\delta}{2^{k+1}}\right) = B\left(x, \frac{\delta}{2^k}\right) - B\left(x, \frac{\delta}{2^{k+1}}\right),$$

and note that

$$\begin{aligned}\int_{B(x,\delta)} \frac{|f(y)|dy}{|x-y|^{n-\alpha}} &= \sum_{k=0}^{\infty} \int_{\mathcal{A}(x,\frac{\delta}{2^k},\frac{\delta}{2^{k+1}})} \frac{|f(y)|dy}{|x-y|^{n-\alpha}} \\ &\le \sum_{k=0}^{\infty} \left(\frac{\delta}{2^{k+1}}\right)^{\alpha-n} \int_{B(x,\frac{\delta}{2^k})} |f|dx \\ &= \alpha(n) \sum_{k=0}^{\infty} \left(\frac{1}{2}\right)^{\alpha-n} \left(\frac{\delta}{2^k}\right)^{\alpha} \fint_{B(x,\frac{\delta}{2^k})} |f|dx \\ &\le C\delta^{\alpha} M(f)(x),\end{aligned}$$

where $\alpha(n)$ denotes the volume of the unit n-ball. This proves (i). □

We now will see that the Sobolev inequality for Riesz potentials is an easy consequence of the above results.

2.8.4. Theorem. *Let $\alpha > 0$, $1 < p < \infty$, and $\alpha p < n$. Then, there is a constant $C = C(n,p)$ such that*

$$\|I_\alpha(f)\|_{p^*} \le C\|f\|_p, \quad p^* = \frac{np}{n-\alpha p},$$

whenever $f \in L^p(R^n)$.

Proof. For $\delta > 0$, Hölder's inequality implies that

$$\int_{R^n - B(x,\delta)} \frac{|f(y)|}{|x-y|^{n-\alpha}} dy \le \omega(n-1)\|f\|_p \left(\int_\delta^\infty r^{n-1-p'(n-\alpha)} dr\right)^{1/p'}$$

where $r = |x-y|$. The integral on the right is dominated by $\delta^{\alpha-(n/p)}$ since $\alpha p < n$, and therefore, by Lemma 2.8.3(i),

$$|I_\alpha(f)(x)| \le C\left[\delta^\alpha M(f)(x) + \delta^{\alpha-(n/p)}\|f\|_p\right]. \tag{2.8.4}$$

If we choose

$$\delta = \left(\frac{M(f)(x)}{\|f\|_p}\right)^{-p/n},$$

then (2.8.4) becomes

$$|I_\alpha(f)(x)| \le C[M(f)(x)]^{1-(\alpha p/n)}\|f\|_p^{\alpha/n}$$

or,

$$|I_\alpha(f)(x)|^{p^*} \le C[M(f)(x)]^p\|f\|_p^{(\alpha p/n)p^*}.$$

An application of Theorem 2.8.2 now yields the desired conclusion. □

2.8.5. Remark. If we are willing to settle for a slightly weaker result in Theorem 2.8.4, an easy proof is available that also provides an estimate of the constant C that appears on the right-hand side of the inequality. Thus, if Ω is a domain with finite measure, $f \in L^p(\Omega)$, and $p \le q < p^*$, we can obtain a bound on $\|I_\alpha(f)\|_q$ by a method that essentially depends only on Hölder's inequality.

For this purpose, let $\frac{1}{r} = 1 - (\frac{1}{p} - \frac{1}{q})$ and note that because $q < p^*$,

$$|x-y|^{\alpha-n} \in L^r(\Omega) \tag{2.8.5}$$

for each fixed $x \in R^n$. As in the proof of (2.4.7), if $|B(x,R)| = |\Omega|$, then

$$\begin{aligned}\int_\Omega |x-y|^{(\alpha-n)r}dy &\le \int_{B(x,R)} |x-y|^{(\alpha-n)r}dy = \frac{\omega(n-1)R^{(\alpha-n)r+n}}{(\alpha-n)r+n}\\ &= \frac{\omega(n-1)|\Omega|^\gamma}{[(\alpha-n)r+n]\alpha(n)^\gamma} \equiv C(\alpha,r,\Omega)\end{aligned} \tag{2.8.6}$$

where $\gamma = ((\alpha-n)r)/n + 1$. For each fixed x, observe that

$$\begin{aligned}|x-y|^{(\alpha-n)}|f(y)| &= \left(|x-y|^{(\alpha-n)r}|f(y)|^p\right)^{1/q}\\ &\cdot\left(|x-y|^{(\alpha-n)r/p'}\right)\cdot|f(y)|^{p\delta}\end{aligned} \tag{2.8.7}$$

where $\delta = \frac{1}{p} - \frac{1}{q}$. Because $\frac{1}{p'} + \frac{1}{q} + \delta = 1$, we may apply Hölder's inequality to the three factors on the right side of (2.8.7) to obtain

$$\begin{aligned}|I_\alpha(f)(x)| &\le \left(\int_\Omega |x-y|^{(\alpha-n)r}|f(y)|^p dy\right)^{1/q}\\ &\cdot\left(\int_\Omega |x-y|^{(\alpha-n)r}dy\right)^{1/p'}\left(\int_\Omega |f(y)|^p dy\right)^\delta.\end{aligned}$$

Therefore, by Fubini's theorem and (2.8.6),

$$\begin{aligned}\int_\Omega (I_\alpha f)^q dx &\le \int_\Omega\int_\Omega |x-y|^{(\alpha-n)r}|f(y)|^p dxdy\\ &\quad\cdot C(\alpha,\delta,\Omega)\frac{q}{p'}\cdot\|f\|_p^{pq\delta}\\ &\le C(\alpha,\delta,\Omega)\cdot\|f\|_p^p\cdot C(\alpha,\delta,\Omega)\frac{q}{p'}\cdot\|f\|_p^{pq\delta}.\end{aligned}$$

Thus,

$$\begin{aligned}\|I_\alpha f\|_q &\le C(\alpha,\delta,\Omega)^{(1/q)+(1/p')}\|f\|_p\\ &\le C(\alpha,\delta,\Omega)^{1/r}\|f\|_p.\end{aligned}$$

2.8.6. Remark. It is an easy matter to see that Theorem 2.8.4 provides another proof of Theorem 2.4.2. Indeed, if $u \in C_0^k(R^n)$, recall from (2.4.5) that for every $x \in R^n$,

$$|u(x)| \le C(n) I_1(|Du|). \tag{2.8.8}$$

In fact, if we employ the Riesz composition formula which states that

$$I_\alpha * I_\beta = I_{\alpha+\beta}, \quad \alpha+\beta < n,$$

an application of (2.8.8) to the derivatives of u gives the estimate

$$|u(x)| \le C(n,k) I_k(|D^k u|).$$

From Theorem 2.8.4 we have

$$\|I_k(|D^k u|)\|_{p^*} \le C\|D^k u\|_p$$

if $kp < n$. Thus,

$$\|u\|_{p^*} \le C\|D^k u\|_p \le C\|u\|_{k,p}$$

which is the conclusion in (2.4.10) when $\Omega = R^n$.

Of course, one could also employ Theorem 2.6.1 which states that each $u \in W^{k,p}(R^n)$ can be represented as $u = g_k * f$ for some $f \in L^p(R^n)$, where $\|f\|_p \sim \|u\|_{k,p;R^n}$. Then, in view of the fact that $g_k \le C I_k$, (2.4.10) follows from Theorem 2.8.4.

2.9 Limiting Cases of the Sobolev Inequality

In previous sections all Sobolev-type inequalities were established under the restriction $kp \ne n$. We now treat the case $kp = n$ in the context of Riesz potentials and since the Riesz kernel I_α is defined for all positive numbers α, we will therefore replace the integer k by α.

When $\alpha p = n$, one might hope that $I_\alpha * f$ is bounded because $p^* \to \infty$ as $\alpha p \to n$. However, while boundedness is trivially true when $n = 1$, it is false when $n > 1$. As an example, consider $u(x) = |\log|x||^{1-2/(n-1)}$; clearly $u \in W^{1,n}(B(0,r))$ for $r < 1$, but $u \notin L^\infty(B(0,r))$. Although an L^∞ estimate cannot, in general, be obtained it is possible to obtain results that provide a good substitute. Our first result below offers exponential integrability as a substitute for boundedness. We begin with a simple and elegant proof of this fact which follows easily from the estimate discussed in Remark 2.8.5.

2.9.1. Theorem. *Let $f \in L^p(\Omega)$, $p > 1$, and define*

$$g = I_{n/p} * f.$$

Then there are constants C_1 and C_2 depending only on p and n such that

$$\fint_\Omega \exp\left[\frac{g}{C_1\|f\|_p}\right]^{p'} dx \leq C_2. \tag{2.9.1}$$

Proof. Let $p \leq q < \infty$ and recall from Remark 2.8.5 the estimate

$$\|I_\alpha f\|_q \leq C(\alpha, \delta, \Omega)^{1/r}\|f\|_p, \tag{2.9.2}$$

where $\frac{1}{r} = 1 - (\frac{1}{p} - \frac{1}{q})$,

$$C(\alpha, \delta, \Omega) = \frac{\omega(n-1)|\Omega|^\gamma}{[(\alpha - n)r + n]\alpha(n)^\gamma},$$

and

$$\gamma = \frac{(\alpha - n)r}{n} + 1.$$

In the present situation, $\alpha p = n$, and therefore

$$(\alpha - n)r + n = \frac{np}{pq + (p - q)}.$$

Thus, we can write

$$C(\alpha, \delta, \Omega) \equiv C = K_1|\Omega|^\gamma \left(\frac{pq + (p - q)}{np}\right) \leq K|\Omega|^\gamma q$$

where K_1 and K are constants that depend only on p and n. Thus, since $\gamma q/r = 1$, from (2.9.2) we have

$$\begin{aligned}\int_\Omega |g|^q dx &\leq C^{q/r}\|f\|_p^q \\ &\leq (qK)^{q/r}|\Omega|\,\|f\|_p^q = (qK)^{1+(q/p')}\cdot\|\Omega|\,\|f\|_p^q.\end{aligned}$$

Now replacement of q by $p'q$ (which requires that $q > p - 1$) yields

$$\int_\Omega |g|^{p'q}dx \leq (p'qK)^{1+q}|\Omega|\,\|f\|_p^{p'q}.$$

In preparation for an expression involving an infinite series, substitute an integer k, $k > p - 1$, for q to obtain

$$\int_\Omega \frac{1}{k'}\left(\frac{|g|}{C\|f\|_p}\right)^{p'k} dx \leq p'K\frac{k^k}{(k-1)!}|\Omega|\left(\frac{Kp'}{C^{p'}}\right)^k$$

for any constant $C > 0$. Consequently,

$$\int_\Omega \sum_{k=k_0}^{\infty} \frac{1}{k!}\left(\frac{|g|}{C\|f\|_p}\right)^{p'k} dx \le p'K|\Omega| \sum_{k=k_0}^{\infty} \frac{k^k}{(k-1)!}\left(\frac{Kp'}{C^{p'}}\right)^k$$

where $k_0 = [p]$. The series on the right converges if $C^{p'} > eKp'$ and thus the result follows from (1.5.12) when applied to the terms involving $k < k_0$ and the monotone convergence theorem. □

By appealing to a different method, we will give another proof of exponential integrability that gives a slightly stronger result than the one just obtained.

2.9.2. Theorem. *Let $f \in L^p(R^n)$, $\operatorname{spt} f \subset B$ where B is a ball of radius R, and let $p = n/\alpha > 1$. Then, for any $\varepsilon > 0$, there is a constant $C = C(\varepsilon, n, p)$ such that*

$$⨍_B \exp\left[\frac{n}{\omega_{n-1}}\left|\frac{I_{n/p}(f)(x)}{\|f\|_p} - \varepsilon\right|^{p'}\right] dx \le C. \tag{2.9.3}$$

Proof. Clearly, we may assume that $\|f\|_p = 1$. Then,

$$I_\alpha(f)(x) = \int_{B(x,\delta)} f(y)|x-y|^{\alpha-n} dy + \int_{B-B(x,\delta)} f(y)|x-y|^{\alpha-n} dy$$

where $x \in B$ and $0 < \delta \le R$. By Lemma 2.8.3(i), the first integral on the right is dominated by $C\delta^\alpha M(f)(x)$. By Hölder's inequality and the fact that $\|f\|_p = 1$, the second integral on the right can be estimated as follows: if $r = |x-y|$, then

$$\begin{aligned}\int_{B-B(x,\delta)} f(y)|x-y|^{\alpha-n} dy &\le \left[\omega(n-1)\int_\delta^R r^{(\alpha-n)p'+n-1} dr\right]^{1/p'} \\ &= [\omega(n-1)\log(R/\delta)]^{1/p'}.\end{aligned}$$

Thus

$$|I_\alpha(f)(x)| \le C\delta^\alpha M(f)(x) + (\omega(n-1)\log(R/\delta))^{1/p'}.$$

If we choose

$$\delta^\alpha = \min(\varepsilon C^{-1}[M(f)(x)]^{-1}, R^\alpha),$$

then we have

$$|I_\alpha(f)(x)| \le \varepsilon + \left[\omega(n-1)\log^+(R\varepsilon^{-1/\alpha}C^{1/\alpha}M(f)(x)^{1/\alpha})\right]^{1/p'},$$

or

$$(I_\alpha(f)(x) - \varepsilon)^{+p'} \le \omega(n-1)n^{-1}\log^+(R^n\varepsilon^{-p}C^p M(f)(x)^p)$$

since $\alpha p = n$. Because $\|f\|_p = 1$, the conclusion now follows immediately from Theorem 2.8.2. □

2.9.3. Remark. Inequality (2.9.3) clearly implies that if $\beta < n/\omega(n-1)$, then there is a constant $C = C(\beta, n, p)$ such that

$$\fint_B \exp\left[\beta\left|\frac{I_{n/p}(f)(x)}{\|f\|_p}\right|^{p'}\right] dx \leq C, \tag{2.9.4}$$

thus recovering inequality (2.9.1).

Although it is of independent mathematical interest to determine the best possible constants in inequalities, in some applications the sharpness of the constant can play a critical role.

The sharpness of the Sobolev imbedding theorem in the case of critical indices has had many different approaches. For example, in [HMT], it was shown that the space $W_0^{1,n}(\Omega)$ could not be embedded in the Orlicz space $L_\varphi(\Omega)$ where $\varphi(t) = \exp(|t|^{n/(n-1)} - 1)$. On the other hand, with this Sobolev space, it was shown by Moser [MOS] that (2.9.4) remains valid for $\beta = n/\omega(n-1)$; that is, ε can be taken to be zero in (2.9.3). Recently, Adams [AD8] has shown that (2.9.4) is valid for $\beta = n/\omega(n-1)$ with no restriction on α.

Theorems 2.9.1 and 2.9.2 give one version of a substitute for boundedness in the case $\alpha p = n$. We now present a second version which was developed by Brezis and Wainger [BW].

For this, recall the definition of the Bessel kernel, g_α, introduced in (2.6.1) by means of its Fourier transform:

$$\hat{g}_\alpha(x) = 2\pi^{-n/2}(1 + |x|^2)^{-\alpha/2}.$$

Also, recall that the space of Bessel potentials, $L^{\alpha,p}(R^n)$, is defined as all functions u such that $u = g_\alpha * f$ where $f \in L^p(R^n)$. The norm in this space is defined as $\|u\|_{\alpha,p} = \|f\|_p$. Also, referring to Theorem 2.6.1, we have in the case α is a positive integer, that this norm is equivalent to the Sobolev norm of u.

For the development of the next result, we will assume that the reader is familiar with the fundamental properties of the Fourier transform.

2.9.4. Theorem. *Let $u \in L^{\ell,q}(R^n)$ with $\ell q > n$, $1 \leq q \leq \infty$ and let $\alpha p = n$, $1 < p < \infty$. If $\|u\|_{\alpha,p} \leq 1$, then*

$$\|u\|_\infty \leq C\left[1 + \log^{1/p'}(1 + \|u\|_{\ell,q})\right]. \tag{2.9.5}$$

Proof. Because $C_0^\infty(R^n)$ is dense in $L^{\ell,q}(R^n)$ relative to its norm and also in the topology induced by uniform convergence on compact sets, it is sufficient to establish (2.9.5) for $u \in C_0^\infty(R^n)$.

Let $\varphi, \eta \in C^\infty(R^n)$ be functions with $\operatorname{spt}\varphi$ compact, $\varphi \equiv (2\pi)^{-n/2}$ on some neighborhood of the origin, and $\varphi + \eta \equiv (2\pi)^{-n/2}$ on R^n. Since $u \in C_0^\infty(R^n)$, u may be written in terms of the inverse Fourier transform as

$$\begin{aligned} u(x) &= \int e^{ix\cdot y}\hat{u}(y)\varphi(y/R)dy + \int e^{ix\cdot y}\hat{u}(y)\eta(y/R)dy \\ &\equiv u_1(x) + u_2(x), \end{aligned} \tag{2.9.6}$$

where $R \geq 2$ is a positive constant to be determined later.

The proof will be divided into two parts. In Part 1, the following inequality will be established,

$$\|u_1\|_\infty \leq C(\log R)^{1/p'}$$

while in Part 2, it will be shown that

$$\|u_2\|_\infty \leq CR^{-\delta}\|u\|_{\ell,q}$$

for some $\delta > 0$. The conclusion of the theorem will then follow by taking

$$R = \max(2, \|u\|_{\ell,q}^{1/\delta}).$$

Proof of Part 1. We proceed to estimate u_1 as follows:

$$\begin{aligned} u_1(x) &= (2\pi)^{-n/2}\int e^{ix\cdot y}(2\pi)^{n/2}(1+|y|^2)^{\alpha/2}\hat{u}(y)\frac{\varphi(y/R)}{(1+|y|^2)^{\alpha/2}}dy \\ &= f * K_R(x) \end{aligned}$$

where

$$\hat{f}(y) = (2\pi)^{n/2}(1+|y|^2)^{\alpha/2}\hat{u}(y)$$

and

$$\hat{K}_R(y) = \frac{\varphi(y/R)}{(1+|y|^2)^{\alpha/2}}. \tag{2.9.7}$$

Note that $u = g_\alpha * f$ (see Section 2.6) and therefore

$$\|f\|_p = \|u\|_{\alpha,p} \leq 1.$$

Consequently, in order to establish Part 1 it will be sufficient to show that

$$\|K_R\|_{p'} \leq C(\log R)^{1/p'}, \quad R > 2.$$

We now define a function L such that $\hat{L} = \varphi$. Note that L is a rapidly decreasing function and thus, in particular, $L \in L^1(R^n) \cap C^\infty(R^n)$. Let

$$L_R(x) = R^n L(Rx).$$

From (2.9.7) we have that

$$(2\pi)^{-n/2}\hat{K}_R(y) = \hat{L}_R(y)\cdot\hat{g}_\alpha(y)$$

and

$$(2\pi)^{-n/2}K_R = L_R * g_\alpha.$$

Let $B(R)$ be the ball of radius R centered at the origin. Define two functions G_α^1 and G_α^2 by

$$G_\alpha^1(x) = g_\alpha(x)\chi_{B(R^{-1})}(x)$$
$$G_\alpha^2(x) = g_\alpha(x) - G_\alpha^1(x).$$

Then

$$(2\pi)^{-n/2}K_R(x) = L_R * G_\alpha^1(x) + L_R * G_\alpha^2(x).$$

An application of Young's inequality yields

$$\|L_R * G_\alpha^1\|_{p'} \le \|L_R\|_{p'}\cdot\|G_\alpha^1\|_1 \tag{2.9.8}$$

and it is easily verified that

$$\|L_R\|_{p'} = CR^{n/p}$$

while from (2.6.3) and $\alpha p = n$, it follows that

$$\|G_\alpha^1\|_1 \le CR^{-n/p}.$$

Similarly, from (2.6.3) we see that

$$g_\alpha(x) \le C_1|x|^{\alpha-n}e^{-C_2|x|}$$

and therefore

$$\begin{aligned}\|G_k^2\|_{p'} &\le C\left(\int_{1/R<|x|<1} C_1|x|^{(\alpha-n)p'}\,dx\right)^{1/p'}\\ &\quad + C\left(\int_{1\le|x|<\infty} e^{-C_2|x|p'}\,dx\right)^{1/p'}\\ &\le C(\log R)^{1/p'} + C\\ &\le C(\log R)^{1/p'}\end{aligned}$$

since $R > 2$. Hence,

$$\|L_R * G_k^2\|_{p'} \le C(\log R)^{1/p'} \tag{2.9.9}$$

because

$$\|L_R\|_1 = \|L\|_1 = C < \infty.$$

Thus, from (2.9.8), and (2.9.9) we have

$$\|K_R\|_{p'} \le C(\log R)^{1/p'}, \tag{2.9.10}$$

thereby establishing Part 1.

Proof of Part 2. We write u_2 as follows:

$$\begin{aligned} u_2(x) &= (2\pi)^{-n/2} \int e^{ix\cdot y} \hat{u}(y)(2\pi)^{n/2}(1+|y|^2)^{\ell/2} \frac{\eta(y/R)}{(1+|y|)^{\ell/2}} dy \\ &= g * K_R(x) \end{aligned} \tag{2.9.11}$$

where

$$\hat{K}_R(y) = \frac{\eta(y/R)}{(1+|y|^2)^{\ell/2}} \tag{2.9.12}$$

and

$$\hat{g}(y) = \hat{u}(y)(2\pi)^{n/2}(1+|y|^2)^{\ell/2}.$$

By assumption, $u \in L^{\ell,q}(R^n)$, and therefore it follows from definition that $g \in L^q(R^n)$ with $u = g_\ell * g$. In order to establish Part 2, it suffices to show that

$$\|K_R\|_{q'} \le CR^{-\delta} \quad \text{for some} \quad \delta > 0.$$

First, consider the case $q = 1$. Since $\ell q > n$ by assumption, we have $\ell > n$. Now write

$$|K_R(x)| = \left| \int e^{ix\cdot y} \frac{\eta(y/R)dy}{(1+|y|^2)^{\ell/2}} \right| \le \int \frac{|\eta(y/R)|dy}{|y|^\ell}.$$

Recall that η vanishes in some neighborhood of the origin, say for all y such that $|y| < \varepsilon R$. Thus, for all x,

$$\begin{aligned} |K_R(x)| &\le \int_{R^n - B(0,\varepsilon R)} \frac{|\eta(y/R)|dy}{|y|^\ell} \\ &\le C \int_{\varepsilon R}^\infty r^{n-1-\ell} dr \\ &\le CR^{n-\ell} \end{aligned}$$

since $\ell > n$. Thus, Part 2 is established if $q = 1$.

Now consider $q > 1$, so that $q' < \infty$ and without loss of generality, let $\ell < n$. Since $\varphi + \eta \equiv (2\pi)^{-n/2}$ on R^n, (2.9.12) can be written as

$$(2\pi)^{n/2} \hat{K}_R(y) = \frac{1}{(1+|y|^2)^{\ell/2}} - \frac{\varphi(y/R)}{(1+|y|^2)^{\ell/2}}.$$

Thus, we have

$$K_R(x) = g_\ell(x) - g_\ell * L_R(x) \tag{2.9.13}$$

where

$$L_R(x) = R^n L(Rx), \quad \hat{L}(y) = \varphi(y).$$

We can rewrite (2.9.13) as

$$K_R(x) = \int [g_\ell(x) - g_\ell(x-y)] L_R(y)dy \tag{2.9.14}$$

because

$$(2\pi)^{-n/2} = \varphi(0) = (2\pi)^{-n/2} \int_{R^n} e^{-i0\cdot y} L(y)dy.$$

To estimate $\int |g_\ell(x-y) - g_\ell(x)|^{q'} dx$ we write

$$\begin{aligned}\int_{R^n} |g_\ell(x-y) - g_\ell(x)|^{q'} dx &= \int_{|x|\le 2|y|} |g_\ell(x-y) - g_\ell(x)|^{q'} dx \\ &\quad + \int_{|x|>2|y|} |g_\ell(x-y) - g_\ell(x)|^{q'} dx \\ &= I_1 + I_2.\end{aligned}$$

Now

$$\begin{aligned} I_1 &\le C\left(\int_{|x|\le 2|y|} |g_\ell(x-y)|^{q'} dx + \int_{|x|\le 2|y|} |g_\ell(x)|^{q'} dx\right) \\ &\le C\left(\int_{|x-y|\le 3|y|} |g_\ell(x-y)|^{q'} dx + \int_{|x|\le 2|y|} |g_\ell(x)|^{q'} dx\right) \\ &\le C|y|^{(\ell-n)q'+n},\end{aligned}$$

by (2.6.3). To estimate I_2, note that g_ℓ is smooth away from the origin, and therefore we may write $|g_\ell(x-y) - g_\ell(x)| \le |Dg_\ell(z)| \cdot |y|$ where $z = t(x-y) + (1-t)x = x - ty$ for some $t \in [0,1]$. Since, $|z| \ge 1/(2|x|)$ when $|x| > 2|y|$ we have, with the help of (2.6.4),

$$\begin{aligned}\int_{|x|>2|y|} |g_\ell(x-y) - g_\ell(x)|^{q'} dx &\le C \int_{|x|>2|y|} e^{-C|x|} |x|^{(\ell-n-1)q'} |y|^{q'} dx \\ &\le |y|^{q'} \int_{R^n} e^{-C|x|} |x|^{(\ell-n-1)q'} dx \\ &\le C|y|^{q'}.\end{aligned}$$

Consequently, combining the estimates for I_1 and I_2, we have

$$\|g_\ell(x-y) - g_\ell(x)\|_{q'} \le C|y|^\delta \tag{2.9.15}$$

where $\delta = [(\ell + 1 - n)q' + n]/q' > 0$. Referring to (2.9.14), we estimate $\|K_R\|_{q'}$ with the aid of Minkowski's inequality and (2.9.15) as follows:

$$\left(\int |K_R(x)|^{q'} dx\right)^{1/q'} \le \int \left(\int |g_\ell(x-y) - g_\ell(x)|^{q'} dx\right)^{1/q'} |L_R(y)|dy$$

$$
\begin{aligned}
&\le C \int |L_R(y)|\, |y|^\delta dy \\
&= CR^{n-\delta} \int |L(Ry)|\, |Ry|^\delta dy \\
&\le CR^{-\delta}.
\end{aligned}
$$

The integral $\int |L(z)|\,|z|^\delta dz$ is finite because $\hat{L} = \varphi$ and thus L is rapidly decreasing. The proof of Part 2 is now complete and the combination of Parts 1 and 2 completes the proof of the theorem. □

2.10 Lorentz Spaces, A Slight Improvement

In this section we turn to the subject of Lorentz spaces which was introduced in Chapter 1, Section 8. We will show that the Sobolev inequality for Riesz potentials (Theorem 2.8.4) as well as the development in Chapter 2, Section 9, can be improved by considering Lorentz spaces instead of L^p spaces.

We begin by proving a result that is similar to Young's inequality for convolutions.

2.10.1. Theorem. *If* $h = f * g$, *where*

$$f \in L(p_1, q_1),\ g \in L(p_2, q_2),\ \text{and}\ \frac{1}{p_1} + \frac{1}{p_2} > 1,$$

then $h \in L(r, s)$ *where*

$$\frac{1}{p_1} + \frac{1}{p_2} - 1 = \frac{1}{r}$$

and $s \ge 1$ *is any number such that*

$$\frac{1}{q_1} + \frac{1}{q_2} \ge \frac{1}{s}.$$

Moreover,

$$\|h\|_{(r,s)} \le 3r \|f\|_{(p_1,q_1)} \|g\|_{(p_2,q_2)}.$$

Proof. Let us suppose that q_1, q_2, s are all different from ∞. Then, by Lemma 1.8.9,

$$
\begin{aligned}
(\|h\|_{(r,s)})^s &= \int_0^\infty [x^{1/r} h^{**}(x)]^s \frac{dx}{x} \le \int_0^\infty \left[x^{1/r} \int_x^\infty f^{**}(t) g^{**}(t)dt \right]^s \frac{dx}{x} \\
&= \int_0^\infty \left[\frac{1}{y^{1/r}} \int_0^y f^{**}\left(\frac{1}{u}\right) g^{**}\left(\frac{1}{u}\right) \frac{du}{u^2} \right]^s \frac{dy}{y}. \qquad (2.10.1)
\end{aligned}
$$

The last equality is by the change of variables $x = 1/y$, $t = 1/u$. Now use Hardy's inequality (Lemma 1.8.11) to obtain,

$$\begin{aligned}\int_0^\infty \left[\frac{1}{y^{1/r}}\int_0^y f^{**}\left(\frac{1}{u}\right) g^{**}\left(\frac{1}{u}\right)\frac{du}{u^2}\right]^s \frac{dy}{y} \\ \le r^s \int_0^\infty \left[y^{1-(1/r)}\frac{f^{**}(1/y)g^{**}(1/y)}{y^2}\right]^s \frac{dy}{y}; \\ = r^s \int_0^\infty [x^{1+(1/r)} f^{**}(x)g^{**}(x)]^s \frac{dx}{x},\end{aligned}$$

by letting $y = 1/x$.

Since $s/q_1 + s/q_2 \ge 1$, we may find positive numbers m_1, m_2 such that

$$\frac{1}{m_1} + \frac{1}{m_2} = 1 \quad \text{and} \quad \frac{1}{m_1} \le \frac{s}{q_1},\ \frac{1}{m_2} \le \frac{s}{q_2}.$$

Therefore $q_1 \le sm_1$, $q_2 \le sm_2$. An application of Hölder's inequality with indices m_1, m_2, yields

$$\begin{aligned}(\|h\|_{(r,s)})^s &\le r^s \int_0^\infty \frac{[x^{1/p_1} f^{**}(x)]^s}{x^{1/m_1}} \frac{[x^{1/p_2} g^{**}(x)]^s}{x^{1/m_2}} dx \\ &\le r^s \left[\int_0^\infty [x^{1/p_1} f^{**}(x)]^{sm_1} \frac{dx}{x}\right]^{1/m_1} \\ &\quad \cdot \left[\int_0^\infty [x^{1/p_2} g^{**}(x)]^{sm_2} \frac{dx}{x}\right] \\ &= r^s(\|f\|_{(p_1,sm_1)})^s(\|g\|_{(p_2,sm_2)})^s.\end{aligned}$$

Thus, by Lemma 1.8.13

$$\begin{aligned}\|h\|_{(r,s)} &\le r\|f\|_{(p_1,sm_1)}\|g\|_{p_2,sm_2)} \\ &\le e^{1/e}e^{1/e}r\|f\|_{(p_1,q_1)}\|g\|_{(p_2,q_2)} \\ &\le 3r\|f\|_{(p_1,q_1)}\|g\|_{(p_2,q_2)}.\end{aligned}$$

Similar reasoning leads to the desired result in case one or more of q_1, q_2, s are ∞. □

As an application of Theorem 2.10.1, consider the kernel $I_\alpha(x) \equiv |x|^{\alpha-n}$ which is a constant multiple of the Riesz kernel that was introduced in Chapter 2, Section 6. For simplicity of notation in this discussion, we omit the constant $\gamma(\alpha)^{-1}$ that appears in the definition of $I_\alpha(x)$. Observe that the distribution function of I_α is given by

$$\begin{aligned}\alpha_{I_\alpha}(t) &= |\{x : |x|^{\alpha-n} > t\}| \\ &= |\{x : |x| < t^{1/(\alpha-n)}\}| \\ &= \alpha(n)t^{n/(\alpha-n)}\end{aligned}$$

and because I_α^* is the inverse of the distribution function, we have $I_\alpha^*(t) = (\frac{t}{\alpha(n)})^{(\alpha-n)/n}$. It follows immediately from definition that

$$I_\alpha^{**}(t) = \left(\frac{n}{\alpha(n)\alpha}\right) t^{(\alpha-n)/n}$$

and therefore that $I_\alpha \in L(n/(n-\alpha), \infty)$. If we form the convolution $I_\alpha * f$ where $f \in L^p = L(p,p)$, then Theorem 2.10.1 states that

$$I_\alpha * f \in L(q,p)$$

where

$$\frac{1}{q} = \frac{1}{p} + \frac{n-\alpha}{n} - 1 = \frac{1}{p} - \frac{\alpha}{n}.$$

Moreover, it follows from Lemma 1.8.13 that $L(q,p) \subset L(q,q)$ and thus we have an improvement of Theorem 2.8.4 which allows us to conclude only that $I_\alpha * f \in L^q$. As a consequence of Theorem 2.10.1, we have the following result that is analogous to Theorem 2.8.4.

2.10.2. Theorem. *If $f \in L(p,q)$ and $0 < \alpha < n/p$, then*

$$I_\alpha * f \in L(r,q)$$

and

$$\begin{aligned} \|I_\alpha * f\|_{(r,q)} &\le \|I_\alpha\|_{(n/(n-\alpha),\infty)} \|f\|_{(p,q)} \\ &= C\|f\|_{(p,q)} \end{aligned}$$

where

$$\frac{1}{r} = \frac{1}{p} - \frac{\alpha}{n}.$$

We now consider the limiting case of $1/p_1 + 1/p_2 = 1$ in Theorem 2.10.1. In preparation for this, we first need the following lemma.

2.10.3. Lemma. *Let φ be a measurable function defined on $(0,1)$ such that $t\varphi(t) \in L^p(0,1;dt/t)$, $p > 1$. Then,*

$$\|(1 + |\log t|)^{-1} \int_t^1 \varphi(s)ds\|_{L^p(0,1;dt/t)} \le \frac{p}{p-1} \|t\varphi(t)\|_{L^p(0,1;dt/t)}.$$

Proof. By standard limit procedures, we may assume without loss of generality that $\varphi \in L^1(0,1)$ is non-negative and bounded. Let

$$I = \int_0^1 (1 + |\log t|)^{-p} \left(\int_t^1 \varphi(s)ds\right)^p \frac{dt}{t}.$$

Since

$$I = \frac{1}{(p-1)} \int_0^1 \left(\int_t^1 \varphi(s)ds \right)^p d(1-\log t)^{-p+1}$$

integration by parts yields

$$I = \frac{p}{p-1} \int_0^1 \left(\int_t^1 \varphi(s)ds \right)^{p-1} (1-\log t)^{-p+1} \varphi(t)dt,$$

and Hölder's inequality implies

$$I \leq \frac{p}{p-1} \|(1+|\log t|)^{-1} \int_t^1 \varphi(s)ds\|^{p-1}_{L^p(0,1;dt/t)} \|t\varphi(t)\|_{L^p(0,1;dt/t)},$$

from which the conclusion follows. □

2.10.4. Remark. Before proving the next theorem, let us recall the following elementary proof concerning convolutions. If $f, g \in L^1(R^n)$ we may conclude that

$$\begin{aligned} \int_{R^n} \int_{R^n} |f(x-y)g(y)| dx dy &= \int_{R^n} |g(y)| \cdot \int_{R^n} |f(x-y)| dx dy \\ &= \int_{R^n} |g(y)| \cdot \|f\|_1 dy \\ &= \|f\|_1 \cdot \|g\|_1 < \infty. \end{aligned}$$

Thus, the mapping $y \to f(x-y)g(y) \in L^1(R^n)$ for almost all $x \in R^n$ and $f * g \in L^1(R^n)$.

In the event that one of the functions, say f, is assumed only to be an element of $L(p,q)$, $p > 1$, $q \geq 1$, while $g \in L^1(R^n)$, then the convolution need not belong to $L^1(R^n)$, but it will at least be defined. To see this, let

$$f_1(x) = \begin{cases} 1 & \text{if } f(x) > 1 \\ f(x) & \text{if } -1 \leq f(x) \leq 1 \\ -1 & \text{if } f(x) < -1 \end{cases}$$

and let $f_2 = f - f_1$. Then $f_1 * g$ is defined because f_1 is bounded. We will now show that $f_2 \in L^1(R^n)$ thus implying that $f_2 * g$ is defined and therefore, similarly for $f * g$. In order to see that $f_2 \in L^1(R^n)$ let

$$\tilde{f}_2(x) = \begin{cases} f(x) & \text{if } |f(x)| > 1 \\ 0 & \text{if } |f(x)| \leq 1. \end{cases}$$

Clearly $\alpha_{\tilde{f}_2}(s) = \alpha_f(1)$ if $0 < s < 1$ and $\alpha_{\tilde{f}_2}(s) = \alpha_f(s)$ if $s \geq 1$. Consequently

$$\tilde{f}_2^*(\alpha_f(1)) = \inf\{s : \alpha_{\tilde{f}_2}(s) \leq \alpha_f(1)\} = 0.$$

Thus, since $\tilde{f}_2^*$ is non-increasing, $\tilde{f}_2^*(t)$ vanishes for all $t \geq \alpha_f(1)$ and it is easy to see that $f^*(t) = \tilde{f}_2^*(t)$ for all $t < \alpha_f(1)$. We may assume that $q > p$ for, if $q \leq p$, then Lemmas 1.8.13 and 1.8.10 imply that $f \in L(p,q) \subset L(p,p)$. Consequently, by Young's inequality for convolutions, $f * g$ is defined. In fact, $f * g \in L^p$. With $q > p$ and $a = \alpha_f(1)$, we have

$$\begin{aligned}\infty > \int_0^\infty \frac{f^{**}(t)^q}{t^{1-(q/p)}}dt &\geq \int_0^\infty \frac{f^*(t)^q}{t^{1-(q/p)}}dt \geq \int_0^a \frac{f^*(t)^q}{t^{1-(q/p)}}dt \\ &\geq \overline{a}^{(q/p)-1} \int_0^a f^*(t)^q dt \\ &= \overline{a}^{(q/p)-1} \int_0^a \tilde{f}_2^*(t)^q dt \\ &= \overline{a}^{(q/p)-1} \int_0^\infty \tilde{f}_2^*(t)^q dt\end{aligned}$$

where $\overline{a} \in [0,a]$. Since $\tilde{f}_2^*(t)$ vanishes for $t \geq a$ whereas $\tilde{f}_2^*(t) \geq 1$ for $t < a$, it therefore follows that

$$\int_0^\infty \tilde{f}_2^*(t)dt \leq \int_0^\infty \tilde{f}_2^*(t)^q dt < \infty,$$

thus showing that $\tilde{f}_2$ is integrable because $\tilde{f}_2$ and $\tilde{f}_2^*$ have the same distribution function. Therefore f_2 is integrable.

2.10.5. Lemma. *Let $1 < p < \infty$, $1 \leq q_1 \leq \infty$, $1 \leq q_2 \leq \infty$ be such that $1/q_1 + 1/q_2 < 1$ and set $1/r = 1/q_1 + 1/q_2$. Assume $f \in L(p,q_1)$ and $g \in (p', q_2) \cap L^1(R^n)$ and let $u = f * g$. Then*

$$\left[\int_0^1 \left[\frac{u^*(t)}{1+|\log t|}\right]^r \frac{dt}{t}\right]^{1/r} \leq C\|f\|_{(p,q_1)} \cdot (\|g\|_{(p',q_2)} + \|g\|_1)$$

where C depends only on p, q_1, and q_2.

Proof. Note from the preceeding remark, that u is defined. For simplicity we set $\|f\|_{(p,q_1)}$ and $\|g\| = \|g\|_{(p',q_2)} + \|g\|_1$. Also, for notational convenience in this discussion, we will insert a factor of $(q/p)^{1/q}$ in the definition of the $\|f\|_{(p,q)}$; thus,

$$\|f\|_{(p,q)} = \begin{cases} \left(\dfrac{q}{p}\right)^{1/q} \left(\displaystyle\int_0^\infty [t^{1/p} f^{**}(t)]^q \frac{dt}{t}\right)^{1/q}, & 1 \leq p < \infty, 0 < q < \infty \\ \left(\dfrac{q}{p}\right)^{1/q} \displaystyle\sup_{t>0} t^{1/p} f^{**}(t), & 1 \leq p < \infty, q = \infty. \end{cases}$$

We distinguish two cases:

(i) $r < \infty$

(ii) $r = \infty$ (i.e. $q_1 = q_2 = \infty$).

(i) The case $r < \infty$. Recall from Lemma 1.8.8 that for every $t > 0$,

$$u^{**}(t) \le t f^{**}(t) g^{**}(t) + \int_t^\infty f^*(s) g^*(s) ds. \tag{2.10.2}$$

Clearly, the following inequalities hold for every $s > 0$:

$$f^*(s) \le f^{**}(s) \le \frac{1}{s^{1/p}} \|f\|_{(p,q_1)}, \tag{2.10.3}$$

$$g^*(s) \le g^{**}(s) \le \frac{1}{s^{1/p'}} \|g\|_{(p',q_2)}, \tag{2.10.4}$$

$$g^*(s) \le g^{**}(s) \le \frac{1}{s} \|g\|_1. \tag{2.10.5}$$

For $t < 1$, it follows from (2.10.2), (2.10.3), and (2.10.4) that,

$$u^{**}(t) \le t \frac{1}{t^{1/p}} \frac{1}{t^{1/p'}} \|f\|_{(p,q_1)} \|g\|_{(p',q_2)} + \int_t^1 f^*(s) g^*(s) ds$$
$$+ \int_1^\infty f^*(s) g^*(s) ds. \tag{2.10.6}$$

From (2.10.3) and (2.10.5) we have that

$$\int_1^\infty f^*(s) g^*(s) ds \le p[\|f\|_{(p,q_1)} \cdot \|g\|_1].$$

This in conjunction with (2.10.6) yields

$$u^{**}(t) \le p\|f\| \, \|g\| + \int_t^1 f^*(s) g^*(s) ds. \tag{2.10.7}$$

By Lemma 2.10.3, (2.10.7), and (2.10.4) we have

$$\begin{aligned} \|(1 + |\log t|)^{-1} u^{**}(t)\|_{L^r(0,1;dt/t)} &\le C\|f\| \, \|g\| + C\|t f^*(t) g^*(t)\|_{L^r(0,1;dt/t)} \\ &= C\|f\| \, \|g\| + C\|t^{1/p} f^*(t) t^{1/p'} \\ &\quad \cdot g^*(t)\|_{L^r(0,1;dt/t)} \\ &\le C\|f\| \, \|g\| + C\|f\|_{(p,q_1)} \|g\|_{(p',q_2)}. \end{aligned}$$

(ii) The case $r = \infty$. By (2.10.7), (2.10.3), and (2.10.4) we have, for $t < 1$,

$$u^{**}(t) \le p\|f\| \, \|g\| + |\log t| \, \|f\|_{(p,\infty)} \|g\|_{(p',\infty)}$$

and therefore

$$\|(1 + |\log t|)^{-1} u^{**}(t)\|_{L^\infty(0,1)} \le C\|f\| \, \|g\|. \qquad \square$$

2.10.6. Theorem. *Let* $1 < p < \infty$, $1 \le q_1 \le \infty$, $1 \le q_2 \le \infty$ *be such that* $1/q_1 + 1/q_2 < 1$ *and set* $1/r = 1/q_1 + 1/q_2$. *Assume* $f \in L(p, q_1)$ *and* $g \in L(p', q_1) \cap L^1$ *and let* $u = f * g$. *Then*

(i) *if* $r < \infty$, $e^{\lambda|u|^{r'}} \in L^1_{\text{loc}}(R^n)$ *for every* $\lambda > 0$.

(ii) *if* $r \leq \infty$, *there exists positive numbers* $C = C(p, q_1, q_2)$ *and* $M = M(|Q|)$ *such that*

$$\int_Q e^{C|u|^{r'}}\, dx \leq M$$

for every f *and* g *with* $\|f\|_{(p,q_1)} \leq 1$ *and* $\|g\|_{(p',q_2)} + \|g\|_1 \leq 1$.

Before proceeding with the proof, let us see how this result extends the analogous one established in Theorem 2.9.1. To make the comparison, take one of the functions in the above statement of the theorem, say f, as the Riesz kernel, $I_{n/p}$. As we have seen from the discussion preceeding Theorem 2.10.2, $f \in L(p', \infty)$. The other function g is assumed to be an element of $L^p(\Omega)$ where Ω is a bounded set. Thus, by Lemma 1.8.10, $g \in L(p,p) \cap L_1(R^n)$. In this context, $q_1 = \infty$ and $q_2 = p > 1$ thus proving that this result extends Theorem 2.9.1.

Proof. Consider part (i) first. Because $u^*(t)$ is non-increasing we have that

$$|u^*(r)|^r \int_0^t (1 - \log s)^{-r} \frac{ds}{s} \leq \int_0^t u^*(s)^r (1 - \log s)^{-r} \frac{ds}{s}$$

for every $t < 1$. The first integral equals $(1 - \log t)^{-r+1}/r - 1$ with $r > 1$ and Lemma 2.10.5 implies that $I(t) \to 0$ as $t \to 0$ where $I(t)$ denotes the second integral. Note that there exists a constant $K = K(r)$ such that

$$|u^*(t)|^{r'} \leq K(1 + |\log t|) I(t)^{1/(r-1)}, \quad 0 < t < 1. \tag{2.10.8}$$

With $Q \subset R^n$ any bounded measurable set, we have with the help of Lemma 1.5.1,

$$\int_Q \exp(C|u(x)|)^{r'} dx = \int_0^{|Q|} \exp(C|u^*(t)|)^{r'} dt = \int_0^{t_0} \exp(C|u^*(t)|)^{r'} dt$$

$$+ \int_{t_0}^{|Q|} \exp(C|u^*(t)|)^{r'} dt \tag{2.10.9}$$

where $0 < t_0 < |Q|$. Because u^* is non-increasing, it is only necessary to show that the first integral is finite. For this purpose, choose $t_0 < 1$ so that $C^{r'} K I(t_0)^{1/(r-1)} < 1$. Then, from (2.10.8),

$$\exp(C|u^*(t)|)^{r'} \leq (e/t)^{\alpha}$$

where $\alpha = C^{r'} K I(t_0)^{1/(r-1)}$. Thus, part (i) of the theorem is established.

For the proof of (ii), Lemma 2.10.5 and the fact that $u^*(t)$ is non-increasing allow us to conclude that for $0 < t < 1$,

$$|u^*(t)|^r \int_0^t (1 - \log s)^{-r} \frac{ds}{s} \leq C\|f\|^r \cdot \|g\|^r,$$

where $\|f\| = \|f\|_{(p,q_1)}$ and $\|g\| = \|g\|_{(p',q_2)} + \|g\|_1$. Therefore,

$$|u^*(t)|^{r'} \leq K(1 - \log t)\|f\|^{r'}\|g\|^{r'}.$$

Similar to part (i), the proof of (ii) now follows from (2.10.9) by choosing $KC < 1$. □

Exercises

2.1. Prove that $W^{k,p}(R^n) = W_0^{k,p}(R^n)$.

2.2. If f and g are integrable functions defined on R^n such that

$$\int f\varphi \, dx = \int g\varphi \, dx$$

for every function $\varphi \in C_0^\infty(R^n)$, prove that $f = g$ almost everywhere on R^n.

2.3. Prove the following extension of the Rellich–Kondrachov compactness theorem. If Ω is a domain having the extension property, then

$$W^{k+m,p}(\Omega) \to W^{k,q}(\Omega)$$

is a compact imbedding if $mp < n$, $1 \leq q \leq np/(n - mp)$ and m a non-negative integer.

2.4. Verify the following equivalent formulation of Bessel capacity:

$$\begin{aligned} B_{\alpha,p}(E) &= \inf_f \left\{ \int_{x \in E} g_\alpha * f(x) \right\}^{-p} \\ &= \left\{ \sup_f \int_{x \in E} g_\alpha * f(x) \right\}^{-p} \end{aligned}$$

where $f \in L^p(R^n)$, $f \geq 0$, and $\|f\|_p \geq 1$.

2.5. Prove that the Riesz and Bessels capacities have the same null sets; that is, $R_{\alpha,p}(E) = 0$ if and only if $B_{\alpha,p}(E) = 0$ for every set $E \subset R^n$.

2.6. Show that there is a constant $C = C(\alpha, p, n)$ such that

$$B_{\alpha,p}(E) \geq C$$

provided $\alpha p > n$ and E is non-empty.

2.7. As an extension of Corollary 2.1.9, prove that if Ω is connected and $u \in W^{k,p}(\Omega)$ has the property that $D^\alpha u = 0$ almost everywhere on Ω, for all $|\alpha| = k$, then u is a polynomial of degree at most $k-1$.

2.8. Let $1 < p$, $kp < n$. If $K \subset R^n$ is a compact set, let

$$\gamma_{k,p}(K) = \inf\{\|u\|_{k,p}^p : u \geq 1 \text{ on a neighborhood of } K,$$
$$u \in C_0^\infty(R^n)\}.$$

With the aid of Theorem 2.6.1, prove that there exists a constant $C = C(p, n)$ such that

$$C^{-1} B_{k,p}(K) \leq \gamma_{k,p}(K) \leq C B_{k,p}(K).$$

2.9. Show that for each compact set $K \subset R^n$,

$$\inf\left\{\int_{R^n} |Du|^n dx : u \geq 1 \text{ on } K, \; u \in C_0^\infty(R^n)\right\} = 0.$$

2.10. Prove that there exists a sequence of piecewise linear maps

$$L_k : R^n \to R^1$$

satisfying (2.7.7) and (2.7.8). See the discussion in Exercise 5.2.

2.11. Suppose that $u: R^n \to R^1$ is Lipschitz with $\|Du\|_{1;R^n} < \infty$. Define $\alpha_u : R^1 \to R^1$ by $\alpha_u(t) = |\{x : u(x) > t\}|$. Since α_u is non-increasing, it is differentiable almost everywhere.

(a) Prove that for almost all t,

$$-\alpha_u'(t) \geq \int_{u^{-1}(t)} |Du|^{-1} dH^{n-1}.$$

(b) Prove that equality holds in (a) if

$$|\{x : Du(x) = 0\}| = 0.$$

2.12. Theorem 2.8.4 gives the potential theoretic version of Theorem 2.4.2, but observe that the latter is true for $p = 1$ whereas the former is false in this case. To see this, choose $f_i \geq 0$ with $\int_{R^n} f_i dx = 1$ and spt $f_i \subset B(0, 1/i)$. Prove that $I_\alpha * f_i \to I_\alpha$ uniformly on $R^n - B(0,r)$ for every $r > 0$. Thus conclude that

$$\int_{R^n} (I_\alpha * f_i)^{n/(n-\alpha)} dx \geq \int_{\tilde{B}(0,r)} (I_\alpha * f_i)^{n/(n-\alpha)} dx.$$

The right side tends to

$$\int_{\tilde{B}(0,r)} (I_\alpha)^{n/(n-\alpha)} dx = C \int_{\tilde{B}(0,r)} |x|^{-n} dx.$$

But

$$\int_{\tilde{B}(0,r)} |x|^{-n} dx \to \infty \quad \text{as} \quad r \to 0.$$

2.13. Show that Theorem 2.8.4 is false when $\alpha p = n$. For this consider

$$f(x) = \begin{cases} |x|^{-\alpha} \left(\log \frac{1}{|x|}\right)^{-\alpha(1+\varepsilon)/n} & |x| \leq 1 \\ 0 & |x| > 1 \end{cases}$$

where $\varepsilon > 0$. Then $f \in L^p$ since $\alpha p = n$ but $I_\alpha * f(0) = \infty$ whenever $\alpha(1+\varepsilon)/n \leq 1$.

2.14. Prove the following extension of Theorem 2.8.2. Suppose $|f| \log(2 + |f|)$ is integrable over the unit ball B. Then $Mf \in L^1(B)$. To prove this, note that (with the notation of Theorem 2.8.2)

$$\int_B Mf\, dx \leq |B| + \int_{A_1} M\, dx \leq |B| + \int_1^\infty |A_t| dt + |A_1|.$$

Now use (2.8.3) and Exercise 1.3.

2.15. There is a variety of methods available to treat Theorem 2.6.16. Here is one that shows that $B_{1,p}(K) = 0$ if $H^{n-p} < \infty$, $1 < p < n$.

STEP 1. Use Exercise 2.8 to replace $B_{1,p}(K)$ by $\gamma_{1,p}(K)$.

STEP 2. There exists $C = C(n,k)$ such that for any open set $U \supset K$, there exist an open set $V \supset K$ and $u \in W_0^{1,p}(R^n)$ such that

(i) $u \geq 0$

(ii) $\operatorname{spt} u \subset U$

(iii) $K \subset V \subset \{x : u(x) = 1\}$

(iv) $\int_{R^n} |Du|^p dx \leq C$.

To prove Step 2, first observe that $H_S^{n-p}(K) < \infty$ (see Exercise 1.10). Since K is compact, there exists a finite sequence of open balls $\{B(r_i)\}$, $i = 1, 2, \ldots, m$, such that

$$K \subset \bigcup_{i=1}^{m} B(r_i) \subset \bigcup_{i=1}^{m} B(2r_i) \subset U$$

and

$$\sum_{i=1}^{\infty} \alpha(n-p) r_i^{n-p} \leq H_S^{n-p}(K) + 1.$$

Let $V = \cup_{i=1}^{m} B(r_i)$ and define u_i to be that piecewise linear function such that $u_i = 1$ on $\overline{B}(r_i)$, $u_i = 0$ on $R^n - B(2r_i)$. Let

$$u = \max\{u_i : i = 1, 2, \ldots, m\}$$

to establish Step 2.

STEP 3. For each positive integer k, let

$$U_k = \left\{ x : d(x, K) < \frac{1}{k} \right\}.$$

Employ Step 2 to find corresponding u_k such that

$$\int_{R^n} |Du_k|^p dx \leq C$$

for $k = 1, 2, \ldots$.

STEP 4. Use Theorem 2.5.1 to find a subsequence $\{u_k\}$ and $u \in W_0^{1,p}(R^n)$ such that $u_k \to u$ weakly in $W_0^{1,p}(R^n)$ and $u_k \to u$ strongly in L^p. Hence, conclude that $u = 1$ almost everywhere on K and that $u = 0$ on $R^n - K$.

STEP 5. Conclude from Theorem 2.1.4 that $|K| = 0$ and therefore that $u = 0$.

STEP 6. Use the Banach–Saks theorem to find a subsequence $\{u_k\}$ such that

$$v_j = \frac{1}{j} \sum_{k=1}^{j} u_k$$

converges strongly to u in $W_0^{1,p}(R^n)$. Thus, $\|Dv_j\|_p \to 0$ as $j \to \infty$. But

$$\gamma_{1,p}(K) \leq \int_{R^n} |Dv_j|^p dx$$

for each $j = 1, 2, \ldots$.

2.16. In this problem we sketch a proof of the fact that $\gamma_{1,p}(K) = 0$ implies $H^{n-p+\varepsilon}(K) = 0$ whenever $\varepsilon > 0$. The proof requires some elementary results found in subsequent chapters.

STEP 1. For each positive integer i, there exists $u_i \in C_0^\infty(R^n)$ such that $u_i \geq 1$ on a neighborhood of K and

$$\int_{R^n} |Du_i|^p dx \leq \frac{1}{2^i}.$$

Let $v = \sum_{i=1}^\infty u_i$ and conclude that $v \in W^{1,p}(R^n)$. Also note that $K \subset$ interior $\{x : v(x) \geq k\}$ whenever $k \geq 1$. Therefore, for $x \in K$,

$$\liminf_{r \to 0} \overline{v}(x,r) = \infty$$

where

$$\overline{v}(x,r) = \fint_{B(x,r)} v(y)dy.$$

STEP 2. For all $x \in K$ and $\varepsilon > 0$,

$$\limsup_{r \to 0} r^{p-n-\varepsilon} \int_{B(x,r)} |Dv|^p dy = \infty.$$

If this were not true, there would exist $k < \infty$ such that

$$r^{p-n-\varepsilon} \int_{B(x,r)} |Dv|^p dy \leq k$$

for all small $r > 0$. For all such r, it follows from a classical version of the Poincaré inequality (Theorem 4.2.2) that

$$\fint_{B(x,r)} |v(y) - \overline{v}(x,r)|^p dy \leq Cr^{p-n} \int_{B(x,r)} |Dv|^p dy \leq Cr^\varepsilon.$$

Thus conclude that

$$|\overline{v}(x,r/2) - \overline{v}(x,r)| \leq Cr^{\varepsilon/p}$$

for all small $r > 0$. Therefore, the sequence $\{\overline{v}(x, 1/2^j)\}$ has a finite limit, contradicting the conclusion of Step 1.

STEP 3. Use Lemma 3.2.1 to reach the desired conclusion.

2.17. At the end of Section 2.3 we refer to [AR2] for the result that $C^\infty(\overline{\Omega})$ is dense in $W^{k,p}(\Omega)$ provided Ω possesses the segment property. Prove this result directly if the boundary of Ω can be locally represented as the graph of a Lipschitz function.

2.18. Show that $C^{0,1}(\Omega) = W^{1,\infty}(\Omega)$ whenever Ω is a domain in R^n.

2.19. This problem addresses the issue raised in (2.6.6). If $f \in L^p(R^n)$ and $\alpha p < n$, then Theorem 2.8.4 states that

$$\|I_\alpha * f\|_q \leq C\|f\|_p$$

where $q = p^*$. A simple homogeneity argument shows that in order for this inequality to hold for all $f \in L^p$, it is necessary for $q = p^*$. For $\delta > 0$, let $\tau_\delta f(x) = f(\delta x)$. Then

$$\|I_\alpha * (\tau_\delta f)\|_q \leq \delta^{-n/p}\|f\|_p$$

and

$$I_\alpha * (\tau_\delta f) = \delta^{-\alpha}\tau_\delta(I_\alpha * f).$$

Hence,

$$\|I_\alpha * (\tau_\delta f)\|_q = \delta^{-\alpha}\delta^{-n/q}\|I_\alpha * f\|_q,$$

thus requiring

$$\frac{1}{q} = \frac{1}{p} - \frac{\alpha}{n}.$$

Historical Notes

2.1. It is customary to refer to the spaces of weakly differentiable functions as Sobolev spaces, although various notions of weak differentiability were used before Sobolev's work, [SO2]; see also [SO1], [SO3]. Beppo Levi in 1906 and Tonelli [TO] both used the class of functions that are absolutely continuous on almost all lines parallel to the coordinate axes, the property that essentially characterizes Sobolev functions (Theorem 2.1.4). Along with Sobolev, Calkin [CA] and Morrey [MO1] developed many of the properties of Sobolev functions that are used today. Although many authors contributed to the theory of Sobolev spaces, special note should be made of the efforts of Aronszajn and Smith, [ARS1], [ARS2], who made a detailed study of the pointwise behavior of Sobolev functions through their investigations of Bessel potentials.

2.2. Theorem 2.2.1 was originally proved by Rademacher [RA]. The proof that is given is attributed to C.B. Morrey [MO1, Theorem 3.1.6]; the proof we give appears in [S]. In our development, Rademacher's theorem was used to show that Sobolev functions remain invariant under composition with bi-Lipschitzian transformations. However, it is possible to obtain a stronger result by using different techniques as shown in [Z3]. Suppose $T : R^n \to R^n$ is a bi-measurable homeomorphism with the property that it and its inverse are in $W^{1,p}_{\text{loc}}(R^n, R^n)$, $p > n - 1$. If $u \in W^{1,\bar{p}}_{\text{loc}}(R^n)$ where $\bar{p} = p[p - (n-1)]^{-1}$, then $u \circ T \in W^{1,1}_{\text{loc}}(R^n)$. With this it is possible to show that if $u \in W^{1,n}_{\text{loc}}(R^n)$ and T is a K-quasiconformal mapping, then $u \circ T \in W^{1,n}_{\text{loc}}(R^n)$.

2.3. Theorem 2.3.2 is due to Meyers and Serrin, [MSE].

2.4. Theorem 2.4.1 is the classical Sobolev inequality [SO1], [SO2], which was also developed by Gagliardo [GA1], Morrey [MO1], and Nirenberg [NI2]. The proof of Theorem 2.4.1 for the case $p < n$ is due to Nirenberg [NI2].

2.5. Theorem 2.5.1 originated in a paper by Rellich [RE] in the case $p = 2$ and by Kondrachov [KN] in the general case. Generally, compactness theorems are of importance in analysis, but this one is of fundamental importance, especially in the calculus of variations and partial differential equations. There are variations of the Rellich–Kondrachov result that yield a slightly stronger conclusion. For example, we have the following result due to Frehse [FRE]: Let $\Omega \subset R^n$ be a bounded domain and suppose $u_i \in W^{1,p}(\Omega)$, $1 \leq p < n$, is a bounded sequence of functions with the property that for each $i = 1, 2, \ldots,$

$$\int_{R^n} |Du_i|^{p-2} Du_i \cdot D\varphi \, dx \leq M' \|\varphi\|_\infty$$

for all $\varphi \in W^{1,p}(\Omega) \cap L^\infty(\Omega)$. Then there exist $u \in W^{1,p}(\Omega)$ and a subsequence such that $u_i \to u$ strongly in $W^{1,q}(\Omega)$, whenever $q < p$.

2.6. Potential theory is an area of mathematics whose origins can be traced to the 18th century when Lagrange in 1773 noted that gravitational forces derive from a function. This function was labeled a potential function by Green in 1828 and simply a potential by Gauss in 1840. In 1782 Laplace showed that in a mass free region, this function satisfies what is now known as Laplace's equation. The fundamental principles of this theory were developed during the 19th century through the efforts of Gauss, Dirichlet, Riemann, Schwarz, Poincaré, Kellogg, and many others, and they constitute today classical potential theory. Much of the theory is directed to the understanding of boundary value problems for the Laplace operator and its linear counterparts. With the work of H. Cartan [CAR1], [CAR2], in the early 1940s, began an important new phase in the development of potential theory with an approach based on a Hilbert-space structure of sets of measures of finite energy. Later, J. Deny [DE] enriched the theory further with the concepts and techniques of distributions. At about the same time, potential theory and a general theory of capacities were being developed from the point of view of an abstract structure based on a set of fundamental axioms. Among those who made many contributions in this direction were Brelot [BRT], Choquet [CH], Deny, Hervé, Ninomiya, and Ohtsuka. The abstract theory of capacities is compatible with the recent development of capacities associated with non-linear potential theory which, among other applications, is used to study questions related to non-linear partial differential equations. The first comprehensive treatment of non-linear potential theory and its associated Bessel capacity was developed by Meyers [ME1],

Havin and Maz'ya [HM], and Rešetnjak [RES]. Most of the material in Section 2.6 has been adopted from [ME1].

2.7. The co-area formula as stated in Theorem 2.7.3 was proved by Federer in [F1]. In the case $m = 2$, $k = 1$, Kronrod [KR] used the right side of (2.7.1) to define the variation of a function of two variables. Fleming and Rishel [FR] established a version of Theorem 2.7.1 for BV functions. Another version resembling the statement in Theorem 2.7.3 for BV functions appears in [F4, Section 4.5.9].

The proof of the best possible constant in the Sobolev inequality (Theorem 2.7.4) is due to Fleming and Federer [FF]. Their result can be stated as follows:

$$n\alpha(n) = \sup \frac{\|Du\|}{\|u\|_{n/(n-1)}}$$

where the supremum is taken over all $u \in C_0^\infty(R^n)$. Talenti [TA] extended this result to the case $p > 1$ by determining the constant $C(n,p)$ defined by

$$C(n,p) = \sup \frac{\|Du\|_p}{\|u\|_{p^*}}.$$

He showed that

$$C(n,p) = \pi^{-1/2} n^{-1/2} \left(\frac{p-1}{n-p}\right)^{1-(1/p)} \left[\frac{\Gamma(1+\frac{n}{2})\Gamma(n)}{\Gamma(\frac{n}{p})\Gamma(1+n-\frac{n}{p})}\right]^{1/n}.$$

He also showed that if the supremum is taken over all functions which decay rapidly at infinity, the function u that attains the supremum in the definition of $C(n,p)$ is of the form

$$u(x) = (a + b|x|^{p/(p-1)})^{1-n/p}$$

where a and b are positive constants. This leads to the following observation: in view of the form of the extremal function, it follows that if Ω is a bounded domain and if $u \in W_0^{1,p}(\Omega)$ has compact support, then by extending u to be zero outside of Ω, we have

$$\|u\|_{p^*} < C(n,p)\|Du\|_p.$$

Brezis and Lieb [BL] provide a lower bound for the difference of the two sides of this inequality for $p = 2$. They show that there is a constant $C(\Omega, n)$ such that

$$C(\Omega,n)\|u\|^2_{(q,\infty)} + \|u\|^2_{2^*} \leq C(n,p)\|Du\|_2^2$$

where $q = n/(n-2)$ and $\|u\|_{(q,\infty)}$ denotes the weak L^q-norm of u (see Definition 1.8.6).

2.8. The maximal theorem 2.8.2 was initially proved by Hardy–Littlewood [HL] for $n = 1$ and for arbitrary n by Wiener [WI]. The proofs of Theorem 2.8.4 and its preceding lemma are due to Hedberg [HE1].

2.9. The proof of exponential integrability in Theorem 2.9.1 is taken from [GT] while the improved version that appears in Theorem 2.9.2 was proved by Hedberg [HE1]. The question concerning sharpness of this inequality has an interesting history. Trudinger [TR] proved (2.9.1) for Sobolev functions in $W^{k,p}$, $n/p = k$, with the power p' replaced by n'. However, when $n/p > 1$, Strichartz [STR] noted that Trudinger's result could be improved with the appearance of the larger power p'. The reason why Trudinger's proof did not obtain the optimal power is that the case of $k > 1$ was reduced to the case of $k = 1$ by using the result that if $u \in W^{k,p}$, $k \geq 2$, $kp = n$, then $u \in W^{1,n}$. However, in this reduction argument, some information is lost because if $u \in W^{k,p}$ then u is actually in a better space than $W^{1,n}$. In fact, by appealing to Theorem 2.10.3, we find that the first derivatives are in the Lorentz space $L(n,p) \subset L^n$. This motivated Brezis and Wainger to pursue the matter further in [BW] where Theorem 2.9.4 and other interesting results are proved. The sharpness of the Sobolev imbedding theorem in the case of critical indices was also considered in [HMT], where it was shown that the space $W_0^{1,n}(\Omega)$ could not be imbedded in the Orlicz space $L_\varphi(\Omega)$ where $\varphi(t) = \exp(|t|^{n/(n-1)} - 1)$.

The other question of sharpness of the inequality pertains to the constant β that appears in (2.9.4). It was shown in [MOS] that (2.9.4) remains valid for $\beta = n/\omega(n-1)$ in the case of Sobolev functions that vanish on the boundary of a domain. The optimal result has recently been proved by Adams [AD8] where (2.9.4) has been established for $\beta = n/\omega(n-1)$ and all $\alpha > 0$.

2.10. Most of the material in this section was developed by Brezis and Wainger [BW] although Theorems 2.10.1 and 2.10.2 and due to O'Neil [O].

3

Pointwise Behavior of Sobolev Functions

In this chapter the pointwise behavior of Sobolev functions is investigated. Since the definition of a function $u \in W^{k,p}(\Omega)$ requires that the kth-order distributional derivatives of u belong to $L^p(\Omega)$, it is therefore natural to inquire whether the function u possesses some type of regularity (smoothness) in the classical sense. The main purpose of this chapter is to show that this question can be answered in the affirmative if interpreted appropriately. Although it is evident that Sobolev functions do not possess smoothness properties in the usual classical sense, it will be shown that if $u \in W^{k,p}(R^n)$, then u has derivatives of order k when computed in the metric induced by the L^p-norm. That is, it will be shown for all points x in the complement of some exceptional set, there is a polynomial P_x of degree k such that the L^p-norm of the integral average of the remainder $|u - P_x|$ over a ball $B(x,r)$ is $o(r^k)$. Of course, if u were of class C^k, then the L^p-norm could be replaced by the sup norm.

We will also investigate to what extent the converse of this statement is true. To this end, it will be shown that if u has derivatives of order k in the L^p-sense at all points in an open set Ω, and if the derivatives are in $L^p(\Omega)$, then $u \in W^{k,p}(\Omega)$. This is analogous to the classical fact that if a function u defined on a bounded interval is differentiable at each point and if u' is integrable, then u is absolutely continuous. In order to further pursue the question of regularity, it will be established that u can be approximated in a strong sense by functions of class C^ℓ, $\ell \leq k$. The approximants will have the property that they are close to u in the Sobolev norm and that they agree pointwise with u on large sets. That is, the sets on which they do not agree will have small capacity, thus establishing a Lusin-type approximation for Sobolev functions.

3.1 Limits of Integral Averages of Sobolev Functions

In this and the next two sections, it will be shown that a Sobolev function $u \in W^{k,p}(\Omega)$ can be defined everywhere, except for a set of capacity zero, in terms of its integral averages. This result is analogous to the one that holds for integrable functions, namely, if $u \in L^1$, then

$$\fint_{B(x,r)} |u(y) - u(x)| dy \to 0 \quad \text{as} \quad r \to 0$$

for almost all $x \in R^n$. Since our result deals with Sobolev functions, the proof obviously will require knowledge of the behavior of the partial derivatives of u. The development we present here is neither the most efficient nor elegant. These qualities have been sacrificed in order give a presentation that is essentially self-contained and clearly demonstrates the critical role played by the gradient of u in order to establish the main result, Theorem 3.3.3. Later, in Section 3.10, we will return to the subject of Lebesgue points and prove a result (Theorem 3.10.2) that extends Theorem 3.3.3. Its proof will employ the representation of Sobolev functions as Bessel potentials (Theorem 2.6.1) and the Hardy–Littlewood maximal theorem (Theorem 2.8.2).

In this first section, it will be shown that the limit of integral averages of Sobolev functions exist at all points except possibly for a set of capacity zero. We begin by proving a lemma that relates the integral average of u over two concentric balls in terms of the integral of the gradient.

3.1.1. Lemma. *Let $u \in W^{1,p}[B(x_0, r)]$, $p \geq 1$, where $x_0 \in R^n$ and $r > 0$. Let $0 < \delta < r$. Then*

$$\begin{aligned} r^{-n} \int_{B(x_0,r)} u(y) dy - \delta^{-n} \int_{B(x_0,\delta)} u(y) dy &= \frac{1}{n} r^{-n} \int_{B(x_0,r)} [Du(y) \cdot (y - x_0)] dy \\ &- \frac{1}{n} \delta^{-n} \int_{B(x_0,\delta)} [Du(y) \cdot (y - x_0)] dy \\ &- \frac{1}{n} \int_{B(x_0,r) - B(x_0,\delta)} |y - x_0|^{-n} [Du(y) \cdot (y - x_0)] dy. \end{aligned} \tag{3.1.1}$$

Proof. Define μ on R^1 by

$$\mu(t) = \begin{cases} \delta^{-n} - \rho^{-n} & t \leq \delta \\ t^{-n} - \rho^{-n} & \delta < t \leq r \\ 0 & t > r. \end{cases}$$

Define a vector field V by $V(y) = \mu(|y - x_0|)(y - x_0)$. Since u is the strong limit of smooth functions defined on $B(x_0, r)$ (Theorem 2.3.2), an application of the Gauss–Green theorem implies

$$\int_{B(x_0,r)} u(y) \operatorname{div} V(y) dy = - \int_{B(x_0,r)} Du(y) \cdot (y - x_0) \mu(|y - x_0|) dy. \tag{3.1.2}$$

An easy calculation of $\operatorname{div} V$ establishes equation (3.1.1). □

3.1.2. Lemma. *Let ℓ be a positive real number such that $\ell p < n$, $p \geq 1$, and let $u \in W^{1,p}(R^n)$. Then*

$$\int_{R^n} |y-x|^{\ell-n}|u(y)|dy \leq \frac{1}{\ell}\int_{R^n} |y-x|^{\ell-n+1}|Du(y)|dy \tag{3.1.3}$$

for all $x \in R^n$.

Proof. (i) We suppose first of all that u vanishes outside a bounded set. Let $x \in R^n$ and for each positive integer j, define a C^∞ vector field V_j on R^n by

$$V_j(y) = \left[\frac{1}{j} + |y-x|^2\right]^{(1/2)(\ell-n)} (y-x).$$

Since $|u| \in W^{1,p}(R^n)$ (by Corollary 2.1.8), $|u|$ is therefore the strong limit of smooth functions with compact support. Therefore, by the Gauss–Green theorem,

$$\int_{R^n} \operatorname{div} V_j(y)|u(y)|dy = -\int_{R^n} V_j(y) \cdot D(|u|)(y)]dy.$$

Moreover, since $|D(|u|)| = |Du|$ a.e.,

$$\int_{R^n} \operatorname{div} V_j(y)|u(y)|dy \leq \int_{R^n} V_j(y)|Du(y)|dy. \tag{3.1.4}$$

By calculating the divergence on the left-hand side of (3.1.4) one obtains

$$\int_{R^n} \left[\frac{1}{j} + |y-x|^2\right]^{(1/2)(\ell-n-2)} \left[\ell|y-x|^2 + \frac{n}{j}\right] |u(y)|dy$$

$$\leq \int_{R^n} \left[\frac{1}{j} + |y-x^2|\right]^{(1/2)(\ell-n)} |y-x|\,|Du(y)|dy.$$

The inequality (3.1.3) now follows, in this case, when $j \to \infty$.

(ii) The general case. Let η be a C^∞ function on R, such that $0 \leq \eta \leq 1$, $\eta(t) = 1$ when $t \leq 1$ and $\eta(t) = 0$ when $t \geq 2$. Define

$$u_j(y) = u(y)\eta(j^{-1}|y|)$$

for $y \in R^n$. By applying (i) to u_j and then letting $j \to \infty$, one can verify (3.1.3) in the general case. □

3.1.3. Lemma. *Let ℓ be a positive real number and k a positive integer such that $(k+\ell-1)p < n$. Then there exists a constant $C = C(n,k,\ell)$ such that*

$$\int_{R^n} |y-x|^{\ell-n}|u(y)|dy \leq C\sum_{|\alpha|=k}\int_{R^n} |y-x|^{\ell-n+k}|D^\alpha u(y)|dy$$

for all $x \in R^n$ and all $u \in W^{k,p}(R^n)$.

This follows from Lemma 3.1.2 by mathematical induction. □

We are now in a position to prove the main theorem of this section concerning the existence of integral averages of Sobolev functions.

3.1.4. Theorem. *Let k be a positive integer such that $kp < n$, $p > 1$, let Ω be a non-empty open subset of R^n and let $u \in W^{k,p}(\Omega)$. Then there exists a subset E of Ω, such that*

$$B_{k,p}(E) = 0$$

and

$$\lim_{\delta \to 0+} ⨍_{B(x,\delta)} u(y)dy \tag{3.1.5}$$

exists for all $x \in \Omega - E$.

Proof. (i) We suppose first of all that $\Omega = R^n$. Define

$$g(y) = \sum_{|\alpha|=k} |D^\alpha u(y)| \tag{3.1.6}$$

for $y \in R^n$. Then $g \in L^p(R^n)$. Let E be the set of all those points x of R^n for which

$$(I_k * g)(x) = \infty. \tag{3.1.7}$$

Then, from the definition of Riesz capacity (Definition 2.6.2),

$$R_{k,p}(E) = 0,$$

and therefore from (2.6.7),

$$B_{k,p}(E) = 0.$$

Consider $x \in R^n \sim E$. By (3.1.1),

$$\begin{aligned} ⨍_{B(x,1)} u(y)dy - \delta^{-n} ⨍_{B(x,\delta)} u(y)dy &= \frac{1}{n} ⨍_{B(x,1)} [Du(y) \cdot (y-x)]dy \\ &- \frac{1}{n}\delta^{-n} ⨍_{B(x,\delta)} [Du(y) \cdot (y-x)]dy \\ &- \frac{1}{n} ⨍_{\delta<|y-x|<1} |y-x|^{-n}[Du(y) \cdot (y-x)]dy. \end{aligned} \tag{3.1.8}$$

When $k = 1$, it follows from (3.1.6) and (3.1.7) that

$$⨍_{B(x,1)} |y-x|^{1-n}|Du(y)|dy < \infty. \tag{3.1.9}$$

When $k > 1$, it follows from Lemma 3.1.3 with $\ell = 1$ and $k-1$ substituted for k, that

$$\int_{R^n} |y-x|^{1-n}|D_i u(y)|dy \leq C \sum_{|\alpha|=k-1} \int_{R^n} |y-x|^{k-n}|D^\alpha[D_i u(y)]|dy,$$

which, by (3.1.6) and (3.1.7), is finite. Thus (3.1.9) still holds when $k > 1$.
By (3.1.9)

$$\lim_{\delta\to 0+} \int_{\delta<|y-x|<1} |y-x|^{-n}[Du(y)\cdot(y-x)]dy \tag{3.1.10}$$

exists. It also follows from (3.1.9) that

$$\lim_{\delta\to 0+} \int_{B(x,\delta)} |y-x|^{1-n}|Du(y)|dy = 0,$$

hence

$$\delta^{-n} \int_{B(x,\delta)} [Du(y)\cdot(y-x)]dy \to 0 \tag{3.1.11}$$

as $\delta \to 0+$. It now follows from (3.1.8), (3.1.10), and (3.1.11) that the limit in (3.1.5) exists.

(ii) The general case. Let Ω be an open set of R^n. There exists an increasing sequence $\{\varphi_j\}$ of non-negative C^∞ functions on R^n, with compact supports, and $\operatorname{spt}\varphi_j \subset \Omega$ for all j such that the interiors of the sets

$$\{x : x \in R^n \quad \text{and} \quad \varphi_j(x) = 1\}$$

tend to Ω as $j \to \infty$. Define

$$u_j(x) = \begin{cases} \varphi_j(x)\cdot u(x) & x \in \Omega \\ 0 & x \notin \Omega. \end{cases}$$

By applying (i) to each of the functions u_j, one can easily prove the theorem in this case. □

3.2 Densities of Measures

Here some basic results concerning the densities of arbitrary measures are established that will be used later in the development of Lebesgue points for Sobolev functions.

3.2.1. Lemma. *Let $\mu \geq 0$ be a Radon measure on R^n. Let $0 < \lambda < \infty$ and $0 < \alpha \leq n$. Suppose for an arbitrary Borel set $A \subset R^n$ that*

$$\limsup_{r\to 0} \frac{\mu[B(x,r)]}{r^\alpha} > \lambda$$

for each $x \in A$. Then there is a constant $C = C(\alpha, n)$ such that

$$\mu(A) \geq C\lambda H^\alpha(A).$$

Proof. Assume $\mu(A) < \infty$ and choose $\varepsilon > 0$. Let $U \supset A$ be an open set with $\mu(U) < \infty$. Let $\mathcal{G}$ be the family of all closed balls $B(x, r) \subset U$ such that

$$x \in A, \quad 0 < r < \varepsilon/2, \quad \frac{\mu[B(x,r)]}{r^\alpha} > \lambda.$$

Clearly, $\mathcal{G}$ covers A finely and thus, by Corollary 1.3.3, there is a disjoint subfamily $\mathcal{F} \subset \mathcal{G}$ such that

$$A \subset [\cup\{B : B \in \mathcal{F}^*\}] \cup [\cup\{\hat{B} : B \in \mathcal{F} - \mathcal{F}^*\}]$$

whenever $\mathcal{F}^*$ is a finite subfamily of $\mathcal{F}$. Thus, by Definition 1.4.1,

$$H^\alpha_{5\varepsilon}(A) \leq C \sum_{B \in \mathcal{F}^*} \left(\frac{\delta(B)}{2}\right)^\alpha + C5^\alpha \sum_{B \in \mathcal{F} - \mathcal{F}^*} \left(\frac{\delta(B)}{2}\right)^\alpha$$

where $\delta(B)$ denotes the diameter of the ball B. Since $\mathcal{F} \subset \mathcal{G}$ and $\mathcal{F}$ is disjoint, we have

$$\begin{aligned} \sum_{B \in \mathcal{F}} \left(\frac{\delta(B)}{2}\right)^\alpha &\leq C\lambda^{-1} \sum_{B \in \mathcal{F}} \mu(B) \\ &\leq C\lambda^{-1}\mu(U) < \infty. \end{aligned}$$

Since

$$C5^\alpha \sum_{B \in \mathcal{F} - \mathcal{F}^*} \left(\frac{\delta(B)}{2}\right)^\alpha$$

can be made arbitrarily small with an appropriate choice of $\mathcal{F}^*$, we conclude

$$H^\alpha_{5\varepsilon}(A) \leq C\lambda^{-1}\mu(U).$$

Since μ is a Radon measure, we have that $\mu(A) = \inf\{\mu(U) : U \supset A,\ U \text{ open}\}$. Thus, letting $\varepsilon \to 0$, we obtain the desired result. □

3.2.2. Lemma. *Let $\mu \geq 0$ be a Radon measure on R^n that is absolutely continuous with respect to Lebesgue measure. Let*

$$A = R^n \cap \left\{x : \limsup_{r \to 0} \frac{\mu[B(x,r)]}{r^\alpha} > 0\right\}.$$

Then, $H^\alpha(A) = 0$ whenever $0 \leq \alpha < n$.

Proof. The result is obvious for $\alpha = 0$, so choose $0 < \alpha < n$. For each positive integer i let

$$A_i = R^n \cap \left\{x : |x| < i, \limsup_{r \to 0} \frac{\mu[B(x,r)]}{r^\alpha} > i^{-1}\right\}$$

and conclude from the preceding lemma that

$$\mu(A_i) \geq Ci^{-1}H^\alpha(A_i). \tag{3.2.1}$$

Since A_i is bounded, $\mu(A_i) < \infty$. Therefore $H^\alpha(A_i) < \infty$ from (3.2.1). Since $\alpha < n$, $H^n(A_i) = 0$ and therefore $|A_i| = 0$ from Theorem 1.4.2. The absolute continuity of μ implies $\mu(A_i) = 0$ and consequently $H^\alpha(A_i) = 0$ from (3.2.1). But $A = \cup_{i=1}^\infty A_i$, and the result follows. □

3.2.3. Corollary. *Suppose* $u \in L^p(R^n)$, $1 \leq p < \infty$, *and let* $0 \leq \alpha < n$. *If* E *is defined by*

$$E = \left\{ x : \limsup_{r \to 0} r^{-\alpha} \int_{B(x,r)} |u(x)|^p dx > 0 \right\},$$

then $H^\alpha(E) = 0$.

Proof. This follows directly from Lemma 3.2.2 by defining a measure μ as

$$\mu(A) = \int_A |u|^p dx.$$ □

3.3 Lebesgue Points for Sobolev Functions

We will now prove the principal result of the first three sections (Theorem 3.3.3) which is concerned with the existence of Lebesgue points for Sobolev functions. We will show that if $u \in W^{k,p}(R^n)$, then

$$\lim_{r \to 0} \fint_{B(x,r)} |u(y) - u(x)|^p dx = 0$$

for $B_{k,p}$-q.e. $x \in R^n$. This is stronger than the conclusion reached in Theorem 3.1.4, which only asserts the existence of the limit of integral averages. However, in case $u \in W^{1,p}(R^n)$, the existence of the limit of integral averages implies the one above concerning Lebesgue points. In this case, we can use the fact that $|u - \rho| \in W^{1,p}_{\text{loc}}(R^n)$ for each real number ρ and then apply Theorem 3.1.4 to conclude that

$$\lim_{r \to 0} \fint_{B(x,r)} |u(y) - \rho| dy$$

exists for $B_{1,p}$-q.e. $x \in R^n$. Of course, the exceptional set here depends on ρ. The object of Exercise 3.1 is to complete this argument. This approach fails to work if $u \in W^{k,p}(R^n)$ since it is not true in general that $|u| \in W^{k,p}(R^n)$, cf. Remark 2.1.10.

3.3.1. Lemma. *Let k be a non-negative integer and λ, p real numbers such that $p > 1$, $kp < n$, and $k < \lambda < n/p$. If*

$$u \in W^{k,p}(R^n), \tag{3.3.1}$$

then

$$\delta^{\lambda-k} \fint_{B(x,\delta)} u(y)dy \to 0 \tag{3.3.2}$$

as $\delta \to 0+$, for all $x \in R^n$ except for a set E with $B_{\lambda,p}(E) = 0$.

Theorem 3.1.4 states that the integral averages converge to a finite value at all points in the complement of a $B_{k,p}$-null set. This lemma offers a slight variation in that the integral averages when multiplied by the factor $\delta^{\lambda-k}$ converge to 0 on a larger set, the complement of a $B_{\lambda,p}$-null set. At some points of this larger set, the integral averages may converge to infinity, but at a rate no faster than $\delta^{k-\lambda}$.

Proof of Lemma 3.3.1. (i) Suppose $k = 0$. It follows from Corollary 3.2.3 that

$$\delta^{p\lambda-n} \int_{B(x,\delta)} |u(y)|^p dy \to 0 \tag{3.3.3}$$

as $\delta \to 0+$, for all $x \in R^n$ except for a set E with $H^{n-\lambda p}(E) = 0$. From the definition of Hausdorff measure, for $\varepsilon > 0$ there is a countable number of sets $\{E_i\}$ such that $E \subset \cup E_i$ and $\Sigma(\operatorname{diam} E_i)^{n-\lambda p} < \varepsilon$. Each E_i is contained in a ball B_i of radius r_i where $r_i = \operatorname{diam} E_i$. Therefore, with the aid of Theorem 2.6.13,

$$B_{\lambda,p}(E) \le \sum_{i=1}^{\infty} B_{\lambda,p}(E_i) \le C \sum_{i=1}^{\infty} r_i^{n-\lambda p} \le C\varepsilon.$$

Since ε is arbitrary, we have that $B_{\lambda,p}(E) = 0$.

Now consider $x \in R^n - E$. From Hölder's inequality, there is a constant $C = C(n,p)$ such that

$$\delta^{\lambda-n} \left| \int_{B(x,\delta)} u(y)dy \right| \le C \left[\delta^{\lambda p-n} \int_{B(x,\delta)} |u(y)|^p dy \right]^{1/p}. \tag{3.3.4}$$

(3.3.2) now follows from (3.3.3) and (3.3.4), and (i) is established.

(ii) Now suppose $k > 0$. Let E be the set of all x for which

$$\int_{R^n} |y-x|^{\lambda-n} \left[\sum_{|\alpha|=k} |D^\alpha u(y)| \right] dy = \infty. \tag{3.3.5}$$

Then $R_{\lambda,p}(E) = 0$ and therefore $B_{\lambda,p}(E) = 0$ by (2.6.8).

Consider $x \in R^n - E$. When $k = 1$, it follows from (3.3.5) that

$$\int_{B(x,1)} |y-x|^{\lambda-k+1-n}|Du(y)|dy < \infty. \tag{3.3.6}$$

When $k > 1$, we replace ℓ by $\lambda - k + 1$ and k by $k - 1$ in Lemma 3.1.3 and again derive (3.3.6) from (3.3.5). For $x \in R^n - E$, we now show that

$$\delta^{\lambda-k} \int_{\delta<|y-x|<1} |y-x|^{1-n}|Du(y)|dy \to 0, \tag{3.3.7}$$

as $\delta \downarrow 0$. Let $r \in (0,1)$ be arbitrary. Clearly

$$\delta^{\lambda-k} \int_{r<|y-x|<1} |y-x|^{1-n}|Du(y)|dy \to 0 \tag{3.3.8}$$

as $\delta \downarrow 0$. When $0 < \delta < r$ we have

$$\delta^{\lambda-k} \int_{\delta<|y-x|<r} |y-x|^{1-n}|Du(y)|dy \le \int_{B(x,r)} |y-x|^{\lambda-k+1-n}|Du(y)|dy. \tag{3.3.9}$$

It follows from (3.3.6) that the right-hand side of (3.3.9) approaches zero as $r \downarrow 0$. (3.3.7) now follows from (3.3.8) and (3.3.9).

Clearly,

$$\int_{B(x,1)} |y-x|\,|Du(y)|dy < \infty. \tag{3.3.10}$$

Since $\lambda - k - n < 0$, it follows that

$$\delta^{\lambda-k-n} \int_{B(x,\delta)} |y-x|\,|Du(y)|dy \le \int_{B(x,\delta)} |y-x|^{\lambda-k+1-n}|Du(y)|dy,$$

so that by (3.3.6),

$$\delta^{\lambda-k-n} \int_{B(x,\delta)} |y-x|\,|Du(y)|dy \to 0 \tag{3.3.11}$$

as $\delta \downarrow 0$. By putting $r = 1$ in Lemma 3.1.1, one can obtain (3.3.2) from Lemma 3.1.1, (3.3.10), (3.3.11), and (3.3.7). □

3.3.2. Theorem. *Let ℓ, k be integers such that $k \ge 1$, $0 \le k \le \ell$ and $\ell p < n$, $p > 1$. Let $u \in W^{k,p}(R^n)$ and for each $x \in R^n$ and $r > 0$ put*

$$u_{x,r} = \fint_{B(x,r)} u(y)dy.$$

Then

$$r^{(\ell-k)p} \fint_{B(x,r)} |u(y) - u_{x,r}|^p dy \to 0 \tag{3.3.12}$$

as $r \downarrow 0$, *for all* $x \in R^n$ *except for a set* E *with* $B_{\ell,p}(E) = 0$.

Proof. We proceed by induction on k. Suppose to begin with that $k = 0$. It follows from Corollary 3.2.3 that

$$r^{\ell p-n} \int_{B(x,r)} |u(y)|^p dy \to 0 \tag{3.3.13}$$

for all $x \in R^n$, except for a set E' with $H^{n-\ell p}(E') = 0$ and therefore

$$B_{\ell,p}(E') = 0. \tag{3.3.14}$$

We now have, for $x \in R^n - E'$,

$$r^{(\ell-(n/p))} \left[\int_{B(x,r)} |u(y) - u_{x,r}|^p dy\right]^{1/p} \le r^{(\ell-(n/p))} \left[\int_{B(x,r)} |u(y)|^p dy\right]^{1/p}$$

$$+ r^{(\ell-(n/p))} |u_{x,r}| \left[\int_{B(x,r)} dy\right]^{1/p}. \tag{3.3.15}$$

But by Lemma 3.3.1,

$$r^{\ell} u_{x,r} \to 0 \tag{3.3.16}$$

as $r \downarrow 0$, for all $x \in R^n$ except for a set E'' with $B_{\ell,p}(E'') = 0$. (3.3.12) now follows from (3.3.13), (3.3.15), and (3.3.16) in the case $k = 0$.

Now suppose that $k > 0$ and that the theorem has been proved for all functions of $W^{k-1,p}(R^n)$. Let $u \in W^{k,p}(R^n)$. By the Poincaré inequality, which we shall prove in a more general setting later in Chapter 4 (for example, see Theorem 4.4.2),

$$r^{(\ell-k)p-n} \int_{B(x,r)} |u(y) - u_{x,r}|^p dy \le C r^{[\ell-(k-1)]p-n} \int_{B(x,r)} |Du(y)|^p dy, \tag{3.3.17}$$

for all $x \in R^n$, where C depends only on n. By the induction assumption, there exists a set F', with

$$B_{\ell,p}(F') = 0 \tag{3.3.18}$$

and

$$r^{[\ell-(k-1)]p-n} \int_{B(x,r)} |D_i u(y) - (D_i u)_{x,r}|^p dy \to 0 \tag{3.3.19}$$

as $r \downarrow 0$, for all $x \in R^n - F'$. But

$$\left[\int_{B(x,r)} |D_i u(y)|^p dy\right]^{1/p} \le \left[\int_{B(x,r)} |D_i u(y) - (D_i u)_{x,r}|^p dy\right]^{1/p}$$

$$+ \ |(D_i u)_{x,r}| \left[\int_{B(x,r)} dy \right]^{1/p}, \tag{3.3.20}$$

and by Lemma 3.3.1

$$r^{\ell-(k-1)}(D_i u)_{x,r} \to 0 \tag{3.3.21}$$

as $r \downarrow 0$, for all $x \in R^n$, except for a set F'' with

$$B_{\ell,p}(F'') = 0.$$

(3.3.12) now follows from (3.3.17), (3.3.19), (3.3.20), and (3.3.21). This completes the proof. □

3.3.3. Theorem. *Let k be a positive integer such that $kp < n$, let Ω be an open set of R^n and let $u \in W^{k,p}(\Omega)$. Then*

$$\fint_{B(x,r)} |u(y) - u(x)|^p dy \to 0 \tag{3.3.22}$$

as $r \downarrow 0$, for all $x \in \Omega$, except for a set E with $B_{k,p}(E) = 0$.

Proof. (i) When $\Omega = R^n$, (3.3.22) follows from Theorem 3.3.2 and Theorem 3.1.4.

(ii) When Ω is arbitrary, the theorem can be derived from (i) as in the proof of Theorem 3.1.4. □

3.3.4. Corollary. *Let k be a positive integer such that $kp < n$, let Ω be an open set of R^n and let $u \in W^{k,p}(\Omega)$. Then*

$$\lim_{r \to 0+} \fint_{B(x,r)} |u(y)|^p dy \quad \textit{exists and} = |u(x)|^p \tag{3.3.23}$$

for all $x \in \Omega$, except for a set E with $B_{k,p}(E) = 0$.

3.3.5. Remark. Theorem 3.3.3 states that on the average, the oscillation of u at x is approximately equal to $u(x)$ at $B_{k,p}$-q.e. $x \in \Omega$. This can also be stated in terms of the classical concept of *approximate continuity,* which will be used extensively in Chapter 5. A function u is said to be approximately continuous at x_0 if there exists a measurable set A such that

$$\lim_{r \to 0} \frac{|B(x_0, r) \cap A|}{|B(x_0, r)|} = 1 \tag{3.3.24}$$

and u is continuous at x_0 relative to A. It is not difficult to show that if u has a Lebesgue point at x_0 then u is approximately continuous at x_0. A proof of this is given in Remark 4.4.5. Thus, in particular, Theorem 3.3.3 implies that $u \in W^{1,p}(R^n)$ is approximately continuous at $B_{1,p}$-q.e. $x \in R^n$.

Approximate continuity is a concept from measure theory. A similar concept taken from potential theory is *fine continuity* and is defined in terms of *thin* sets. A set $A \subset R^n$ is said to be *thin at x_0 relative to the capacity $B_{k,p}$* if

$$\int_0^1 \left[\frac{B_{k,p}[A \cap B(x_0,r)]}{B_{k,p}[B(x_0,r)]}\right]^{1/(p-1)} \frac{dr}{r} < \infty. \tag{3.3.25}$$

A function u is *finely continuous at x_0* if there exists a set A that is thin at x_0 and

$$\lim_{\substack{x \to x_0 \\ x \notin A}} u(x) = u(x_0).$$

It follows from standard arguments in potential theory that A can be taken as a measurable set. In the case of the capacity, $B_{1,2}$, which is equivalent to Newtonian capacity, these definitions are in agreement with those found in classical potential theory. In view of the fact that

$$|A| \le C[B_{k,p}(A)]^{n/(n-kp)}$$

for any set $A \subset R^n$, it follows that (3.3.25) implies

$$\lim_{r \to 0} \frac{|B(x_0,r) \cap (R^n - A)|}{|B(x_0,r)|} = 1,$$

and therefore fine continuity implies approximate continuity.

We now will show that the approximate continuity property of Sobolev functions can be replaced by fine continuity. First, we need the following lemma.

Lemma. *If $\{A_i\}$ is a sequence of sets each of which is thin at x_0, then there exists a sequence of real numbers $\{r_i\}$ such that*

$$\bigcup_{i=1}^{\infty} A_i \cap B(x_0, r_i)$$

is thin at x_0.

Proof. Because A_i is thin at x_0, it follows that there exists a sequence $\{r_i\} \to 0$ such that

$$\frac{B_{k,p}[A_i \cap B(x_0,r_i)]}{B_{k,p}[B(x_0,r_i)]} \to 0 \quad \text{as} \quad i \to \infty.$$

We may assume the r_i to have been chosen so that

$$\int_0^{r_i} \left[\frac{B_{k,p}[A_i \cap B(x_0,r)]}{B_{k,p}[B(x_0,r)]}\right]^{1/(p-1)} \frac{dr}{r} < 2^{-(i+1)}.$$

Then,

$$
\begin{aligned}
&\int_0^1 \left[\frac{B_{k,p}[A_i \cap B(x_0,r) \cap B(x_0,r_i)]}{B_{k,p}[B(x_0,r)]}\right]^{1/(p-1)} \frac{dr}{r} \\
&= \int_0^{r_i} \left[\frac{B_{k,p}[A_i \cap B(x_0,r) \cap B(x_0,r_i)]}{B_{k,p}[B(x_0,r)]}\right]^{1/(p-1)} \frac{dr}{r} \\
&\quad + \int_{r_i}^1 \left[\frac{B_{k,p}[A_i \cap B(x_0,r) \cap B(x_0,r_i)]}{B_{k,p}[B(x_0,r)]}\right]^{1/(p-1)} \frac{dr}{r} \\
&< 2^{-(i+1)} + B_{k,p}[A_i \cap B(x_0,r_i)]^{1/(p-1)} \int_{r_i}^1 \left[\frac{1}{Cr^{n-kp}}\right]^{1/(p-1)} \frac{dr}{r} \\
&< 2^{-(i+1)} + B_{k,p}[A_i \cap B(x_0,r_i)]^{1/(p-1)} C_1 \left[1 - \frac{1}{B_{k,p}[B(x_0,r_i)]^{1/(p-1)}}\right] \\
&< 2^{-i} \quad \text{for } r_i \text{ sufficiently small.}
\end{aligned}
$$

Since capacity is countably subadditive, the result easily follows. □

For ease of exposition, we now restrict our attention to $u \in W^{1,p}(R^n)$. Again, we see the important role played by the growth of the gradient in order to obtain some regularity at a given point.

Theorem. *Let $x_0 \in R^n$, $p > 1$, and suppose $u \in W^{1,p}(R^n)$ has the property that*

$$\int_0^1 \left[r^{p-n} \int_{B(x_0,r)} |Du|^p dx\right]^{1/(p-1)} \frac{dr}{r} < \infty.$$

Suppose also that

$$\lim_{r\to 0} \fint_{B(x_0,r)} u(y)dy = u(x_0).$$

Then u is finely continuous at x_0.

Proof. For each $\varepsilon > 0$, let

$$A(x_0,\varepsilon) = R^n \cap \{x : |u(x) - u(x_0)| > \varepsilon\}.$$

For $r > 0$, let

$$v_r(x) = \varphi_r(x)[u(x) - \overline{u}(2r)]$$

where φ_r is a smooth function such that $\varphi_r \equiv 1$ on $B(x_0,r)$, $\operatorname{spt}\varphi_r \subset B(x_0,2r)$, $|D\varphi_r| \leq Cr^{-1}$ and where

$$\overline{u}(2r) = \fint_{B(x_0,2r)} u\, dx.$$

Because of the assumption $\overline{u}(2r) \longrightarrow u(x_0)$ as $r \to 0$, note that for all sufficiently small r,

$$v_r(x) \geq \varepsilon/2 \quad \text{for} \quad x \in A(x_0, \varepsilon) \cap B(x_0, r).$$

Therefore, by appealing to Exercise 2.8, which allows $B_{1,p}$ to be expressed in terms of a variational integral, there exists $C = C(p, n)$ such that

$$\begin{aligned} B_{1,p}[A(x_0,\varepsilon) \cap B(x_0,r)] &\leq C(2\varepsilon^{-1})^p \int_{B(x_0,2r)} |Dv_r|^p dx \\ &\leq C(2\varepsilon^{-1})^p \int_{B(x_0,2r)} |Du|^p dx \\ &\quad + (C2\varepsilon^{-1}r^{-1})^p \int_{B(x_0,2r)} |u - \overline{u}(2r)|^p dx. \quad (3.3.26) \end{aligned}$$

An application of Poincaré's inequality (cf. Theorem 4.4.2) yields

$$\int_{B(x_0,2r)} |u - \overline{u}(2r)|^p dx \leq Cr^p \int_{B(x_0,2r)} |Du|^p dx$$

and therefore (3.3.26) can be written as

$$\frac{B_{1,p}[A(x_0,\varepsilon) \cap B(x_0,r)]}{r^{n-p}} \leq C\varepsilon^{-1} r^{p-n} \int_{B(x_0,2r)} |Du|^p dx,$$

which directly implies that $A(x_0, \varepsilon)$ is thin at x_0. Now let ε_j be a sequence tending to 0. By the preceding lemma, there is a decreasing sequence $r_j \to 0$ such that

$$A = \bigcup_{j=1}^{\infty} [A(x_0, \varepsilon_j) \cap B(x_0, r_j)]$$

is thin at x_0. Clearly

$$\lim_{\substack{x \to x_0 \\ x \in R^n - A}} u(x) = u(x_0)$$

and the theorem is established. □

It can be shown that

$$\int_0^1 \left[r^{p-n} \int_{B(x_0,r)} |Du|^p dx \right]^{1/(p-1)} \frac{dr}{r} < \infty \qquad (3.3.27)$$

for $B_{1,p}$-q.e. $x_0 \in R^n$ cf. [ME3]. Therefore, with Theorem 3.3.3, we obtain the following.

Corollary. *If $u \in W^{1,p}(R^n)$ then u finely continuous at all points except for a set of $B_{1,p}$ capacity zero.*

Observe that Corollary 3.2.3 implies

$$\limsup_{r\to 0} r^{p-n} \int_{B(x_0,r)} |Du|^p dx = 0 \tag{3.3.28}$$

for H^{n-p}-a.e. $x_0 \in R^n$. Although (3.3.27) implies (3.3.28) for each x_0, the exceptional set for the former is larger than that for the latter.

3.4 L^p-Derivatives for Sobolev Functions

In the previous three sections, the continuity properties of Sobolev functions were explored through an investigation of Lebesgue points and fine continuity. We now proceed to analyze their differentiability properties. We begin by proving that Sobolev functions can be expanded in a finite Taylor series such that for all points in the complement of an exceptional set, the integral average of the remainder term tends to 0, (Theorem 3.4.2). In keeping with the spirit of this subject, it will be seen that the exceptional set has zero capacity. Observe that Theorem 3.3.3 provides the first step in this direction if we interpret the associated polynomial as one of degree 0 and the remainder at x as $|u(y) - u(x)|$.

When k, m are integers such that $0 \le m \le k$, $(k-m)p < n$ and $u \in W^{k,p}(R^n)$, it follows from Theorem 3.1.4 that there exists a subset E of R^n such that

$$B_{k-m,p}(E) = 0 \tag{3.4.1}$$

and

$$\lim_{r\to 0+} \fint_{B(x,r)} D^\alpha u(y)dy \tag{3.4.2}$$

exists for all $x \in R^n - E$ and for each multi-index α with $0 \le |\alpha| \le m$. Thus, for all such x, we are able to define the Taylor polynomial $P_x^{(m)}$ in the usual way:

$$P_x^{(m)}(y) = \sum_{0\le|\alpha|\le m} \frac{1}{\alpha!} D^\alpha u(x)(y-x)^\alpha. \tag{3.4.3}$$

(Recall the notation introduced in Section 1.1.) Observe that when u is a C^m function on R^n, Taylor's theorem can be expressed in the form

$$\begin{aligned} u(y) = P_x^{(m-1)}(y) + m \sum_{|\alpha|=m} \frac{1}{\alpha!} \Big[\int_0^1 (1-t)^{m-1} \\ \cdot D^\alpha u[(1-t)x + ty]dt \Big] (y-x)^\alpha. \end{aligned} \tag{3.4.4}$$

3.4.1. Theorem. *Let $1 \le m \le k$ and suppose $(k-m)p < n$. Let $u \in$*

$W^{k,p}(R^n)$ and E be the set described in (3.4.1) and (3.4.2). Then

$$\left[\int_{B(x,r)} |u(y) - P_x^{(m)}(y)|^p dy\right]^{1/p} \leq r^m \sum_{|\alpha|=m} \frac{m}{\alpha!} \int_0^1 (1-t)^{m-1}$$

$$\cdot \left[t^{-n} \int_{B(x,tr)} |D^\alpha u(y) - D^\alpha u(x)|^p dy\right]^{1/p} dt \qquad (3.4.5)$$

and

$$\left[\int_{B(x,r)} |u(y) - P_x^{(m-1)}(y)|^p dy\right]^{1/p} \leq r^m \sum_{|\alpha|=m} \frac{m}{\alpha!} \int_0^1 (1-t)^{m-1}$$

$$\cdot \left[t^{-n} \int_{B(x,tr)} |D^\alpha u(y)|^p dy\right]^{1/p} dt, \qquad (3.4.6)$$

for all $x \in R^n$ except for a set $E' \supset E$ with $B_{k-m,p}(E') = 0$.

Proof. (i) Suppose first of all that u is a C^m function on R^n. Let $x \in R^n$, $r > 0$ and put $B = B(x,r)$. Let φ be a function of $L^{p'}(B)$ with $\|\varphi\|_{p'} \leq 1$ where p' is the conjugate of p. By (3.4.3) and (3.4.4),

$$\int_B [u(y) - P_x^{(m)}(y)]\varphi(y)dy = \sum_{|\alpha|=m} \frac{m}{\alpha!}(1-t)^{m-1}$$

$$\cdot \left[\int_B \{D^\alpha u((1-t)x + ty) - D^\alpha u(x)\}(y-x)^\alpha \varphi(y)dy\right] dt.$$

Hence, by Hölder's inequality,

$$\left|\int_B [u(y) - P_x^{(m)}(y)]\varphi(y)dy\right| \leq r^m \sum_{|\alpha|=m} \frac{m}{\alpha!} \int_0^1 (1-t)^{m-1}$$

$$\cdot \left[\int_B |D^\alpha u((1-t)x + ty) - D^\alpha u(x)|^p dy\right]^{1/p} dt.$$

By making the substitution $z = x + t(y-x)$ in the right-hand side and then taking the supremum over all φ, one obtains (3.4.5).

The inequality (3.4.6) can be derived similarly.

(ii) Now let u be an arbitrary function of $W^{k,p}(R^n)$. By Theorem 3.3.3, there exists a set $E' \supset E$, with $B_{k-m,p}(E') = 0$ such that

$$\lim_{\delta \to 0+} \fint_{B(x,\delta)} |D^\alpha u(y) - D^\alpha u(x)|^p dy = 0 \qquad (3.4.7)$$

when $0 \le |\alpha| \le m$ and $x \in R^n - E'$.

Consider $x \in R^n - E'$. There exists a constant M (depending on x), such that

$$\fint_{B(x,\delta)} |D^\alpha u(y)|^p dy \le M \tag{3.4.8}$$

for all $|\alpha| = m$ and all $\delta > 0$. Let $\{\varphi_\varepsilon\}$ be a sequence of regularizers as discussed in Section 1.6. Thus, $\varphi_\varepsilon \in C_0^\infty(R^n)$,

$$\int_{R^n} \varphi_\varepsilon(x)dx = 1, \tag{3.4.9}$$

spt $\varphi_\varepsilon \subset B(0,\varepsilon)$ and

$$\sup_{x\in R^n} \varphi_\varepsilon(x) \le C\varepsilon^{-n} \tag{3.4.10}$$

for all ε (where C depends only on n), while

$$(\varphi_\varepsilon * D^\alpha u)(x) \to D^\alpha u(x) \tag{3.4.11}$$

as $\varepsilon \downarrow 0$, for $0 \le |\alpha| \le m$ and $x \in R^n - E'$. Put $u_\varepsilon = \varphi_\varepsilon * u$. Each $u_\varepsilon \in C^\infty(R^n) \cap W^{k,p}(R^n)$. Let us denote by $(3.4.5)_\varepsilon$ and $(3.4.6)_\varepsilon$ the inequalities (3.4.5) and (3.4.6) with u replaced by u_ε. Since u_ε is smooth we know that $(3.4.5)_\varepsilon$ and $(3.4.6)_\varepsilon$ are valid. By (3.4.11) and Fatou's lemma, the lower-limit as $\varepsilon \downarrow 0$ of the left-hand side of $(3.4.5)_\varepsilon$ and $(3.4.6)_\varepsilon$ is greater than or equal to the left-hand side of (3.4.5) and (3.4.6). The result of the theorem will thus follow from Theorem 1.6.1(ii) and Lebesgue's Dominated Convergence theorem when we show for each α with $|\alpha| = m$ and $r > 0$ fixed, that the following function of t, $0 \le t \le 1$, is bounded; that is,

$$t^{-n} \int_{B(x,tr)} |D^\alpha u_\varepsilon(y)|^p dy \le Mr^n \tag{3.4.12}$$

where M is independent of ε.

We now proceed to establish (3.4.12). For any measurable subset E of R^n, we have (when $|\alpha| = m$)

$$\int_E |D^\alpha u_\varepsilon(y)|^p dy = \int_E \left| \int_{R^n} \varphi_\varepsilon(y-z) D^\alpha u(z) dz \right|^p dy,$$

hence by (3.4.10)

$$\int_E |D^\alpha u_\varepsilon(y)|^p dy \le C^p \varepsilon^{-(np)} \int_E \left[\int_{B(y,\varepsilon)} |D^\alpha u(z)| dz \right]^p dy. \tag{3.4.13}$$

Thus, when $p > 1$, we have by Hölder's inequality

$$\int_E |D^\alpha u_\varepsilon(y)|^p dy \le C^p \varepsilon^{-(np)} \int_E \left[\int_{B(y,\varepsilon)} |D^\alpha u(z)|^p dz \right] \left[\int_{B(y,\varepsilon)} dz \right]^{p-1} dy,$$

so that

$$\int_E |D^\alpha u_\varepsilon(y)|^p dy \le C\varepsilon^{-n} \int_E \int_{B(y,\varepsilon)} |D^\alpha u(z)|^p dz dy, \tag{3.4.14}$$

where C depends only on n and p. When $p = 1$, (3.4.14) follows from (3.4.13).

When $tr \le 3\varepsilon$, we let E be the ball with center x and radius tr. Since $B(y,\varepsilon) \subset B(x, 4\varepsilon)$ when $y \in B(x,tr)$, (3.4.14) implies that

$$\int_{B(x,tr)} |D^\alpha u_\varepsilon(y)|^p dy \le C\varepsilon^{-n} \int_{B(x,tr)} \int_{B(x,4\varepsilon)} |D^\alpha u(z)|^p dz dy.$$

It now follows from (3.4.8) and (3.4.12) holds in the case where $tr \le 3\varepsilon$.

When $tr > 3\varepsilon$, we have

$$\int_{B(x,tr)} |D^\alpha u_\varepsilon(y)|^p dy = \int_{B(x,3\varepsilon)} |D^\alpha u_\varepsilon(y)|^p dy + \int_{3\varepsilon \le |y-x| < tr} |D^\alpha u_\varepsilon(y)|^p dy$$

and a double application of (3.4.14) yields

$$\begin{aligned}\int_{B(x,tr)} |D^\alpha u_\varepsilon(y)|^p dy \le C\varepsilon^{-n} &\int_{B(x,3\varepsilon)} \int_{B(x,4\varepsilon)} |D^\alpha u(z)|^p dz dy \\ &+ C\varepsilon^{-n} \int_{3\varepsilon \le |y-x| < tr} \int_{B(y,\varepsilon)} |D^\alpha u(z)|^p dz dy\end{aligned}$$

and by (3.4.8)

$$\begin{aligned}&\le C' t^n r^n + C\varepsilon^{-n} \int_{3\varepsilon \le |y-x| < tr} \int_{B(0,\varepsilon)} |D^\alpha u(w+y)|^p dw dy \\ &\le C' t^n r^n + C\varepsilon^{-n} \int_{B(0,\varepsilon)} \left[\int_{2\varepsilon \le |y-x| < tr+\varepsilon} |D^\alpha u(y)|^p dy \right] dw\end{aligned}$$

so that

$$\int_{B(x,tr)} |D^\alpha u_\varepsilon(y)|^p dy \le C'' t^n r^n.$$

Thus (3.4.12) is established. □

3.4.2. Theorem. *Let* $0 \le m \le k$ *and suppose* $(k-m)p < n$. *Let* $u \in W^{k,p}(R^n)$. *Then,*

$$r^{-m} \left[\fint_{B(x,r)} |u(y) - P_x^{(m)}(y)|^p dy \right]^{1/p} \to 0$$

as $r \downarrow 0$, *for all* $x \in R^n$, *except for a set* F *with*

$$B_{k-m,p}(F) = 0.$$

This is the main result of this section. In particular, it states that the integral average over a ball of radius r of the remainder term involving the formal Taylor polynomial of degree k tends to 0 as $r \to 0$ at a speed greater than r^k at almost every point. If a Taylor polynomial of smaller degree is considered, the integral average tends to 0 at perhaps a slower speed, but on a larger set.

Proof of Theorem 3.4.2. When $m = 0$, the theorem reduces to Theorem 3.3.3. Suppose $m > 0$. By Theorem 3.3.3,

$$⨍_{B(x,r)} |D^\alpha u(y) - D^\alpha u(x)|^p dy \to 0 \tag{3.4.15}$$

as $r \downarrow 0$, for all $|\alpha| = m$ for all $x \in R^n$, except for a set F with

$$B_{k-m,p}(F) = 0.$$

Consider $x \in R^n - F$ and an α with $|\alpha| = m$. Define

$$\eta(r) = \left[r^{-n} \int_{B(x,r)} |D^\alpha u(y) - D^\alpha u(x)|^p dy \right]^{1/p} \tag{3.4.16}$$

for $r > 0$. By (3.4.15), $\eta(r) \to 0$ as $r \downarrow 0$, hence

$$\int_0^1 (1-t)^{m-1} \eta(tr) dt \to 0 \tag{3.4.17}$$

$r \downarrow 0$. The required result now follows from (3.4.16), (3.4.17), and Theorem 3.4.1. □

3.5 Properties of L^p-Derivatives

In this section we consider arbitrary functions that possess formal Taylor series expansions and investigate their relationship with those functions that have Taylor series expansions in the metric of L^p, such as those discussed in the previous section.

3.5.1. Definition. Let $E \subset R^n$. A bounded function u defined on E belongs to $T^k(E)$, $k > 0$, if there is a positive number M and for each $x \in E$ there is a polynomial $P_x(\cdot)$ of degree less than k of the form

$$P_x(y) = \sum_{|\alpha| \geq 0} \frac{u_\alpha(x)}{\alpha!} (y-x)^\alpha, \quad (u_0 = u) \tag{3.5.1}$$

whose coefficients u_α satisfy

$$|u_\alpha(x)| \leq M \quad \text{for} \quad x \in E, \ 0 \leq |\alpha| < k,$$

and

$$u_\alpha(y) = D^\alpha P_x(y) + R_\alpha(x, y)$$

whenever $x, y \in E$ and where $R_\alpha(x, y) \leq C|y - x|^{k-|\alpha|}$, $0 \leq |\alpha| < k$.

The class $t^k(E)$ is defined as all functions u on E such that for each $x \in E$ there is a polynomial $P_x(\cdot)$ of degree less than or equal to k of the form (3.5.1) such that for $0 \leq |\alpha| \leq k$,

$$D^\alpha P_y(y) = D^\alpha P_x(y) + R_\alpha(x, y),$$

whenever $x, y \in E$ with $|R_\alpha(x, y)| \leq C|y - x|^{k-|\alpha|}$ and

$$\lim_{y \to x} \frac{R_\alpha(x, y)}{|y - x|^{k-|\alpha|}} = 0$$

uniformly on E.

As a mnemonic, $T^k(E)$ and $t^k(E)$ may be considered as classes of functions that possess formal Taylor series expansions relative to E whose remainder terms tend to 0 "big O" or "little O," respectively.

3.5.2. Remark. Clearly, if $u \in T^k(E)$ then u_α is locally Lipschitz on E, $0 \leq |\alpha| < k$. If E is an open set, note that the derivatives $D^\alpha u$ exist on E, $0 \leq |\alpha| < k$, and that

$$D^\alpha u(x) = D^\alpha P_x(x) = u_\alpha(x) \quad \text{for} \quad x \in E.$$

Since $|D^\alpha P_x(x)| \leq M$ for $x \in E$, it follows that $u \in W^{k-1,p}_{\text{loc}}(E)$ for every $p \geq 1$. The space $t^k(E)$ may be considered as the class of functions on E that admit formal Taylor series expansions of degree k. Of course, if E were open and $u \in C^k(E)$, then u would have an expansion as in Definition 3.5.1 with

$$P_x(y) = \sum_{0 \leq |\alpha| \leq k} \frac{1}{\alpha!} D^\alpha u(x)(y - x)^\alpha.$$

Moreover, if $u \in C^k(R^n)$ and $E \subset R^n$, then the restriction of u to E, $u|E$, belongs to $t^k(F)$ for each compact set $F \subset E$. One of the reasons for identifying the class $t^k(E)$ is that it applies directly to the Whitney extension theorem [WH], which we state here without proof. We will provide a different version in Section 3.6.

3.5.3. Whitney Extension Theorem. *Let $E \subset R^n$ be compact. If $u \in t^k(E)$, $k > 0$ an integer, then there exists $\overline{u} \in C^k(R^n)$ such that for $0 \leq |\beta| \leq k$*

$$D^\beta \overline{u}(x) = D^\beta P_x(x) \quad \textit{for all} \quad x \in E.$$

In view of this result, it follows that $u \in t^k(E)$ if and only if u is the restriction to E of a function of class $C^k(R^n)$.

We now introduce another class of functions similar to those introduced in Definition 3.5.1 but different in the respect that the remainder term is required to have suitable decay relative to the L^p-norm instead of the L^∞-norm. The motivation for this definition is provided by the results established in Section 4 concerning Taylor expansions for Sobolev functions.

3.5.4. Definition. For $1 \le p \le \infty$, k a non-negative integer, and $x \in R^n$, $T^{k,p}(x)$ will denote those functions $u \in L^p$ for which there exists a polynomial $P_x(\cdot)$ of degree less than k and a constant $M = M(x, u)$ such that for $0 < r < \infty$

$$\left(\fint_{B(x,r)} |u(y) - P_x(y)|^p dy \right)^{1/p} \le Mr^k. \tag{3.5.2}$$

When $p = \infty$, the left side of (3.5.2) is interpreted to mean $\operatorname{ess\,sup}_{y \in B(x,r)} |u(y) - P_x(y)|$. $T^{k,p}(x)$ is a Banach space if for each $u \in T^{k,p}(x)$ the norm of u, $\|u\|_{T^{k,p}(x)}$, is defined as the sum of $\|u\|_p$, the absolute value of the coefficients of P_x, and the smallest value of M in (3.5.2).

3.5.5. Definition. A function $u \in T^{k,p}(x)$ belongs to $t^{k,p}(x)$ if there is a polynomial of degree less than or equal to k such that

$$\left(\fint_{B(x,r)} |u(y) - P_x(y)|^p dy \right)^{1/p} = o(r^k) \quad \text{as} \quad r \to 0. \tag{3.5.3}$$

Note that if $u \in T^{k,p}(x)$ the polynomial P_x is uniquely determined. To see this write

$$u(y) = P_x(y) + R_x(y)$$

where

$$\left(\fint_{B(x,r)} |R_x(y)|^p dy \right)^{1/p} \le Mr^k.$$

If P_x were not uniquely determined, we would have $u(y) = Q_x(y) + \overline{R}_x(y)$, where $\overline{R}_x$ satisfies an integral inequality similar to that of R_x.

Let $S_x(y) = P_x(y) - Q_x(y)$. In order to show that $S_x \equiv 0$, first note that

$$\fint_{B(x,r)} |S_x(y)| dy \le \left(\fint_{B(x,r)} |S_x(y)|^p dy \right)^{1/p} \le Cr^k, \quad 0 < r < \infty.$$

Now let L_x be the sum of terms of S_x of lowest order and let $M_x = S_x - L_x$. Thus, L_x has the property that for each $\lambda \in R^1$, $L_x(\lambda y + x) = \lambda^a L_x(y + x)$, where a is an integer, $0 < a \le k - 1$. Since M_x is a polynomial of degree at most $k - 1$, we have

$$\fint_{B(x,r)} |M_x(y)| dy \le Cr^{k-1}, \quad 0 < r < \infty.$$

It follows from the inequality $|L_x(y)| \le |S_x(y)| + |M_x(y)|$ that

$$\begin{aligned} r^a \fint_{B(x,1)} |L_x(y)|dy &= \\ &= \fint_{B(x,r)} |L_x(y)|dy \\ &\le Cr^k + Cr^{k-1}, \quad 0 < r < \infty. \end{aligned}$$

This is impossible for all small $r > 0$ if $a < k - 1$ and L_x is non-zero. If $a = k - 1$, then $M_x \equiv 0$ and the term Cr^{k-1} above can be replaced by 0.

A similar argument holds in case $u \in t^{k,p}(x)$.

Obviously, $t^k(E) \subset t^{k,p}(x)$ and $T^k(E) \subset T^{k,p}(x)$ whenever $x \in E$ and $p \ge 1$. We now consider the question of the reverse inclusion. For this purpose, we first need the following lemma.

3.5.6. Lemma. *Let k be a non-negative integer. Then there exists $\varphi \in C_0^\infty(R^n)$ with $\operatorname{spt}\varphi \subset \{|x| \le 1\}$ such that for every polynomial P on R^n of degree $\le k$ and every $\varepsilon > 0$,*

$$\varphi_\varepsilon * P = P$$

where $\varphi_\varepsilon(x) = \varepsilon^{-n}\varphi(x/\varepsilon)$.

Proof. Let $V = C_0^\infty(B)$ where B is the closed unit ball centered at the origin and let W denote the vector space of all m-tuples $\{y_\alpha\}$ whose components are indexed by multi-indices $\alpha = (\alpha_1, \alpha_2, \ldots, \alpha_n)$ with $0 \le |\alpha| \le k$. The number m is determined by k and n. Define a linear map $T: V \to W$ by

$$T(\varphi) = \left\{ \int_{R^n} \varphi(x) x^\alpha dx \right\};$$

thus,

$$y_\alpha = \int_{R^n} \varphi(x) x^\alpha dx$$

where $0 \le |\alpha| \le k$ and $x^\alpha = x_1^{\alpha_1} x_2^{\alpha_2} \cdots x_n^{\alpha_n}$.

Note that vector space, range T, has the property that range $T = W$ for if not, there would exist a vector, $a = \{a_\alpha\}$ orthogonal to range T. That is,

$$\Sigma\, a_\alpha y_\alpha = 0 \quad \text{whenever} \quad y = \{y_\alpha\} \in \operatorname{range} T.$$

This implies,

$$\int_{R^n} \varphi(x) \Sigma\, a_\alpha x^\alpha dx = 0 \quad \text{whenever} \quad \varphi \in V.$$

Select $\eta \in V$ such that $\eta > 0$ in $\{x : |x| < 1\}$. Now define ψ by $\psi = \Sigma\, a_\alpha x^\alpha \eta$ and note that $\psi \in V$. Therefore,

$$\int_{R^n} (\Sigma\, a_\alpha x^\alpha)^2\, \eta(x) dx = 0,$$

which implies $\Sigma\, a_\alpha x^\alpha = 0$ whenever $|x| < 1$. But this implies that all m numbers $a_\alpha = 0$, a contradiction. Thus, range $T = W$. In particular, this implies there is $\varphi \in V$ such that

$$\int_{R^n} \varphi(x) dx = 1, \quad \int_{R^n} \varphi(x) x^\alpha dx = 0, \quad 0 < |\alpha| \le k.$$

Since any polynomial Q of degree no greater than k is of the form

$$Q(z) = \sum_{0 \le |\alpha| \le k} b_\alpha z^\alpha,$$

it follows that

$$\int_{R^n} \varphi(z) Q(z) dz = Q(0).$$

Given a polynomial $P = P(x)$ as in the statement of the lemma, let $z = (x - y)/\varepsilon$ and set $Q(z) = P(x - \varepsilon z)$ to obtain the desired result. □

The next theorem is the main result of this section. Roughly speaking, it states that if a function possesses a finite Taylor expansion in the L^p-sense at all points of a compact set E, then it has a Taylor expansion in the classical sense on E. It is rather interesting that we are able to deduce a L^∞-conclusion from a L^p-hypothesis. A critical role is played by the existence of a smoothing kernel φ that leaves all polynomials of a given degree invariant under the action of convolution.

3.5.7. Theorem. *Let $E \subset R^n$ be closed and suppose $u \in T^{k,p}(x)$, $1 \le p \le \infty$, $k > 0$, with $\|u\|_{T^{k,p}(x)} \le M$ for all $x \in E$. Then $u \in T^k(E)$. Also, if E is compact and if $u \in t^{k,p}(x)$ for all $x \in E$ with (3.5.3) holding uniformly on E, then $u \in t^k(E)$.*

In view of Whitney's Extension theorem (Theorem 3.5.3), note that a function satisfying the second part of the theorem is necessarily the restriction of a function of class $C^k(R^n)$. In the next section, we will investigate Whitney's theorem in the context of L^p.

Proof of Theorem 3.5.7. Let $\varphi \in C_0^\infty(R^n)$ be the function obtained in Lemma 3.5.6 such that

$$\varphi_\varepsilon * P(x) = P(x) \tag{3.5.4}$$

whenever P is a polynomial of degree less than k, $\varepsilon > 0$, and $x \in R^n$. Note that (3.5.4) implies

$$D^\alpha \varphi_\varepsilon * P(x) = \varphi_\varepsilon * D^\alpha P(x) = D^\alpha P(x). \tag{3.5.5}$$

Since $u \in T^{k,p}(x_0)$ for all $x_0 \in E$, we have for x_0 and $x \in E$,

$$u(y) = P_{x_0}(y) + R(x_0, y) \tag{3.5.6}$$

and

$$u(y) = P_x(y) + R(x, y) \tag{3.5.7}$$

where

$$\left(⨍_{B(x^*,r)} |R(x^*, y)|^p dy \right)^{1/p} \leq Mr^k, \tag{3.5.8}$$

with x^* either x_0 or x. Now let $\varepsilon = |x - x_0|$ and for $0 \leq |\beta| < k$ consider

$$I = D^\beta \varphi_\varepsilon * u(x).$$

For each fixed $z \in R^n$, define R_z as $R_z(x) = R(z, x)$ whenever $x \in R^n$. From (3.5.6) and (3.5.5) it follows that

$$\begin{aligned} I &= D^\beta \varphi_\varepsilon * P_{x_0}(x) + D^\beta \varphi_\varepsilon * R_{x_0}(x) \\ &= D^\beta P_{x_0}(x) + D^\beta \varphi_\varepsilon * R_{x_0}(x). \end{aligned}$$

Similarly, using (3.5.7) and (3.5.5), we have

$$\begin{aligned} I &= D^\beta P_x(x) + D^\beta \varphi_\varepsilon * R_x(x) \\ &= u_\beta(x) + D^\beta \varphi_\varepsilon * R_x(x). \end{aligned}$$

Therefore,

$$\begin{aligned} D^\beta P_x(x) &= D^\beta P_{x_0}(x) + [D^\beta \varphi_\varepsilon * (R_{x_0} - R_x)](x) \\ &= D^\beta P_{x_0}(x) \\ &\quad + \int \varepsilon^{-(n+|\beta|)} D^\beta \varphi \left[\frac{(x-y)}{\varepsilon} \right] [R(x_0, y) - R(x, y)] dy. \end{aligned}$$

Because $\varphi \equiv 0$ on $|x| > 1$, the last integral is taken over $B(x, \varepsilon)$. Since $B(x, \varepsilon) \subset B(x_0, 2\varepsilon)$, the integral is dominated by

$$C \left[⨍_{B(x_0,2\varepsilon)} |R(x_0, y)| dy + ⨍_{B(x,\varepsilon)} |R(x, y)| dy \right] \varepsilon^{-|\beta|} \tag{3.5.9}$$

where C depends on an upper bound for $|D^\beta \varphi|$. Jensen's inequality and (3.5.8) implies that (3.5.9) is bounded by $CM\varepsilon^{k-|\beta|}$, thus proving $u \in T^k(E)$.

A similar proof establishes the second assertion of the theorem. Indeed, as before we obtain

$$\begin{aligned} D^\beta P_x(x) &= D^\beta P_{x_0}(x) + [D^\beta \varphi_\varepsilon * (R_{x_0} - R_x)](x) \\ &= D^\beta P_{x_0}(x) \\ &\quad + \int \varepsilon^{-(n+|\beta|)} D^\beta \varphi \left[\frac{(x-y)}{\varepsilon}\right] [R(x_0, y) - R(x,y)]dy \\ &\le C \left[\fint_{B(x_0, 2\varepsilon)} |R(x_0, y)|dy + \fint_{B(x,\varepsilon)} |R(x,y)|dy \right] \varepsilon^{-|\beta|}. \end{aligned}$$

Since (3.5.3) is assumed to hold uniformly on E, for $\eta > 0$ arbitrary, the last expression is dominated by $\eta\varepsilon^{k-|\beta|} = \eta|x - x_0|^{k-|\beta|}$ provided $|x - x_0|$ is sufficiently small. The compactness of E is used in this case to ensure that $|R_\beta(x,y)| \le C|x-y|^{k-|\beta|}$ whenever $x, y \in E$. □

3.6 An L^p-Version of the Whitney Extension Theorem

We now return to the Whitney Extension Theorem (Theorem 3.5.3) that was stated without proof in the previous section. It states that for a compact set $E \subset R^n$, a function u is an element of $t^k(E)$ if and only if it is the restriction to E of a function of class $C^k(R^n)$. The result we establish here, which was first proved in [CZ], is slightly stronger in that the full strength of the hypothesis $u \in t^k(E)$ is not required. Instead, our hypothesis requires that $u \in t^{k,p}(x)$ for all $x \in E$ with (3.5.3) holding uniformly on E.

We begin by proving a lemma that establishes the existence of a smooth function which is comparable to the distance function to an arbitrary closed set.

3.6.1. Lemma. *Let $A \subset R^n$ be closed and for $x \in R^n$ let $d(x) = d(x, A)$ denote the distance from x to A. Let $U = \{x : d(x) < 1\}$. Then there is a function $\delta \in C^\infty(U - A)$ and a positive number $M = M(n)$ such that*

$$M^{-1}d(x) \le \delta(x) \le Md(x), \quad x \in U - A,$$

$$|D^\alpha \delta(x)| \le C(\alpha) d(x)^{1-|\alpha|}, \quad x \in U - A, \ |\alpha| \ge 0.$$

Proof. Let $h(x) = \frac{1}{20} d(x)$, $x \in U - A$, and consider a cover of $U - A$ by closed balls $\{\overline{B}(x, h(x))\}$, with center x and radius $h(x)$, $x \in U - A$. From Theorem 1.3.1 there is a countable set $S \subset U - A$ such that $\{\overline{B}(s, h(s)) : s \in S\}$ is disjointed and

$$R^n - A \supset \{\cup \overline{B}(s, 5h(s)) : s \in S\} \supset U - A.$$

With $\alpha = \beta = 10$ and $\lambda = \frac{1}{20}$, we infer from Lemma 1.3.4 that

$$\frac{1}{3} \leq h(x)/h(s) \leq 3 \quad \text{for} \quad s \in S_x. \tag{3.6.1}$$

Let $\theta(x) = H^o(S_x) \leq C(n)$ and let $\eta : R^1 \to [0,1]$ be of class C^∞ with

$$\eta(t) = 1 \text{ for } t \leq 1, \quad \eta(t) = 0 \text{ for } t \geq 2.$$

Now define $\psi \in C^\infty(R^n)$ by $\psi(x) = \eta(|x|)$ and $v_s \in C^\infty(U)$ by

$$v_s(x) = h(s)\psi\left[\frac{(x-s)}{5h(s)}\right] \quad \text{for} \quad s \in S, \ x \in U.$$

Note that spt $v_s \subset \overline{B}(s, 10h(s))$, $v_s \equiv h(s)$ on $\overline{B}(s, 5h(s))$ and from (3.6.1) that

$$\begin{aligned} |D^\alpha v_s(x)| &\leq h(s)N(\alpha)[5h(s)]^{-|\alpha|} \\ &\leq 5^{-|\alpha|}3^{|\alpha|-1}N(\alpha)h(x)^{1-|\alpha|} \quad \text{for } s \in S_x, \end{aligned}$$

where $N(\alpha)$ is a bound for $|D^\beta\psi|$, $|\beta| \leq |\alpha|$. Now define

$$\delta(x) = \sum_{s\in S} v_s(x) = \sum_{s\in S_x} v_s(x) \quad \text{for} \quad x \in U.$$

Clearly,

$$\frac{d(x)}{60} = \frac{h(x)}{3} \leq \delta(x) \leq 3\theta(x)h(x) = \frac{3}{20}\theta(x)d(x)$$

and

$$|D^\alpha\delta(x)| \leq 5^{-|\alpha|}3^{|\alpha|-1}\theta(x)N(\alpha)h(x)^{1-|\alpha|}, \quad \text{for} \quad x \in U - A. \qquad \square$$

The following is only a prelude to the L^p-version of the Whitney extension theorem, although its proof supplies all of the necessary ingredients. Its hypothesis only invokes information pertaining to the spaces $T^{k,p}(x)$ (bounded difference quotients) and not the spaces $t^{k,p}(x)$ (differentiability). In particular, the theorem states that if u is Lipschitz on A (the case when $k = 1$) then u can be extended to a Lipschitz function on an open set containing A. This fact is also contained in the statement of Theorem 3.5.7.

3.6.2. Theorem. *Let $A \subset R^n$ be closed and let $U = \{x : d(x, A) < 1\}$. If $u \in L^p(U)$, $1 \leq p \leq \infty$, and there is a positive constant M such that $\|u\|_{T^{k,p}(x)} \leq M$ for all $x \in A$, where k is a non-negative integer, then there exists $\overline{u} \in C^{k-1,1}(U)$ such that $D^\beta\overline{u}(x) = D^\beta P_x(x)$ for $x \in A$, $0 \leq |\beta| < k$.*

Proof. Let δ denote the function determined in Lemma 3.6.1. Define $\overline{u} = u$ on A and for $x \in U - A$ let

$$\overline{u}(x) = \varphi_{\delta(x)} * u(x) \tag{3.6.2}$$

where φ is the function determined by Lemma 3.5.6 and where

$$\varphi_{\delta(x)}(y) = \delta(x)^{-n}\varphi\left(\frac{y}{\delta(x)}\right).$$

Thus, $\overline{u}$ is defined at x as the convolution of $\varphi_{\delta(x)}$ and u evaluated at x.

Because both φ and δ are of class C^∞ it is easily verified that $\overline{u} \in C^\infty(U - A)$. For $x \in U$, let x^* be a point in A such that $|x - x^*| = d(x) = d(x, A)$. Because $u \in T^{k,p}(x)$ we may write

$$u(x) = P_{x^*}(x) + R_{x^*}(x) \tag{3.6.3}$$

where

$$\left(\fint_{B(x^*,r)} |R_{x^*}(x)|^p dx\right)^{1/p} \leq Mr^k.$$

By substituting this expression into (3.6.2), we obtain

$$\begin{aligned} D^\beta \overline{u}(x) &= D^\beta\left[\varphi_{\delta(x)} * P_{x^*}(x)\right] + D^\beta\left[\varphi_{\delta(x)} * R_{x^*}(x)\right] \\ &= (D^\beta \varphi_{\delta(x)} * P_{x^*}(x) + \int R_\beta(x,y) R_{x^*}(y)dy \\ &= \varphi_{\delta(x)} * (D^\beta P_{x^*})(x) + \int R_\beta(x,y) R_{x^*}(y)dy \end{aligned} \tag{3.6.4}$$

where $R_\beta(x,y) = D^\beta\{\delta(x)^{-n}\varphi[(x-y)\delta(x)^{-1}]\}$. Applying Lemma 3.5.6 to the first term on the right side of (3.6.4) we obtain

$$D^\beta \overline{u}(x) = D^\beta P_{x^*}(x) + \int R_\beta(x,y) R_{x^*}(y)dy. \tag{3.6.5}$$

We wish to estimate the remainder term in (3.6.5) which requires an analysis of $R_\beta(x,y)$. It can be shown that

$$|R_\beta(x,y)| \leq C(\beta)d(x)^{-n-|\beta|}$$

and consequently

$$\left|\int R_\beta(x,y)R_{x^*}(y)dy\right| \leq C(\beta)d(x)^{-n-|\beta|}\int_{\overline{B}(x,\delta)} |R_{x^*}(y)|dy. \tag{3.6.6}$$

Because $\delta(x)$ is comparable to $d(x)$ (Lemma 3.6.1) and $|x - x^*| = d(x)$, it follows that $\overline{B}(x, \delta(x)) \subset \overline{B}(x^*, Kd(x))$ for some $K > 0$. Therefore from (3.6.3) and Hölder's inequality,

$$\int_{B(x^*,Kd(x))} |R_{x^*}(y)|dy \leq M[Kd(x)]^{n+k} \tag{3.6.7}$$

which along with (3.6.5) and (3.6.6) implies,

$$D^\beta \overline{u}(x) - D^\beta P_{x^*}(x) = \tilde{S}_\beta(x^*, x) \tag{3.6.8}$$

where

$$|S_\beta(x^*, x)| \le C(\beta, k)M|x - x^*|^{k-|\beta|}.$$

We emphasize here that for given $x \in U - A$, (3.6.8) is valid only for $x^* \in A$ such that $d(x) = |x - x^*|$. We now proceed to establish the estimate for arbitrary $x^* \in A$.

By assumption $\|u\|_{T^{k,p}(x)} \le M$ for all $x \in A$. Therefore, we may apply Theorem 3.5.7 to conclude that $u \in T^k(A)$. Thus, if $x_1^* \in A$,

$$P_{x^*}(x^*) = u(x^*)$$

and

$$D^\alpha P_{x^*}(x^*) = D^\alpha P_{x_1^*}(x^*) + R_\alpha(x_1^*, x^*), \quad 0 \le |\alpha| < k \tag{3.6.9}$$

where

$$|R_\alpha(x_1^*, x^*)| \le C(\alpha, k)M|x^* - x_1^*|^{k-|\alpha|}.$$

By Taylor's theorem for polynomials, it follows that

$$D^\beta P_{x^*}(x) = \sum_{|\alpha|=0}^{k-1-|\beta|} \frac{1}{\alpha!} D^{\beta+\alpha} P_{x^*}(x^*)(x - x^*)^\alpha.$$

Thus, by (3.6.9) and Taylor's theorem,

$$\begin{aligned} D^\beta P_{x^*}(x) &= \sum_{|\alpha|=0}^{k-1-|\beta|} \frac{1}{\alpha!}[D^{\beta+\alpha} P_{x_1^*}(x^*) + R_{\alpha+\beta}(x_1^*, x^*)](x - x^*)^\alpha \\ &= \sum_{|\alpha|=0}^{k-1-|\beta|} \frac{1}{\alpha!} \left[\sum_{|\gamma|=0}^{k-1-(|\alpha|+|\beta|)} \frac{1}{\gamma!} D^{\beta+\alpha+\gamma} P_{x_1^*}(x_1^*)(x^* - x_1^*)^\gamma \right. \\ &\quad + \left. R_{\alpha+\beta}(x_1^*, x^*) \right] (x - x^*)^\alpha. \end{aligned} \tag{3.6.10}$$

By Taylor's theorem, it follows that

$$D^\beta P_{x_1^*}(x) = \sum_{|\alpha|\ge 0} \frac{1}{\alpha!} D^{\beta+\alpha} P_{x_1^*}(x_1^*)(x - x_1^*)^\alpha.$$

Therefore, since

$$|x - x^*| \le |x - x_1^*| \text{ and } |x^* - x_1^*| \le |x^* - x| + |x - x_1^*| \le 2|x - x_1^*|,$$

(3.6.10) becomes (after some algebraic simplification)

$$D^\beta P_{x^*}(x) - \sum_{|\alpha|=0}^{(k-1)-|\beta|} \frac{1}{\alpha!} D^{\beta+\alpha} P_{x_1^*}(x_1^*)(x-x_1^*)^\alpha = O(|x-x_1^*|^{k-|\beta|}).$$

It follows from (3.6.8), that

$$D^\beta \overline{u}(x) - \sum_{|\alpha|=0}^{(k-1)-|\beta|} \frac{1}{\alpha!} D^{\beta+\alpha} P_{x_1^*}(x_1^*)(x-x_1^*)^\alpha = O(|x-x_1^*|^{k-|\beta|})$$

or

$$D^\beta \overline{u}(x) - D^\beta P_{x_1^*}(x) = O(|x-x_1^*|^{k-|\beta|}). \tag{3.6.11}$$

Thus, (3.6.11) holds whenever $x_1^* \in A$ and $x \in U - A$ and Theorem 3.5.7 implies that it also holds with $D^\beta \overline{u}(x)$ replaced by $u_\beta(x)$ whenever $x \in A$. This implies that $D^\beta \overline{u}$ is a continuous extension of u_β and that this extension has a Taylor series expansion about each point in A. Since $\overline{u} \in C^\infty(U - A)$ it now follows that $\overline{u} \in C^{k-1}(U)$.

In order to prove that $\overline{u} \in C^{k-1,1}(U)$ it suffices to show that $D^\beta \overline{u}$ is Lipschitz, $|\beta| = k-1$. We know from (3.6.11) that if $a \in A$, and $|\beta| = k-1$

$$|D^\beta \overline{u}(x) - D^\beta \overline{u}(a)| \leq C(k)M|x-a| \tag{3.6.12}$$

for $x \in U$. Therefore, it is necessary to consider only the case $x, y \in U - A$. First suppose $|x-y| \geq \frac{1}{2}d(y)$ and let $a \in A$ be such that $d(y) = |a-y|$. Then, $|a-y| \leq 2|x-y|$ and

$$|x-a| \leq |x-y| + |y-a| \leq 3|x-y|.$$

Thus, utilizing (3.6.12),

$$\begin{aligned} |D^\beta \overline{u}(x) - D^\beta \overline{u}(y)| &\leq |D^\beta \overline{u}(x) - D^\beta \overline{u}(a)| + |D^\beta \overline{u}(y) - D^\beta \overline{u}(a)| \\ &\leq |D^\beta \overline{u}(x) - D^\beta \overline{u}(a)| + |D^\beta \overline{u}(y) - D^\beta \overline{u}(a)| \\ &\leq C(k)M[|x-a| + |y-a|] \\ &\leq 5C(k)M|x-y|. \end{aligned}$$

Finally, suppose $|x-y| < \frac{1}{2}d(y)$ and $d(y) = |a-y|$. Using (3.6.5) with $|\beta| = k-1$ and the Mean Value theorem, we have

$$\begin{aligned} |D^\beta \overline{u}(x) - D^\beta \overline{u}(y)| &= \int R(a,z)\,[R_\beta(x,z) - R_\beta(y,z)]\,dz \\ &\leq |x-y| \int |D_x R_\beta(x_0,z)|\,|R(a,z)|dz \end{aligned} \tag{3.6.13}$$

where x_0 is a point on the line segment joining x and y. Now spt $R_\beta(x_0, y) \subset \overline{B}(x_0, \delta(x_0))$ and $\delta(x_0) \leq Cd(x_0)$. Thus,

$$|D_x R_\beta(x_0,z)| \leq C(\beta)d(x_0)^{-n-k},$$

($|\beta| = k-1$). Therefore, (3.6.13) implies

$$|D^\beta\overline{u}(x) - D^\beta\overline{u}(y)| \le C(\beta)|x-y|d(x_0)^{-n-k}\int_{B(x_0,Cd(x_0))}|R(a,z)|dz. \tag{3.6.14}$$

Since $\mathrm{Lip}(d) = 1$, we have

$$2d(x_0) \ge d(x) \ge d(y) - |x-y| > \frac{1}{2}d(y) > |x-y|.$$

If $z \in U$, $|z-x_0| \le Cd(x_0)$, then

$$\begin{aligned}|z-a| &\le |z-x_0| + |x_0-a|\\ &\le Cd(x_0) + |x_0-a|\\ &\le Cd(x_0) + |x_0-y| + d(y)\\ &\le Cd(x_0) + |x-y| + d(y)\\ &\le Cd(x_0) + d(x_0) + 2d(x_0).\end{aligned}$$

That is,

$$B(x_0, Cd(x_0)) \subset B(a, (C+3)d(x_0)).$$

Therefore, reference to (3.6.7) implies

$$\int_{B(x_0,Cd(x_0))}|R(a,z)|dz \le C[d(x_0)]^{n+k}$$

and this, along with (3.6.14) completes the proof. □

This proof leads directly to the following which is the Whitney extension theorem in the context of $t^{k,p}(x)$ spaces.

3.6.3. Theorem. *Let $A \subset R^n$ be closed and let $U = \{x : d(x,A) < 1\}$. If $u \in L^p(U)$, $1 \le p \le \infty$, and $u \in t^{k,p}(x)$ for all $x \in A$ with* (3.5.3) *holding uniformly on A, then there exists $\overline{u} \in C^k(U)$ such that $D^\beta\overline{u}(x) = D^\beta P_x(x)$ for $x \in A$, $0 \le |\beta| \le k$.*

Proof. The proof is essentially the same as the one above with only minor changes necessary. For example, the polynomials in (3.6.8) and (3.6.9) are now of degree k and the remainders can be estimated, respectively, by

$$|S_\beta(x^*, x)| \le \mathrm{o}(|x-x^*|^{k-|\beta|})$$

and

$$|R_\alpha(x_1^*, x^*)| \le \mathrm{o}(|x^*-x_1^*|^{k-|\alpha|}),$$

thus allowing (3.6.11) to be replaced by

$$D^\beta\overline{u}(x) - D^\beta P_{x_1^*}(x) = \mathrm{o}(|x-x_1^*|^{k-|\beta|}).$$

The remainder of the argument proceeds as before. □

3.7 An Observation on Differentiation

We address the technicality of showing that $\|u\|_{T^{k,p}(x)}$ is a measurable function of x and then establish a result in differentiation theory that will be needed later in the sequel.

3.7.1. Lemma. *Let $u \in T^{k,p}(x)$ for all x in a measurable set E. Then, $\|u\|_{T^{k,p}(x)}$ is a measurable function of x.*

Proof. Recall that the norm $\|u\|_{T^{k,p}(x)}$ is the sum of the numbers $\|u\|_p$, $|D^\alpha P_x(x)|$, $0 \le |\alpha| \le k-1$, and the pth root of

$$\sup_{r>0} r^{-kp} \fint_{B(x,r)} |u(y) - P_x(y)|^p dy.$$

Also recall that $D^\alpha P_x(x) = u_\alpha(x)$. To show that $D^\alpha P_x(x)$ is measurable in x consider the function φ of Lemma 3.5.6 and define

$$u_\varepsilon(x) = \varphi_\varepsilon * u(x).$$

If we write $u(y) - P_x(y) = R_x(y)$, then

$$\begin{aligned} D^\alpha u_\varepsilon(x) &= D^\alpha(\varphi_\varepsilon * P_x)(x) + D^\alpha(\varphi_\varepsilon * R_x)(x) \\ &= D^\alpha P_x(x) + \int \varepsilon^{-(n+|\alpha|)} D^\alpha \varphi \left[\frac{y}{\varepsilon}\right] R_x(x-y)dy. \end{aligned}$$

The above integral is dominated by

$$\begin{aligned} C\varepsilon^{-(n+|\alpha|)} \int_{B(x,\varepsilon)} |R_x(x-y)|dy &\le C\varepsilon^{-(n+|\alpha|)}\varepsilon^{k+n} \\ &= C\varepsilon^{k-|\alpha|} \to 0 \quad \text{as} \quad \varepsilon \to 0. \end{aligned}$$

This shows that $D^\alpha P_x(x)$ is the limit of smooth functions $D^\alpha u_\varepsilon(x)$ for all $x \in E$, and is therefore measurable. The remainder of the proof is easy to establish. □

3.7.2. Lemma. *Let $u \in L^p(R^n)$, $1 \le p < \infty$, be such that for some C, $a > 0$ and all $r > 0$,*

$$\left(\fint_{B(x,r)} |u(y)|^p dy\right)^{1/p} \le Cr^a,$$

for all x in a measurable set $E \subset R^n$. Then, for almost all $x \in E$,

$$\left(\fint_{B(x,r)} |u(y)|^p dy\right)^{1/p} = o(r^a) \quad \text{as} \quad r \downarrow 0.$$

Proof. Without loss of generality we may assume that E is bounded and that u has compact support. Given $\varepsilon > 0$, let $A \subset E$ be a closed set such that $|E - A| < \varepsilon$. Let U be the open set defined by

$$U = \{x : d(x, A) < 1\}.$$

It will suffice to establish the conclusion for almost all $x \in A$.

First, observe that the hypotheses imply that

$$\lim_{r \to 0} \fint_{B(x,r)} |u(y)| dy = 0$$

for $x \in A$ and therefore, $u = 0$ almost everywhere on A.

Let $h(x) = \frac{1}{10} d(x, A)$. Recall from Theorem 1.3.1 that there is a countable set $S \subset U - A$ such that $\{\overline{B}(s, h(s)) : s \in S\}$ is disjointed and

$$\{\cup \overline{B}(s, 5h(s)) : s \in S\} \supset U - A.$$

Therefore, since $u = 0$ almost everywhere on A,

$$\begin{aligned} \int_A \int_U \frac{|u(y)|}{|x-y|^{n+a}} dy dx &\leq \int_A \int_{U-A} \frac{|u(y)|}{|x-y|^{n+a}} dy dx \\ &\leq \int_A \sum_{s \in S} \int_{B(s,5h(s))} \frac{|u(y)| dy}{|x-y|^{n+a}} dx \\ &= \sum_{s \in S} \int_{B(s,5h(s))} |u(y)| \int_A \frac{dx}{|x-y|^{n+a}} dy. \qquad (3.7.1) \end{aligned}$$

Let $x_s \in A$ be such that $|s - x_s| = d(s, A) = d(s)$. Hence, $B(s, 5h(s)) \subset B(x_s, |s - x_s| + 5h(s))$ and $|s - x_s| = d(s) = 10h(s)$. By Jensen's inequality and the hypothesis of the lemma

$$\fint_{B(s,5h(s))} |u| \leq \left(\fint_{B(s,5h(s))} |u|^p \right)^{1/p} \leq C \left(\fint_{B(x_s,15h(s))} |u|^p \right)^{1/p} \leq C h(s)^a. \qquad (3.7.2)$$

Now for $x \in A$, $y \in B(s, 5h(s))$, we have

$$|x - y| \geq |x - s| - |s - y| \geq d(s) - 5h(s) = 5h(s).$$

Hence, for $y \in B(s, 5h(s))$ we estimate by spherical coordinates with origin at y,

$$\begin{aligned} \int_A \frac{dx}{|x-y|^{n+a}} &\leq C \int_{|5h(s)|}^{\infty} r^{-a-1} dr \\ &\leq C(a) h(s)^{-a}. \end{aligned}$$

This, along with (3.7.2) yields

$$\int_A \int_{B(s,5h(s))} \frac{|u(y)|}{|x-y|^{n+a}} dydx \leq C(a)h(s)^n.$$

Since $\{\overline{B}(s,h(s)) : s \in S\}$ is disjointed, it follows from (3.7.1) that

$$\int_A \int_U \frac{|u(y)|}{|x-y|^{n+a}} dydx \leq C(a) \sum_{s\in S} h(s)^n < \infty$$

and therefore,

$$\int_U \frac{|u(y)|dy}{|x-y|^{n+a}} < \infty$$

for almost every $x \in A$. Clearly,

$$\int_{R^n - U} \frac{|u(y)|dy}{|x-y|^{n+a}} < \infty$$

for all $x \in A$, and therefore

$$\int_{R^n} \frac{|u(y)|dy}{|x-y|^{n+a}} < \infty$$

for almost all $x \in A$.

An analysis of the argument shows that this was established by using only the fact that

$$\fint_{B(x,r)} |u| \leq Cr^a.$$

If we apply the above argument with $v = |u|^p$, our hypothesis becomes

$$\fint_{B(x,r)} |v(y)|dy \leq Cr^{ap}$$

for all $x \in E$ and therefore

$$\infty > \int_{R^n} \frac{|v(y)|dy}{|y-x|^{n+ap}} = \int_{R^n} \frac{|u(y)|^p dy}{|y-x|^{n+ap}}$$

for almost all $x \in E$. But, for all such x, and for $\varepsilon > 0$,

$$\int_{B(x,r)} \frac{|u(y)|^p dy}{|y-x|^{n+pa}} < \varepsilon \quad \text{for all small} \quad r > 0.$$

That is,

$$\left(\fint_{B(x,r)} |u(y)|^p dy \right)^{1/p} < \varepsilon r^a$$

for all small r. □

3.8 Rademacher's Theorem in the L^p-Context

Recall the fundamental result of Rademacher which states that a Lipschitz function defined on R^n has a total differential at almost all points (Theorem 2.2.1). We rephrase this result in terms of the present setting by replacing the hypothesis that u is Lipschitz by $u \in T^{k,p}(x)$ for all x in some set E. If $k = 1$ and $p = \infty$, this yields the usual Rademacher hypothesis. The conclusion we will establish is that $u \in t^{k,p}(x)$ for almost all $x \in E$.

3.8.1. Theorem. *Let $u \in T^{k,p}(x)$ for all $x \in E$, where $E \subset R^n$ is measurable, k a non-negative integer and $1 < p < \infty$. Then $u \in t^{k,p}(x)$ for almost all $x \in E$.*

Proof. By Lemma 3.7.1 and Lusin's theorem we may assume that E is compact and that $\|u\|_{T^{k,p}(x)} \leq M$ for all $x \in E$. Since $u \in T^{k,p}(x)$ for $x \in E$, we may write $u(y) = P_x(y) + R_x(y)$ where P_x is a polynomial of degree less than k and where

$$\left(\fint_{B(x,r)} |R_x(y)|^p dy \right)^{1/p} \leq M r^k, \quad r > 0. \tag{3.8.1}$$

From Theorem 3.6.2 it follows that there exists an open set $U \supset E$ and $\overline{u} \in C^{k-1,1}(U)$ such that

$$D^\beta \overline{u}(x) = D^\beta P_x(x), \quad 0 \leq |\beta| < k. \tag{3.8.2}$$

Because $\overline{u}$ is of class $C^{k-1,1}$ it follows from Theorem 2.1.4 that $\overline{u} \in W^{k,p}_{\text{loc}}(R^n)$ and therefore we may apply Theorem 3.4.2. Thus, for almost all $x \in R^n$, there is a polynomial Q_x of degree at most k such that $\overline{u}(y) = Q_x(y) + \overline{R}_x(y)$ where

$$\left(\fint_{B(x,r)} |\overline{R}_x(y)|^p dy \right)^{1/p} = o(r^k) \quad \text{as} \quad r \downarrow 0. \tag{3.8.3}$$

Because $\overline{u} \in C^{k-1}(R^n)$, the argument following Definition 3.5.5 implies that

$$D^\beta \overline{u}(x) = D^\beta Q_x(x), \quad 0 \leq |\beta| < k. \tag{3.8.4}$$

Therefore, in view of (3.8.1), (3.8.2), and (3.8.3)

$$\left(\fint_{B(x,r)} |u - \overline{u}|^p \right)^{1/p} \leq C r^k$$

for almost all $x \in E$. Appealing to Lemma 3.7.2 we have

$$\left(\fint_{B(x,r)} |u - \overline{u}|^p \right)^{1/p} = o(r^k) \quad \text{as} \quad r \downarrow 0$$

for almost all $x \in E$. Consequently, for all such x,

$$\begin{aligned}\left(\fint_{B(x,r)} |u(y) - Q_x(y)|^p dy\right)^{1/p} &\leq \left(\fint_{B(x,r)} |u(y) - \overline{u}(y)|^p dy\right)^{1/p} \\ &\quad + \left(\fint_{B(x,r)} |\overline{u}(y) - Q_x(y)|^p dy\right)^{1/p} \\ &\leq o(r^k) \quad \text{as} \quad r \to 0,\end{aligned}$$

thus establishing the result. □

3.9 The Implications of Pointwise Differentiability

We have seen in Section 4 of this chapter that Sobolev functions possess L^p-derivatives almost everywhere. This runs parallel to the classical result that an absolutely continuous function f on the real line is differentiable almost everywhere. Of course, the converse is false. However, if it is assumed that f' exists everywhere and that $|f'|$ is integrable, then f is absolutely continuous (Exercise 3.16). It is natural, therefore, to inquire whether this result has a counterpart in the multivariate L^p theory. It will be shown that this question has an affirmative answer. Indeed, we will establish that if a function has an L^p derivative everywhere except for a small exceptional set, and if the coefficients of the associated Taylor polynomial are in L^p, then the function is in a Sobolev space.

We begin the investigation by asking the following question. Suppose $u \in L^P(R^n)$ has L^p-derivatives at $x \in R^n$; that is, suppose $u \in t^{k,p}(x)$ where k is a positive integer. Then, is it possible to relate the distributional derivatives of u (which always exist) to the L^p-derivatives of u? The first step in this direction is given by the following lemma. First, recall that $u \in t^{k,p}(x)$ if there is a polynomial P_x of degree k such that

$$\left(\fint_{B(x,r)} |u(y) - P_x(y)|^p dy\right)^{1/p} = o(r^k) \quad \text{as} \quad r \to 0, \tag{3.9.1}$$

and $u \in T^{k,p}(x)$ if there is a polynomial P_x of degree less than k and a number $M > 0$ such that

$$\left(\fint_{B(x,r)} |u(y) - P_x(y)|^p dy\right)^{1/p} \leq Mr^k, \quad 0 < r < \infty.$$

3.9.1. Lemma. *Suppose* $u \in L^p(R^n)$, $p \geq 1$.

(i) *If* $u \in T^{k,p}(x)$, *then*

$$\liminf_{t \to 0} \varphi_t * D^\alpha u(y) > -\infty,$$

with $|x - y| < t$, *and where* $D^\alpha u$ *denotes the distributional derivative of* u, $0 \leq |\alpha| \leq k$;

(ii) *If* $u \in t^{k,p}(x)$, *then*

$$\limsup_{t \to 0} \varphi_t * D^\alpha u(y) = D^\alpha P_x(x),$$

with $|x - y| < t$, $0 \leq |\alpha| \leq k$.

The function φ_t above is a mollifier as described in Section 1.6. Since $\varphi_t \in C_0^\infty(R^n)$, its convolution with a distribution T is again a smooth function. Moreover, for small t and $|y - x| \leq t$, the quantity $\varphi_t * T(y)$ gives an approximate description of the behavior of T in a neighborhood of x. Indeed, if T is a function, then

$$\limsup_{\substack{t \to 0 \\ |y-x| \leq t}} \varphi_t * T(y) = T(x)$$

whenever x is a Lebesgue point for T. This will be established in the proof of Lemma 3.9.3. Very roughly then, the statement in (ii) of the above lemma states that, on the average, the behavior of the distribution $D^\alpha u$ near x is reflected in the value of the coefficient, $D^\alpha P_x(x)$, of the Taylor series expansion.

Let $F(y, t) = \varphi_t * u(y)$. F is thus a function defined on a subset of R^{n+1}, namely $R^n \times (0, \infty)$ and is smooth in y. The lower and upper limits stated in (i) and (ii) above can be interpreted as non-tangential approach in R^{n+1} of (y, t) to the point $(x, 0)$ whch is located on the hyperplane $t = 0$.

Proof of Lemma 3.9.1. Proof of (ii). Let

$$u(y) = P_x(y) + R_x(y) \quad \text{and} \quad F_t(y) = F(y, t).$$

Then

$$D^\alpha F_t(y) = D^\alpha[\varphi_t * u](y) = D^\alpha \varphi_t * u(y).$$

Therefore

$$\begin{aligned} D^\alpha F_t(x + h) &= \int D^\alpha \varphi_t(x + h - y) u(y) dy \\ &= \int D^\alpha \varphi_t(x + h - y) P_x(y) dy + \int D^\alpha \varphi_t(x + h - y) R_x(y) dy \\ &= (\varphi_t * D^\alpha P_x)(x + h) + \int D^\alpha \varphi_t(x + h - y) R_x(y) dy. \end{aligned} \tag{3.9.2}$$

There is a constant $C = C(|D\varphi|)$ such that

$$|D^\alpha \varphi_t(x + h - y)| \le Ct^{-n-k}$$

for $|\alpha| = k$. Consequently, for $h \in \overline{B}(0,t)$, it follows that

$$\begin{aligned}\left|\int D^\alpha \varphi_t(x+h-y)R_x(y)dy\right| &\le Ct^{-n-k}\int_{B(x+h,t)} |R_x(y)|dy \\ &\le Ct^{-n-k}\int_{B(x,2t)} |R_x(y)|dy \to 0 \quad \text{as} \quad t \downarrow 0,\end{aligned}$$

by (3.9.1). Writing P_x in terms of its Taylor series, we have

$$P_x(y) = \sum_{|\alpha|=0}^{k} \frac{D^\alpha P_x(x)(y-x)^\alpha}{\alpha!},$$

and therefore $D^\alpha P_x(y) = D^\alpha P_x(x)$ for all $y \in R^n$ if $|\alpha| = k$. Hence, $\varphi_t * D^\alpha P_x(x+h) = D^\alpha P_x(x)$, and reference to (3.9.2) yields

$$\limsup_{t\downarrow 0} \varphi_t * D^\alpha u(x+h) = D^\alpha P_x(x), \quad 0 \le |h| \le t, \tag{3.9.3}$$

thus establishing (ii) if $|\alpha| = k$. However, if $0 \le \ell \le k$, then $u \in t^{\ell,p}(x)$ and the associated polynomial is

$$\sum_{|\alpha|=0}^{\ell} \frac{D^\alpha P_x(x)(h-x)^\alpha}{\alpha!}.$$

Thus, applying (3.9.3) to this case leads to the proof of (ii).

The proof of (i) is similar and perhaps simpler. The only difference is that because P_x is of degree at most $k-1$, we have

$$D^\alpha F_t(x+h) = 0 + \int D^\alpha \varphi_t(h-y)R_x(y)dy$$

if $|\alpha| = k$. The integral is estimated as before and its absolute value is seen to be bounded for all $t > 0$, thus establishing (i). □

The next two lemmas, along with the preceding one, will lead to the main result, Theorem 3.9.4.

3.9.2. Lemma. *Let T be a distribution and suppose for all x in an open set $\Omega \subset R^n$ that*

$$\liminf_{\substack{t\to 0\\ t\in S}} \varphi_t * T(y) > -\infty, \quad |x-y| \le t,$$

where $S \subset (0,\infty)$ *is a countable set having* 0 *as its only limit point. Let* C *be a closed set such that* $C \cap \Omega \neq 0$. *Then there exist* $N > 0$ *and an open set* $\Omega_1 \subset \Omega$ *with* $C \cap \Omega_1 \neq \emptyset$ *such that* $\varphi_t * T(y) \geq -N > -\infty$ *whenever* $|y-x| \leq t$, $x \in C \cap \Omega_1$, $t \in S$.

Proof. Let

$$F_*(x) = \inf\{\varphi_t * T(y) : |x-y| \leq t, t \in S\}.$$

Then, $F_*(x) > -\infty$ for $x \in \Omega$ since 0 is the only limit point of S and it is easy to verify that F_* is upper semicontinuous. Thus, the sets

$$C \cap \Omega \cap \{x : F_*(x) \geq -i\}, \quad i = 1, 2, \ldots,$$

are closed relative to $C \cap \Omega$ and their union is $C \cap \Omega$. Since $C \cap \Omega$ is of the second category in itself, the Baire Category theorem implies that one of these sets has a non-empty interior relative to $C \cap \Omega$. □

One of the fundamental results in distribution theory is that a non-negative distribution is a measure. The following lemma provides a generalization of this fact.

3.9.3. Lemma. *Let* $\delta > 0$, $N > 0$, *and suppose* S *is as in Lemma* 3.9.2. *If* T *is a distribution in an open set* Ω *such that*

$$\varphi_t * T(x) > -N > -\infty \quad \textit{for} \quad x \in \Omega,\ t \in S \cap (0,\delta)$$

and

$$\limsup_{\substack{t \downarrow 0 \\ |x-x_0| \leq t}} \varphi_t * T(x) \geq 0 \quad \textit{for almost all} \quad x_0 \in \Omega,$$

then T *is a non-negative measure in* Ω.

Proof. Let $\psi \in \mathscr{D}(\Omega)$, $\psi \geq 0$, and recall from Section 1.7, that the convolution $\varphi_t * T$ is a smooth function defined by

$$\varphi_t * T(x) = T(\tau_x \tilde{\varphi}_t)$$

where $\tilde{\varphi}_t(y) = \varphi_t(-y)$ and $\tau_x \tilde{\varphi}_t(y) = \tilde{\varphi}_t(y-x)$. Then,

$$\begin{aligned} T(\psi * \tilde{\varphi}_t) &= T * (\tilde{\psi} * \varphi_t)(0) \\ &= (T * \varphi_t) * \tilde{\psi}(0) \\ &= \int T * \varphi_t(-y)\tilde{\psi}(y)dy \\ &= \int T * \varphi_t(y)\psi(y)dy. \end{aligned}$$

Now $\psi * \tilde{\varphi}_t \to \psi$ in $\mathscr{D}(\Omega)$ as $t \to 0^+$. Moreover, since ψ is non-negative and $\varphi_t * T(x) \geq -N$ for $x \in \Omega$ and $t \in S \cap (0, \delta)$, it follows with the help of Fatou's lemma, that

$$\begin{aligned} T(\psi) &= \lim_{\substack{t\downarrow 0 \\ t\in S}} T(\psi * \tilde{\varphi}_t) \\ &\geq \liminf_{\substack{t\downarrow 0 \\ t\in S}} \int_\Omega T * \varphi_t(y)\psi(y)dy \\ &\geq \int_\Omega \liminf_{\substack{t\downarrow 0 \\ t\in S}} T * \varphi_t(y)\psi(y)dy \\ &\geq -N \int_\Omega \psi(y)dy. \end{aligned}$$

Thus, the distribution $T + N$ has the property that

$$(T + N)(\psi) \geq 0 \quad \text{for} \quad \psi \in \mathscr{D}(\Omega), \quad \psi \geq 0.$$

That is, $T + N$ is a non-negative measure on Ω, call it μ. Let $\mu = \nu + \sigma$ where ν is absolutely continuous with respect to Lebesgue measure and σ is singular. Clearly, Ω is the union of a countable number of sets of finite ν measure. Thus, by the Radon–Nikodym theorem, there exists $f \in L^1(\Omega)$ such that

$$\nu(E) = \int_E f(x)dx$$

for every measurable set $E \subset \Omega$. Since $T + N = \mu$, it follows that

$$\begin{aligned} \varphi_t * T(x) + N &= \varphi_t * (T + N)(x) = \varphi_t * \mu(x) \\ &= \int_\Omega \varphi_t(x-y)f(y)dy + \int_\Omega \varphi_t(x-y)d\sigma(y), \end{aligned} \tag{3.9.4}$$

for $x \in \Omega$. Because σ is a singular measure, a result from classical differentiation theory states that

$$\lim_{t\to 0} \frac{\sigma[B(x_0, t)]}{|B(x_0, t)|} = 0$$

for almost all $x_0 \in \Omega$, cf. [SA, Lemma 7.1]. Therefore, at all such x_0 with $|x - x_0| \leq t$,

$$\begin{aligned} \int \varphi_t(x-y)d\sigma(y) &= \int_{B(x,t)} \varphi_t(x-y)d\sigma(y) \\ &\leq \int_{B(x_0,2t)} \varphi_t(x-y)d\sigma(y) \\ &\leq C\|\varphi\|_\infty \frac{\sigma[(B(x_0, 2t)]}{|B(x_0, t)|} \\ &\to 0 \ \text{ as } \ t \to 0^+ \ \text{ with } \ |x - x_0| < t. \end{aligned}$$

To treat the other term in (3.9.4), recall that f has a Lebesgue point at almost all $x_0 \in \Omega$. That is,

$$\fint_{B(x_0,r)} |f(y) - f(x_0)|dy \to 0 \quad \text{as} \quad r \to 0^+.$$

Therefore,

$$\int_{B(x,t)} f(y)\varphi_t(x-y)dy - f(x_0) = \int_{B(x,t)} [f(y) - f(x_0)]\varphi_t(x-y)dy$$

$$\leq C\|\varphi\|_\infty \fint_{B(x_0,2t)} |f(y) - f(x_0)|dy \to 0 \text{ as } t \to 0,\ |x - x_0| \leq t.$$

Consequently,

$$N \leq \limsup_{\substack{t\downarrow 0 \\ |x-x_0|\leq t}} \varphi_t * T(x) + N = f(x_0)$$

for almost all $x_0 \in \Omega$. This implies that

$$\nu(E) \geq N|E|$$

for all measurable $E \subset \Omega$. Since $\mu(E) \geq \nu(E)$ it follows that the measure $\mu - N = T$ is non-negative. □

3.9.4. Theorem. *Let T be a distribution in an open set $\Omega \subset R^n$ and let $f \in L^1_{\text{loc}}(\Omega)$. Assume*

$$\limsup_{t\downarrow 0} \varphi_t * T(y) \geq f(x), \quad |x - y| \leq t,$$

for almost all $x \in \Omega$, and

$$\liminf_{\substack{t\downarrow 0 \\ t\in S}} \varphi_t * T(y) > -\infty, \quad |x - y| \leq t,$$

for all $x \in \Omega$. Then $T - f$ is a non-negative measure in Ω.

Proof. We first assume that $f \equiv 0$. Lemma 3.9.2 implies that every open subset of Ω contains an open subset Ω' such that for some $N > 0$, $\varphi_t * T(x) > -N$ for $x \in \Omega'$, $t \in S$. Lemma 3.9.3 implies that T is a measure in Ω'.

Let Ω_1 be the union of all open sets $\Omega' \subset \Omega$ such that T is a non-negative measure on Ω'. From Remark 1.7.2 we know that T is a measure in Ω_1. We wish to show that $\Omega_1 = \Omega$. Suppose not. Applying Lemma 3.9.2 with $C = R^n - \Omega_1$, there is an open set $\Omega' \subset \Omega$ such that $\varphi_t * T(x) \geq -N$ for $y \in C \cap \Omega'$, $|x - y| \leq t$, and $t \in S$. Let $\Omega_2 = \Omega_1 \cup \Omega'$ and note that $\Omega_2 - \Omega_1 = C \cap \Omega'$. Let

$$\Omega_3 = \Omega_2 \cap \{x : d(x, R^n - \Omega_2) > \varepsilon\}$$

for some $\varepsilon > 0$. Take ε sufficiently small so that $\Omega_3 \cap (R^n - \Omega_1) \neq \emptyset$. Consider $\varphi_t * T(x)$ for $x \in \Omega_3$ and $t < \varepsilon$. Now T is a non-negative measure in Ω_1. Therefore, if $d(x, R^n - \Omega_1) > t$, $\varphi_t * T(x) \geq 0$. On the other hand, if $d(x, R^n - \Omega_1) \leq t$, there exists $y \in R^n - \Omega_1$ such that $|x - y| \leq t$. Since $B(x,t) \subset \Omega_2$, it follows that $y \in \Omega_2 - \Omega_1 = C \cap \Omega'$. Consequently, $\varphi_t * T(x) \geq -N$. Hence, $\varphi_t * T(x)$ is bounded below for $x \in \Omega_3$, $t \in S \cap (0, \varepsilon)$, and thus T is a measure in Ω_3 by Lemma 3.9.3. But $\Omega_3 \cap (R^n - \Omega_1) \neq \emptyset$ thus contradicting the definition of Ω_1.

For the case $f \neq 0$, for each $N > 0$ define

$$f_N(x) = \begin{cases} N, & f(x) \geq N \\ f(x), & -N \leq f(x) \leq N \\ -N, & f(x) \leq -N \end{cases}$$

and let R be the distribution defined by $R = T - f_N$. Clearly R satisfies the same conditions as did T when f was assumed to be identically zero. Therefore, R is a non-negative measure in Ω. Thus, for $\psi \in C_0^\infty(\Omega)$, $\psi \geq 0$,

$$R(\psi) = T(\psi) - \int f_N \psi \, dx \geq 0.$$

Letting $N \to \infty$, we have that

$$T(\psi) - \int f \psi \, dx \geq 0.$$

That is, $T - f$ is a non-negative measure in Ω. □

Now that Theorem 3.9.4 is established, we are in a position to consider the implications of a function u with the property that $u \in T^{k,p}(x)$ for every $x \in \Omega$, where Ω is an open subset of R^n. From Theorem 3.8.1 we have that $u \in t^{k,p}(x)$ for almost all $x \in \Omega$. Moreover, in view of Lemma 3.9.1 (ii), it follows that whenever $u \in t^{k,p}(x)$,

$$\limsup_{t \downarrow 0} \varphi_t * D^\alpha u(y) = D^\alpha P_x(x), \quad |x - y| \leq t,$$

for $0 \leq |\alpha| \leq k$. For convenience of notation, let $u_\alpha(x) = D^\alpha P_x(x)$, and assume $u_\alpha \in L^p(\Omega)$. Then Theorem 3.9.4 implies that the distribution $D^\alpha u - u_\alpha$ is a non-negative measure. Similar reasoning applied to the function $-u$ implies that $D^\alpha(-u) - (-u_\alpha)$ is a non-negative measure or equivalently, that $D^\alpha u - u_\alpha$ is a non-positive measure. Thus, we conclude that $D^\alpha u = u_\alpha$ almost everywhere in Ω. That is, the distributional derivatives of u are functions in $L^p(\Omega)$. In summary, we have the following result.

3.9.5. Theorem. *Let $1 \leq p < \infty$ and let k be a non-negative integer. If $u \in T^{k,p}(x)$ for every $x \in \Omega$ and the L^p-derivatives, u_α, belong to $L^p(\Omega)$,*

$0 \leq |\alpha| \leq k$, then $u \in W^{k,p}(\Omega)$.

Clearly, the hypothesis that the L^p-derivatives belong to $L^p(\Omega)$ is necessary. On the other hand, we will be able to strengthen the result slightly by not requiring that $u \in T^{k,p}(x)$ for all $x \in \Omega$. The following allows an exceptional set.

3.9.6. Corollary. *Let $K \subset R^n$ be compact and let $\Omega = R^n - K$. Suppose $H^{n-1}[\pi_i(K)] = 0$ where the $\pi_i : R^n \to R^{n-1}$, $i = 1, 2, \ldots, n$, are n independent orthogonal projections. Assume $u \in T^{k,p}(x)$ for all $x \in \Omega$ and that $u_\alpha \in L^p(\Omega)$, $0 \leq |\alpha| \leq k$. Then $u \in W^{k,p}(R^n)$.*

Proof. Assume initially that the projections π_i are given by

$$\pi_i(x) = (x_1, \ldots, \hat{x}_2, \ldots, x_n)$$

where $(x_1, \ldots, \hat{x}_i, \ldots, x_n)$ denotes the $(n-1)$-tuple with the x_i-component deleted. Theorem 3.9.5 implies that $u \in W^{k,p}(\Omega)$. In view of the assumption on K, reference to Theorem 2.1.4 shows that $u \in W^{1,p}(R^n)$ since u has a representative that is absolutely continuous on almost all lines parallel to the coordinate axes. Now consider $D^\alpha u$, $|\alpha| = 1$. Since $D^\alpha u \in W^{k-1,p}(\Omega)$ a similar argument shows that $D^\alpha u \in W^{1,p}(R^n)$ and therefore that $u \in W^{2,p}(R^n)$. Proceeding inductively, we have that $u \in W^{k,p}(R^n)$.

Recall from Theorem 2.2.2 that $u \in W^{k,p}(R^n)$ remains in the space $W^{k,p}(R^n)$ when subjected to a linear, non-singular change of coordinates. Thus, the initial restriction on the projections π_i is not necessary and the proof is complete. □

In the special case of $k = 1$, it is possible to obtain a similar result that does not require the exceptional set K to be compact. We state the following [BAZ, Theorem 4.5], without proof.

3.9.7. Theorem. *Let $K \subset R^n$ be a Borel set and suppose $H^{n-1}[\pi_i(K)] = 0$ where the $\pi_i : R^n \to R^{n-1}$, $i = 1, 2, \ldots, n$, are n independent orthogonal projections. Let $\Omega = R^n - K$ and assume $u \in L^p_{\text{loc}}(\Omega)$ has the property that its partial derivatives exist at each point of Ω and that they are in $L^p_{\text{loc}}(\Omega)$. Then $u \in W^{1,p}_{\text{loc}}(R^n)$.*

3.10 A Lusin-Type Approximation for Sobolev Functions

Lusin's Theorem states that a measurable function on a compact interval agrees with a continuous function except perhaps for a closed set of arbitrarily small measure. By analogy, it seems plausible that a Sobolev function $u \in W^{k,p}(\Omega)$ should agree with a function of class $C^k(\Omega)$ except for a set of small measure. Moreover, if the requirement concerning the degree of smoothness is lessened, perhaps it could be expected that there is a larger set on which there is agreement. That is, one could hope that u agrees with a function of class $C^\ell(\Omega)$, $0 \leq \ell < k$, except for a set of small $B_{k-\ell,p}$-capacity. Finally, because Sobolev functions can be approximated in norm by functions of class $C^k(\Omega)$, it is also plausible that the Lusin-type approximant could be chosen arbitrarily close to u in norm. The purpose of this and the next section is to show that all of this is possible.

In this section, we begin by showing that if $u \in W^{k,p}(R^n)$, then u agrees with a function, v, of class C^ℓ on the complement of an open set of arbitrarily small $B_{k-\ell,p}$-capacity. In the next section, it will also be shown that $\|u - v\|_{\ell,p}$ can be made small. The outline of the proof of the existence of v is as follows. If $u \in W^{k,p}(R^n)$ and $0 \leq \ell \leq k$, then Theorem 3.4.2 implies that $u \in t^{\ell,p}(x)$ for all x except for a set of $B_{k-\ell,p}$-capacity 0. This means that the remainder terms tends to 0 (with appropriate speed) at $B_{k-\ell,p}$-q.e. $x \in R^n$. We have already established that if a function u has an L^p-derivative of order ℓ at all points of a closed set A (that is, if $u \in t^{\ell,p}(x)$ for each $x \in A$) and if the remainder term tends to 0 in L^p uniformly on A, then there exists a function $v \in C^\ell(R^n)$ which agrees with u on A (Theorem 3.6.3). Thus, to establish our result, we need to strengthen Theorem 3.4.2 by showing that the remainder tends uniformly to 0 on the complement of sets of arbitrarily small capacity. This will be accomplished in Theorem 3.10.4 below.

In the following, we will adopt the notation

$$M_{p,R}u(x) = \sup_{0<r<R} \left(\fint_{B(x,r)} |u(y)|^p dy \right)^{1/p}$$

whenever $u \in L^p(R^n)$, $1 < p < \infty$, and $0 < R < \infty$.

3.10.1. Theorem. *If $1 < p < \infty$ and k is a non-negative integer such that $kp \leq n$, then there is a constant $C = C(k, p, n)$ such that*

$$B_{k,p}[\{x : M_{p,R}u(x) > t\}] \leq \frac{C}{t^p} \|u\|_{k,p}^p \tag{3.10.1}$$

whenever $u \in W^{k,p}(R^n)$ and $R < 1$.

Proof. We use Theorem 2.6.1 to represent u as $u = g_k * f$ where $f \in L^p(R^n)$ and $\|u\|_{k,p} \sim \|f\|_p$. Thus, it is sufficient to establish (3.10.1) with $\|u\|_{k,p}$ replaced by $\|f\|_p$. Since $|u| \leq g_k * |f|$, we may assume $f \geq 0$. Let

$$E_t = \{x : M_{p,R}u(x) > t\}$$

and choose $x \in E_t$. For notational convenience, we will assume that $x = 0$

and denote $B(0,r) = B(r)$. Thus, there exists $0 < r < R < 1$ such that

$$\fint_{B(r)} |u(y)|^p dy > t^p$$

or

$$\fint_{B(r)} \left(\int_{R^n} g_k(y-w) f(w) dw \right)^p dy > t^p.$$

Utilizing the simple inequality $(a+b)^p \leq 2^{p-1}(a^p + b^p)$ whenever $a, b \geq 0$, it therefore follows that either

$$\fint_{B(r)} \left(\int_{|w|\leq 2r} g_k(y-w) f(w) dw \right)^p dy > 2^{1-p} t^p \tag{3.10.2}$$

or

$$\fint_{B(r)} \left(\int_{|w|> 2r} g_k(y-w) f(w) dw \right)^p dy > 2^{1-p} t^p. \tag{3.10.3}$$

If $y \in B(r)$, then from Lemma 2.8.3(i) and the fact that $g_k \leq C I_k$, (2.6.3), we obtain

$$\begin{aligned} \int_{|w|<2r} g_k(y-w) f(w) dw &\leq C \int_{|y-w|\leq 3r} \frac{f(w)}{|y-w|^{n-k}} dw \\ &\leq C r^k M f(y), \end{aligned}$$

where $C = C(k, n)$. Thus, in case (3.10.2) holds, we have

$$t^p \leq C r^{kp} \fint_{B(r)} M f(y)^p dy \tag{3.10.4}$$

where $C = C(k, p, n)$.

We will now establish the estimate

$$\int_{|w|>2r} g_k(y-w) f(w) dw \leq C \inf_{y \in B(r)} \int_{|w|>2r} g_k(y-w) f(w) dw \tag{3.10.5}$$

for all $y \in B(r)$. Recall that $r < 1$. Now if y and w are such that $|y| < r < 2r \leq |w| \leq 2$, we have

$$\frac{3}{2}|w| \geq |w| + |y| \geq |w - y| \geq |w| - |y| \geq |w| - \frac{|w|}{2}.$$

Consequently, if y_1 and y_2 are any two points of $B(r)$, refer to (2.6.3) and the inequality preceding it to conclude that for some constant $C = C(k, n)$

$$\begin{aligned} g_k(w - y_1) &\leq \frac{C}{|w - y_1|^{n-k}} \leq \frac{C}{|w|^{n-k}} \\ &\leq \frac{C}{|w - y_2|^{n-k}} \leq C g_k(w - y_2). \end{aligned} \tag{3.10.6}$$

If $|w| > 2$ and $y \in B(r)$, then $\frac{3}{2}|w| > |w| + 1 \geq |w - y| \geq |w| - |y| \geq |w| - 1 > |w|/2$. Therefore, in this case we also have

$$g_k(w - y_1) \leq C g_k(w - y_2). \tag{3.10.7}$$

Our desired estimate (3.10.5) follows from (3.10.6) and (3.10.7). Thus, in case (3.10.3) holds, there is a constant $C = C(k, p, n)$ such that

$$\begin{aligned} t^p &< C \inf_{y \in B(r)} \left(\int_{|w|>2r} g_k(w - y) f(w) dw \right)^p \\ &\leq C \inf_{y \in B(r)} (g_k * f(y))^p. \end{aligned}$$

To summarize the results of our efforts thus far, for each $x \in E_t$ there exists $0 < r < 1$ such that either

$$t^p \leq C r^{kp} \fint_{B(x,r)} Mf(y)^p dy \tag{3.10.8}$$

or

$$t \leq C \inf_{y \in B(x,r)} g_k * f(y). \tag{3.10.9}$$

Let $\mathcal{G}_1$ be the family of all closed balls for which (3.10.8) holds. By Theorem 1.3.1, there exists a disjoint subfamily $\mathcal{F}$ such that

$$\begin{aligned} B_{k,p}[\{\cup B : B \in \mathcal{G}_1\}] &\leq B_{k,p}[\{\cup \hat{B} : B \in \mathcal{F}\}] \\ &\leq \sum_{B \in \mathcal{F}} B_{k,p}(\hat{B}) \\ &\leq C \sum_{B(x,r) \in \mathcal{F}} (5r)^{n-kp} \quad \text{(by Theorem 2.6.13)} \\ &\leq \frac{C}{t^p} \sum_{B \in \mathcal{F}} \int_B Mf(y)^p dy \\ &\leq \frac{C}{t^p} \|f\|_p^p \quad \text{(by Theorem 2.8.2).} \end{aligned} \tag{3.10.10}$$

Let $\mathcal{G}_2$ be the family of closed balls for which (3.10.9) holds, then the definition of Bessel capacity implies that $B_{k,p}[\{\cup B : B \in \mathcal{G}_2\}] \leq (C/t^p)\|f\|_p^p$. Thus

$$B_{k,p}[E_t] \leq B_{k,p}[\{\cup B : B \in \mathcal{G}_1\}] + B_{k,p}[\{\cup B : B \in \mathcal{G}_2\}] \leq \frac{C}{t^p}\|f\|_p^p,$$

which establishes our result. □

We now have the necessary information to prove that integral averages of Sobolev functions can be made uniformly small on the complement of sets of small capacity. This result provides an alternate proof of Theorem 3.3.3, as promised in the introduction to Section 1 of this chapter.

3.10.2. Theorem. *Let $1 < p < \infty$ and k be a non-negative integer such that $kp \leq n$. If $u \in W^{k,p}(R^n)$, then for every $\varepsilon > 0$ there exists an open set $U \subset R^n$ with $B_{k,p}(U) < \varepsilon$ such that*

$$\fint_{B(x,r)} |u(y) - u(x)|^p dy \to 0$$

uniformly on $R^n - U$ as $r \downarrow 0$.

Proof. With the result of Theorem 3.3.3 in mind, we define

$$A_r u(x) = \fint_{B(x,r)} |u(y) - u(x)|^p dy$$

for $x \in R^n$ and $r > 0$. Select $\overline{\varepsilon}$ such that $0 < \overline{\varepsilon} < 1$. Since $u \in W^{k,p}(R^n)$, there exists $g \in C_0^k(R^n)$ such that

$$\|u - g\|_{k,p}^p < \overline{\varepsilon}^{p+1}/2.$$

Set $h = u - g$. Then

$$A_r u(x) \leq 2^{p-1}[A_r g(x) + A_r h(x)],$$
$$A_r h(x) \leq 2^{p-1} \left(\fint_{B(x,r)} |h(y)|^p dy + |h(x)|^p \right),$$

and therefore,

$$A_r u(x) \leq C \left[A_r g(x) + \fint_{B(x,r)} |h(y)|^p dy + |h(x)|^p \right],$$

where $C = C(p)$. Consequently, for each $x \in R^n$,

$$\sup_{0<r<R} A_r u(x) \leq C \left[\sup_{0<r<R} A_r g(x) + M_{p,R}|h|(x) + |h(x)|^p \right].$$

Since g has compact support, it is uniformly continuous on R^n and therefore there exists $0 < R < 1$ such that

$$\sup_{0<r<R} C A_r g(x) < \overline{\varepsilon}$$

whenever $x \in R^n$. Therefore,

$$\left\{ x : \sup_{0<r<R} A_r u(x) > 3\overline{\varepsilon} \right\} \subset \{x : C M_{p,R}|h|(x) > \overline{\varepsilon}\} \cup \{x : C|h(x)|^p > \overline{\varepsilon}\}$$

$$\subset \{x : CM_{p,R}|h|(x) > \bar{\varepsilon}\} \cup \{x : (C\bar{\varepsilon}^{-1})^{1/p}|h(x)| > 1\}.$$

Since $h \in W^{k,p}(R^n)$, by Theorem 2.6.1 we can write $h = g_k * f$, where $\|f\|_p \sim \|h\|_{k,p}$. Now

$$\{x : (C\bar{\varepsilon}^{-1})^{1/p}|h(x)| > 1\} \subset \{x : (C\bar{\varepsilon}^{-1})^{1/p} g_k * |f|(x) > 1\}$$

and therefore, by the preceding theorem and the definition of capacity, we obtain a constant $C = C(k,p,n)$ such that

$$\begin{aligned} B_{k,p}[\{x : \sup_{0<r<R} A_r u(x) > 3\bar{\varepsilon}\}] &\leq C[\bar{\varepsilon}^{-p}\|h\|_{k,p}^p + \bar{\varepsilon}^{-1}\|h\|_{k,p}^p] \\ &\leq C2\bar{\varepsilon}^{-p}\left(\frac{\bar{\varepsilon}^{p+1}}{2}\right) \\ &\leq C\bar{\varepsilon}. \end{aligned}$$

For each positive integer i and ε as in the statement of the theorem, let $\bar{\varepsilon}_i = C^{-1}\varepsilon 2^{-i}$ to obtain $0 < R_i < 1$ such that

$$B_{k,p}\left[\left\{x : \sup_{0<r<R_i} A_r u(x) > 3\bar{\varepsilon}_i\right\}\right] < \varepsilon 2^{-i}.$$

Let

$$U = \bigcup_{i=1}^{\infty}\left\{x : \sup_{0<r<R_i} A_r u(x) > 3\bar{\varepsilon}_i\right\}$$

to establish the conclusion of the theorem. □

3.10.3. Remark. If we are willing to accept a slightly weaker conclusion in Theorem 3.10.2, the proof becomes less complicated. That is, if we require only that

$$\fint_{B(x,r)} |u(y) - u(x)|dy \to 0$$

uniformly on $R^n - U$ as $r \downarrow 0$, rather than

$$\fint_{B(x,r)} |u(y) - u(x)|^p dy \to 0,$$

then an inspection of the proof reveals that it is only necessary to show

$$B_{k,p}[\{x : Mu(x) > t\}] \leq \frac{C}{t^p}\|u\|_{k,p}^p.$$

To prove this, let $u = g_k * f$, where $\|f\|_p \sim \|u\|_{k,p}$ and define

$$\Gamma_r(x) = \begin{cases} \frac{1}{r^n} & \text{if } |x| \leq r \\ 0 & \text{otherwise.} \end{cases}$$

Then

$$\fint_{B(x,r)} |u(y)|dy = \Gamma_r * |u|(x)$$
$$\leq \Gamma_r * (g_k * |f|)(x)$$
$$\leq g_k * M|f|(x),$$

which implies $Mu \leq g_k * M|f|$. From the definition of capacity,

$$\begin{aligned} B_{k,p}[\{x : Mu(x) \geq t\}] &\leq B_{k,p}[\{x : g_k * M|f|(x) \geq t\}] \\ &\leq t^{-p}\|M\,|f|\,\|_p^p \\ &\leq Ct^{-p}\|f\|_p^p, \quad \text{by Theorem 2.8.2,} \\ &\leq Ct^{-p}\|u\|_{k,p}^p. \end{aligned}$$

As an immediate consequence of Theorem 3.10.2 and the proof of Theorem 3.4.2, we obtain the following theorem which states that Sobolev functions are uniformly differentiable on the complement of sets of small capacity.

3.10.4. Theorem. *Let ℓ, k be non-negative integers such that $\ell \leq k$ and $(k-\ell)p < n$. Let $u \in W^{k,p}(R^n)$. Then, for each $\varepsilon > 0$, there exists an open set U with $B_{k-\ell,p}(U) < \varepsilon$ such that*

$$r^{-\ell}\left[\fint_{B(x,r)} |u(y) - P_x^{(\ell)}(y)|^p dy\right]^{1/p} \to 0$$

uniformly on $R^n - U$ as $r \downarrow 0$.

Finally, as a direct consequence of Theorems 3.10.4 and 3.6.3, we have the following.

3.10.5. Theorem. *Let ℓ, k be non-negative integers such that $\ell \leq k$ and $(k-\ell)p < n$. Let $u \in W^{k,p}(R^n)$ and $\varepsilon > 0$. Then there exists an open set $U \subset R^n$ and a C^ℓ function v on R^n, such that*

$$B_{k-\ell,p}(U) < \varepsilon$$

and

$$D^\alpha v(x) = D^\alpha u(x)$$

for all $x \in R^n - U$ and $0 \leq |\alpha| \leq \ell$.

3.11 The Main Approximation

We conclude the approximation procedure by proving that the smooth function v obtained in the previous theorem can be modified so as to be close to u in norm.

In addition to some preliminary lemmas, we will need the following version of the Poincaré inequality which will be proved in Theorem 4.5.1.

3.11.1. Theorem. *Let $\sigma \in (0,1)$, ℓ a positive integer, and $1 \le p < \infty$. Then there exists a constant $C = C(\sigma, \ell, p, n)$ such that for every non-empty bounded convex subset Ω of R^n with diameter ρ and every $u \in W^{\ell,p}(\Omega)$ for which*

$$|\Omega \cap \{x : u(x) = 0\}| \ge \sigma|\Omega|,$$

we have the inequality

$$\int_\Omega |u(x)|^p dx \le C\rho^{\ell p} \sum_{|\alpha|=\ell} \int_\Omega |D^\alpha u(x)|^p dx.$$

3.11.2. Lemma. *Let ℓ be a positive integer and let u be a function $W^{\ell,p}(R^n)$ which vanishes outside a bounded open set U. Let $\delta, \sigma \in (0,1)$ and let*

$$E = \partial U \cap \left\{x : \inf_{0<t\le\delta} \frac{|K(x,t) \cap (R^n - U)|}{t^n} \ge \sigma\right\} \tag{3.11.1}$$

where $K(x,t)$ denotes the closed cube with center x and side-length t. Let m be a positive integer such that $m \le \ell$ and let $\varepsilon > 0$. Then there exists a function $v \in W^{m,p}(R^n)$ and an open set V such that

(i) $\|u - v\|_{m,p} < \varepsilon$;

(ii) *$E \subset V$ and $v(x) = 0$ when $x \in V \cup (R^n - U)$.*

Proof. For $\lambda \in (0,1]$, let $\mathcal{K}_\lambda$ denote the set of all closed cubes of the form

$$[(i_1 - 1)\lambda, i_1\lambda] \times [(i_2 - 1)\lambda, i_2\lambda] \times \cdots \times [(i_n - 1)\lambda, i_n\lambda]$$

where $i_1, i_2, \ldots, i_n$ are arbitrary integers. Let $\lambda \le \frac{1}{3}\delta$ and let

$$K_1, K_2, \ldots, K_r$$

be those cubes of $\mathcal{K}_\lambda$ that intersect E. Let a_i be the center of K_i and let

$$P_i = K(a_i, 4\lambda).$$

Let ζ be a C^∞ function on R^n, such that $0 \le \zeta \le 1$, $\zeta(x) = 0$, when $x \in K(0,1)$ and $\zeta(x) = 1$ when $x \notin K(0, 3/2)$. Define

$$v_\lambda(x) = u(x) \prod_{i=1}^r \zeta\left(\frac{x - a_i}{2\lambda}\right) \tag{3.11.2}$$

for $x \in R^n$. Clearly $v_\lambda(x) = 0$ when $d(x, E) \le \frac{1}{2}\lambda$, so that, for any λ, we can define v by $v = v_\lambda$ and find an open set V satisfying (ii).

We keep i fixed for the moment and estimate

$$\|u - v_\lambda\|_{\ell,p;P_i}. \tag{3.11.3}$$

We observe that there exists a constant τ, depending only on n and such that at most τ of the cubes P_j intersect P_i (including P_i). Denote these by

$$P_{j_1}, P_{j_2}, \ldots, P_{j_s}$$

where $s \leq \tau$. Then, for $x \in P_i$

$$v_\lambda(x) = u(x)w(x), \tag{3.11.4}$$

where

$$w(x) = \prod_{k=1}^{s} \zeta[(x - a_{j_k})/2\lambda]. \tag{3.11.5}$$

Now, for $x \in P_i$ and any multi-index α with $0 \leq |\alpha| \leq \ell$, we have

$$|D^\alpha w(x)| \leq A_1 \lambda^{-|\alpha|},$$

where A_1 depends only on ℓ and n. Hence for almost all $x \in P_i$ and any multi-index γ with $0 \leq |\gamma| \leq \ell$, we have

$$|D^\gamma v_\lambda(x)| \leq A_2 \sum_{r=0}^{|\gamma|} \lambda^{r-|\gamma|} \sum_{|\beta|=r} |D^\beta u(x)|, \tag{3.11.6}$$

where A_2 depends only on ℓ and n.

Let y be a point where K_i intersects E. Clearly, there is a subcube Q_i of P_i with center y and edge length 3λ. By (3.11.1), u and hence its derivatives are zero on a subset Z of Q_i with

$$|Z| \geq \sigma(3\lambda)^n. \tag{3.11.7}$$

By applying the Poincaré inequality to the interior of the convex set P_i we obtain, when $|\beta| < \ell$,

$$\int_{P_i} |D^\beta u(x)|^p dx \leq A_3 \lambda^{p(\ell-|\beta|)} \sum_{|\xi|=\ell} \int_{P_i} |D^\xi u(x)|^p dx \tag{3.11.8}$$

where $A_3 = A_3(\ell, \sigma, p, n)$. But, with a suitable constant A_3, (3.11.8) will still hold when $|\beta| = \ell$. By (3.11.6) and (3.11.8) (since $\lambda \leq 1$)

$$\int_{P_i} |D^\gamma v_\lambda(x)|^p dx \leq A_4 \sum_{|\xi|=\ell} \int_{P_i} |D^\xi u(x)|^p dx \tag{3.11.9}$$

for $0 \le |\gamma| \le \ell$, where $A_4 = A_4(\ell, p, \sigma, n)$. Let

$$X_\lambda = \bigcup_{i=1}^{r} P_i.$$

Then

$$\int_{X_\lambda} |D^\gamma v_\lambda(x)|^p dx \le A_4 \sum_{|\xi|=\ell} \sum_{i=1}^{r} \int_{P_i} |D^\xi u(x)|^p dx$$

for $0 \le |\gamma| \le \ell$. But each point of X_λ belongs to at most τ of the cubes P_i, hence

$$\int_{X_\lambda} |D^\gamma v_\lambda(x)|^p dx \le \tau A_4 \sum_{|\xi|=\ell} \int_{X_\lambda} |D^\xi u(x)|^p dx, \tag{3.11.10}$$

for $0 \le |\gamma| \le \ell$. Now

$$\|u - v_\lambda\|_{\ell,p}^p \le 2^p \sum_{0 \le |\gamma| \le \ell} \left[\int_{X_\lambda} |D^\gamma v_\lambda(x)|^p dx + \int_{X_\lambda} |D^\gamma u(x)|^p dx \right],$$

so that by (3.11.10)

$$\|u - v_\lambda\|_{\ell,p}^p \le A_5 \sum_{0 \le |\gamma| \le \ell} \int_{X_\lambda \cap U} |D^\gamma u(x)|^p dx \tag{3.11.11}$$

where $A_5 = A_5(\ell, p, \sigma, n)$. But

$$X_\lambda \cap U \subset U \cap \{x : d(x, \partial U) < 2\sqrt{n}\lambda\}.$$

Hence $|(X_\lambda \cap U)| \to 0$ as $\lambda \downarrow 0$. Therefore by (3.11.11)

$$\|u - v_\lambda\|_{\ell,p} \to 0$$

as $\lambda \to 0^+$.

The required function v is now obtained by putting $v = v_\lambda$, with sufficiently small λ. □

3.11.3. Lemma. *Let $0 \le \lambda \le n$. Then there exists a constant $C = C(\lambda, n)$ such that*

$$\fint_{B(z,\delta)} |x - y|^{\lambda - n} dx \le C|y - z|^{\lambda - n}, \tag{3.11.12}$$

for all $y, z \in R^n$ and all $\delta > 0$.

Proof. We first show that there exists a constant C, such that (3.11.12) holds when $y = 0$ and z is arbitrary.

When $|z| \geq 3\delta$, we have

$$|z| \leq |x| + |z - x| < |x| + \delta \leq |x| + \frac{1}{3}|z|,$$

so that $|z| \leq \frac{3}{2}|x|$, $|x|^{\lambda-n} \leq A|z|^{\lambda-n}$, and (3.11.12) holds.

When $|z| < 3\delta$,

$$\delta^{-n} \int_{B(z,\delta)} |x|^{\lambda-n} dx \leq \delta^{-n} \int_{B(0,4\delta)} |x|^{\lambda-n} dx = C\delta^{\lambda-n}.$$

Hence, it is clear that (3.11.12) holds with $y = 0$.

Since we have shown that

$$\delta^{-n} \int_{B(z,\delta)} |x|^{\lambda-n} dx \leq C|z|^{\lambda-n},$$

for all $z \in R^n$, the general result follows by a change of variables; that is, replace z by $z - y$. □

Throughout the remainder of this section, it will be more convenient to employ the Riesz capacity, $R_{k,p}$, rather than the Bessel capacity, $B_{k,p}$. This will have no significant effect on the main result, Theorem 3.11.6. See Remark 3.11.7.

3.11.4. Lemma. *Let k be a non-negative real number such that $kp < n$. Let U be a bounded non-empty open subset of R^n and F a subset of ∂U with the property that for each $x \in F$, there is a $t \in (0,1)$ for which*

$$\frac{|U \cap B(x,t)|}{|B(x,t)|} \geq \sigma \tag{3.11.13}$$

where $\sigma \in (0,1)$. Then there exists a constant $C = C(n,p,k)$ such that

$$R_{k,p}(U \cup F) \leq C\sigma^{-p} R_{k,p}(U). \tag{3.11.14}$$

Proof. Let σ, U, and F be as described above. The cases $k = 0$ and $k > 0$ are treated separately.

(i) We consider first the case where $k > 0$. Let ψ be a non-negative function in $L^p(R^n)$ with the property that

$$\frac{1}{\gamma(k)} \int_{R^n} |x - y|^{k-n} \psi(y) dy \geq 1 \tag{3.11.15}$$

for all $x \in U$. Let C_1 be the constant of Lemma 3.11.3. It can be assumed that $C_1 \geq 1$. Consider a point $b \in F$ and let t be such that (3.11.13) holds for $x = b$. By Lemma 3.11.3,

$$C_1 |y - b|^{k-n} \geq \fint_{B(b,t)} |x - y|^{k-n} dx$$

so that

$$C_1 \int_{R^n} |y-b|^{k-n}\psi(y)dy \geq \fint_{B(b,t)} \left[\int_{R^n} |x-y|^{k-n}\psi(y)dy\right] dx$$

and by (3.11.15),

$$\geq \gamma(k)t^{-n}|U \cap B(b,t)|.$$

Hence by (3.11.13),

$$\frac{C_2}{\gamma(k)} \int_{R^n} |y-b|^{k-n}\psi(y)dy \geq \sigma.$$

Put $\eta = C_2\sigma^{-1}\psi$. Then

$$\frac{1}{\gamma(k)} \int_{R^n} |y-x|^{k-n}\eta(y)dy \geq 1$$

for all $x \in F$, and therefore,

$$R_{k,p}(F) \leq \|\eta\|_p^p = \left(\frac{C_2}{\sigma}\right)^p \|\psi\|_p^p.$$

Thus

$$R_{k,p}(F) \leq \left(\frac{C_2}{\sigma}\right)^p R_{k,p}(U).$$

The required inequality now follows.

(ii) Now let $k = 0$, so that $R_{k,p}$ becomes Lebesgue measure. Let $\mathcal{B}$ be the collection of all closed balls B with center in F and radius between 0 and 1 such that

$$\frac{|U \cap B|}{|B|} \geq \sigma. \tag{3.11.16}$$

Hence, by Theorem 1.3.1, there exists sequence $\{B_r\}$, $B_r \in \mathcal{B}$, such that $B_r \cap B_s = 0$ when $r \neq s$ and

$$F \subset \bigcup_{r=1}^{\infty} \hat{B}_r.$$

Thus

$$|F| \leq \sum_{r=1}^{\infty} |\hat{B}_r| = 5^n \sum_{r=1}^{\infty} |B_r|$$

and by (3.11.16)

$$\leq 5^n\sigma^{-1} \sum_{r=1}^{\infty} |U \cap B_r| \leq 5^n\sigma^{-1}|U|.$$

Since $\sigma < 1$, the required inequality follows. □

3.11.5. Lemma. *Let $p > 1$, k a non-negative real number such that $kp < n$ and ℓ a positive integer. There exists a constant $C = C(n, p, k, \ell)$ such that for each bounded non-empty open subset U of R^n, each $u \in W^{\ell,p}(R^n)$ which vanishes outside U and every $\varepsilon > 0$ there exists a C^∞ function v on R^n with the properties*

(i) $\|u - v\|_{\ell,p} < \varepsilon$,

(ii) $R_{k,p}(\operatorname{spt} v) \leq C R_{k,p}(U)$ *and*

(iii) $\operatorname{spt} v \subset V = R^n \cap \{x : d(x, U) < \varepsilon\}$.

Proof. Let U, u, and ε be as described above. Since $U \neq \emptyset$, it follows that $R_{k,p}(U) > 0$. Let

$$E = \partial U \cap \left\{ x : \inf_{0<t\leq 1/2} \frac{|\overline{B}(x,t) - U)|}{t^n} \geq \frac{1}{2} \right\}. \tag{3.11.17}$$

Then E is closed. By Lemma 3.11.2 there exists a function $v_0 \in W^{\ell,p}(R^n)$ and an open set V_0 such that

$$\|u - v_0\|_{\ell,p} < \frac{1}{2}\varepsilon, \tag{3.11.18}$$

$E \subset V_0$ and $v_0(x) = 0$ when $x \in V_0 \cup (R^n - U)$. Set

$$F = \partial U - E.$$

Then, for each $x \in F$ there exists $t \in (0, 1/2]$ such that

$$\frac{|U \cap \overline{B}(x,t)|}{|\overline{B}(x,t)|} \geq \sigma \tag{3.11.19}$$

where $\sigma = 1 - 1/(2\alpha(n))$. Let C_1 be the constant appearing in Lemma 3.11.4. Then

$$R_{k,p}(U \cup F) \leq \frac{1}{2} C R_{k,p}(U), \tag{3.11.20}$$

where $C = 2C_1\sigma^{-p}$. Let

$$B = R^n \cap \{x : v_0(x) \neq 0\}.$$

Then $\overline{B} \subset U \cup F$ and hence $R_{k,p}(\overline{B}) \leq \frac{1}{2} C R_{k,p}(U)$, so that there exists an open set W with $\overline{B} \subset W$ and

$$R_{k,p}(W) < C R_{k,p}(U).$$

By applying a suitable mollifier to v_0 we can obtain a C^∞ function v with $\operatorname{spt} v \subset V \cap W$ and

$$\|v_0 - v\|_{\ell,p} < \frac{1}{2}\varepsilon. \tag{3.11.21}$$

It follows from (3.11.18) and (3.11.21) that v has the required properties. □

We are now in a position to prove the main theorem.

3.11.6. Theorem. *Let ℓ, m be positive integers with $m \leq \ell$, $(\ell - m)p < n$ and let Ω be a non-empty open subset of R^n. Then, for $u \in W^{\ell,p}(\Omega)$ and each $\varepsilon > 0$, there exists a C^m function v on Ω such that if*

$$F = \Omega \cap \{x : u(x) \neq v(x)\},$$

then

$$R_{\ell-m,p}(F) < \varepsilon \quad \text{and} \quad \|u - v\|_{m,p} < \varepsilon.$$

Proof. It can be assumed that the set $A = \Omega \cap \{x : u(x) \neq 0\}$ is not empty. Initially, it will be assumed that $\Omega = R^n$ and A bounded. We will show that there exists a C^m function v on R^n satisfying the conclusion of the theorem and that $\operatorname{spt} v$ is contained in the set $V = R^n \cap \{x : d(x, A) < \varepsilon\}$.

Let C be the constant of Lemma 3.11.5. Let u be defined by its values at Lebesgue points everywhere on Ω except for a set E with $B_{\ell,p}(E) = R_{\ell,p}(E) = 0$. By Theorem 3.10.5 there exists an open set U of R^n and a C^m function h on R^n, such that $U \supset E$,

$$R_{\ell-m,p}(U) < \frac{\varepsilon}{1+C} \tag{3.11.22}$$

and

$$h(x) = u(x)$$

for all $x \in R^n - U$. We may assume that $\operatorname{spt} h \subset V$ and $\overline{U} \subset V$. By substituting $\ell - m$ for k and $u - h$ for u in Lemma 3.11.5, we obtain a C^∞ function φ on R^n such that

$$\|u - h - \varphi\|_{m,p} < \varepsilon, \tag{3.11.23}$$

$$R_{\ell-m,p}(\operatorname{spt} \varphi) \leq C R_{\ell-m,p}(U), \tag{3.11.24}$$

and

$$\operatorname{spt} \varphi \subset V. \tag{3.11.25}$$

Put $v = h + \varphi$. Then the second part of the theorem follows from (3.11.23). Clearly,

$$F \subset R^n \cap [\{x : h(x) \neq u(x)\} \cup \operatorname{spt} \varphi] \subset U \cup \operatorname{spt} \varphi, \tag{3.11.26}$$

so that by (3.11.24)

$$R_{\ell-m,p}(F) \leq (1 + C) R_{\ell-m,p}(U).$$

Thus, the first part of the conclusion follows from (3.11.22). Since $\operatorname{spt} h$ and $\operatorname{spt} \varphi$ are both contained in V, it follows that $\operatorname{spt} v \subset V$.

We now consider the general case when Ω is an arbitrary open subset of R^n. Let $\{C_j\}_{j=0}^{\infty}$ be an infinite sequence of non-empty compact sets, such that

$$C_i \subset \operatorname{Int} C_{i+1} \tag{3.11.27}$$

for i a non-negative integer and

$$\lim_{i\to\infty} C_i = \Omega. \tag{3.11.28}$$

Put $C_{-1} = \emptyset$. For each $i \geq 0$, let φ_i be a C^∞ function on R^n such that $0 \leq \varphi_i \leq 1$,

$$C_i \subset \operatorname{int}\{x : \varphi_i(x) = 1\}, \tag{3.11.29}$$

and

$$\operatorname{spt}\varphi_i \subset \operatorname{Int} C_{i+1}. \tag{3.11.30}$$

Put

$$\psi_0 = \varphi_0 \quad \text{and} \quad \psi_i = \varphi_i - \varphi_{i-1} \tag{3.11.31}$$

when $i \geq 1$. Then each ψ_i is C^∞ on R^n with compact support and

$$\operatorname{spt} \psi_i \subset (\operatorname{Int} C_{i+1}) - C_{i-1}. \tag{3.11.32}$$

Hence, for each $x \in \Omega$, $\psi_i(x) \neq 0$ for at most two values of i. Therefore

$$\sum_{i=0}^{\infty} \psi_i(x) = 1 \tag{3.11.33}$$

for all $x \in \Omega$. For each $i = 0, 1, 2, \ldots$ define

$$u_i(x) = \begin{cases} u(x)\psi_i(x) & \text{when } x \in \Omega \\ 0 & \text{when } x \notin \Omega. \end{cases} \tag{3.11.34}$$

By the conclusion of our theorem proved under the assumption that $\Omega = R^n$, there exists for each $i \geq 0$ a C^m function v_i on R^n with compact support such that

$$\|u_i - v_i\|_{m,p} < \frac{\varepsilon}{2^{i+1}}, \tag{3.11.35}$$

and

$$R_{\ell-m,p}(F_i) < \frac{\varepsilon}{2^{i+1}} \tag{3.11.36}$$

where

$$F_i = R^n \cap \{x : u_i(x) \neq v_i(x)\}.$$

Moreover,

$$\operatorname{spt} v_i \subset (\operatorname{Int} C_{i+1}) - C_{i-1}. \tag{3.11.37}$$

For each $x \in \Omega$, there are at most two values of i for which $v_i(x) \neq 0$. Hence we can define

$$v(x) = \sum_{i=0}^{\infty} v_i(x)$$

for $x \in \Omega$. It is easily seen that $F \subset \cup_{i=0}^{\infty} F_i$, hence

$$R_{\ell-m,p}(F) \leq \varepsilon.$$

Also

$$\|u - v\|_{m,p} \leq \sum_{i=0}^{\infty} \|u_i - v_i\|_{m,p} < \varepsilon. \qquad \square$$

3.11.7. Remark. We have seen from earlier work in Section 2.6, that $R_{k,p} \leq C B_{k,p}$ and that $R_{k,p}$ and $B_{k,p}$ have the same null sets. However, it also can be shown that $B_{k,p} \leq C[R_{k,p} + (R_{k,p})^{n/(n-kp)}]$ for $kp < n$, cf. [A5]. Therefore, the Riesz capacity in the previous theorem can be replaced by Bessel capacity.

Exercises

3.1. Prove that the statement

$$\lim_{r \to 0} \fint_{B(x,r)} u(y)dy = u(x)$$

for $B_{1,p}$-q.e. $x \in R^n$ and any $u \in W^{1,p}(R^n)$ implies the apparently stronger statement

$$\lim_{r \to 0} \fint_{B(x,r)} |u(y) - u(x)|dy = 0$$

for $B_{1,p}$-q.e. $x \in R^n$. See the beginning of Section 3.3.

3.2. It was proved in Theorem 2.1.4 that a function $u \in W^{1,p}(R^n)$ has a representative that is absolutely continuous on almost all line segments parallel to the coordinate axes. If a restriction is placed on p, more information can be obtained. For example, if it is assumed that $p < n - 1$, then u is continuous on almost all hyperplanes parallel to the coordinate planes. To prove this, refer to Theorem 3.10.2 to conclude that there is a sequence of integral averages

$$A_k(x) = \fint_{B(x,\frac{1}{k})} u(y)dy$$

which, for each $\varepsilon > 0$, converges uniformly to u on the complement of an open set U_ε whose $B_{1,p}$-capacity is less than ε. Hence $E =$

$\bigcap_{\varepsilon>0} U_\varepsilon$ is a set of $B_{1,p}$-capacity 0. It follows from Theorem 2.16.6 (or Exercise 2.16) that the projection of E onto a coordinate axis has linear measure 0. Note that u is continuous on $\pi^{-1}(t)$, $t \notin \pi(E)$ where π denotes the projection. Corresponding results for $p > n - k$, k an integer, can be easily stated and proved.

3.3. At the beginning of Section 3.9, an example is given which shows that u need not be bounded when $u \in W^{1,n}[B(0,r)]$, $r < 1$. This example can be easily modified to make the pathology even more striking. Let $u(x) = \log\log(1/|x|)$ for small $|x|$ and otherwise defined so that u is positive, smooth and has compact support. Now let

$$v(x) = \sum_{k=1}^{\infty} 2^{-k} u(x - r_k)$$

where $\{r_k\}$ is dense in R^n. Then $v \in W^{1,n}(R^n)$ and is unbounded in a neighborhood of each point.

3.4. Use (2.4.18) to show that if $u \in W^{1,p}_{\text{loc}}(R^n)$, $p > n$, then u is classically differentiable almost everywhere.

3.5. Verify that $\|u\|_{T^{k,p}(x)}$, which is discussed in Definition 3.5.4, is in fact a norm.

3.6. If $u \in W^{1,p}(R^n)$, the classical Lebesgue point theorem states that

$$\lim_{r\to 0} \fint_{B(x_0,r)} |u(x) - u(x_0)|dx = 0 \tag{$*$}$$

for a.e. x_0. Of course, $u \in L^1(R^n)$ is sufficient to establish this result. Since $u \in W^{1,p}(R^n)$, this result can be improved to the extent that $(*)$ holds for $B_{1,p}$-q.e. $x_0 \in R^n$ (Theorem 3.3.3). Give an example that shows this result is optimal. That is, show that in general it is necessary to omit a $B_{1,p}$-null set for the validity of $(*)$.

3.7. Prove that (3.3.22) can be improved by replacing p by $p^* = np/(n - kp)$.

3.8. A measurable function u is said to have a Lebesgue point at x_0 if

$$\lim_{r\to 0} \fint_{B(x_0,r)} |u(y) - u(x_0)|dy = 0.$$

A closely related concept is that of *approximate continuity*. A measurable function u is said to be approximately continuous at x_0 if there exists a measurable set E with density 1 at x_0 such that u is continuous at x_0 relative to E. Show that if u has a Lebesgue point at x_0, then u is approximately continuous at x_0. See Remark 4.4.5. Show that the converse is true if u is bounded and that it is false without this assumption.

Another definition of approximate continuity is the following. u is approximately continuous at x_0 if for every $\varepsilon > 0$, the set

$$A_\varepsilon = \{x : |u(x) - u(x_0)| \geq \varepsilon\}$$

has density 0 at x_0. A_ε is said to have density 0 at x_0 if

$$\lim_{r \to 0} \frac{|A_\varepsilon \cap B(x_0, r)|}{|B(x_0, r)|} = 0.$$

Prove that the two definitions of approximate continuity are equivalent.

3.9. The definition of an *approximate total differential* is analogous to that of approximate continuity. If u is a real valued function defined on a subset of R^n, we say that a linear function $L : R^n \to R^1$ is an approximate differential of u at x_0 if for every $\varepsilon > 0$ the set

$$A_\varepsilon = \left\{x : \frac{|u(x) - u(x_0) - L(x - x_0)}{|x - x_0|} \geq \varepsilon\right\}$$

has density 0 at x_0. Prove the analog of Exercise 3.8; show that if u is an element of $t^{1,1}(x_0)$, then u has an approximate total differential at x_0.

3.10. The definition of an approximate total differential given in Exercise 3.9 implies that the difference quotient

$$\frac{|u(x) - u(x_0) - L(x - x_0)}{|x - x_0|}$$

approaches 0 as $x \to x_0$ through a set E whose density at x_0 is 1. In some applications, it is necessary to have more information concerning the set E. For example, if $u \in W^{1,p}(R^n)$, $p > n - 1$, then it can be shown that u has a *regular approximate total differential* at almost all points x_0. The definition of this is the same as that for an approximate total differential, except that the set E is required to be the union of boundaries of concentric cubes centered at x_0. To prove this, consider

$$u_{x_0}(t, z) = \frac{u(x_0 + tz) - u(x_0)}{t} - L(z),$$

and define

$$\gamma_{x_0}(t) = \sup\{|u_t(z)| : z \in \partial C\}$$

where C is a cube centered at x_0. Since x_0 is fixed throughout the argument, let $u_t(z) = u_{x_0}(t, z)$.

STEP 1. For each x_0 and each cube C with x_0 as center, observe that $u_t \in W^{1,p}(C)$ for all sufficiently small $t > 0$. With

$$\alpha_{x_0}(t) = \int_C (|u_t|^p + |Du_t|^p)dx$$

prove that $\alpha_{x_0}(t) \to 0$ as $t \to 0$ for almost all x_0.

STEP 2. Show that u has a regular approximate differential at all x_0 that satisfy the conclusion of Step 1 and for which $Du(x_0)$ exists. For this purpose, let $L(z) = Du(x_0) \cdot z$. Since $u_t \in W^{1,p}(C)$, it follows that $u_t \in W^{1,p}(K_r)$ for almost all $r > 0$ where K_r is the boundary of a cube of side length $2r$. Moreover, from Exercise 3.2, we know that u_t is continuous on all such K_r. Let

$$\varphi_t(r) = \int_{K_r} (|u_t|^p + |Du_t|^p)dH^{n-1}.$$

Let $E_t = [1/2, 1] \cap \{r : \varphi_t(r) < \alpha_{x_0}(t)^{1/2}\}$ and conclude that

$$|([1/2, 1] - E_t)| \leq \alpha_{x_0}(t)^{1/2}.$$

STEP 3. Use the Sobolev inequality to prove that for $z \in K_r$, $r \in E_t$, and $\beta = (n-1)/p$

$$\begin{aligned}|u_t(z)| &\leq Mr^{-\beta}\left(\int_{K_r} |u_t|^p dH^{n-1}\right)^{1/p} \\ &\quad + Mr^{1-\beta}\left(\int_{K_r} |Du_t|^p dH^{n-1}\right)^{1/p} \\ &\leq Mr^{-\beta}\varphi_t(r)^{1/p} + Mr^{1-\beta}\varphi_t(r)^{1/p} \\ &\leq [M2^\beta + M]\alpha_{x_0}(t)^{1/2p},\end{aligned}$$

where $M = M(p, n)$.

STEP 4. Thus, for $z \in K_r$ and $r \in E_t$,

$$\begin{aligned}\gamma_{x_0}(t \cdot r) &= r^{-1}\sup\{|u_t(z)| : z \in K_r\} \\ &\leq 2[M2^\beta + 1] \cdot \alpha_{x_0}(t)^{1/2p}.\end{aligned}$$

STEP 5. For each positive integer i, let $t_i = 2^{-i}$ and let E_{t_i} be the associated set as in Step 2. Set $A = \cup_{i=1}^{\infty} E_{t_i}$ and note that 0 is a point of right density for A (Step 1) and that $\gamma_{x_0}(t) \to 0$ as $t \to 0$, $t \in A$.

3.11. Prove that $\overline{u}$ defined in (3.6.2) belongs to $C^\infty(U - A)$.

3.12. Give an example which shows that the uniformity condition in the second part of the statement in Theorem 3.5.7 is necessary.

3.13. In this and the next exercise, it will be shown that a function with minimal differentiability hypotheses agrees with a C^1 function on a set of large measure, thus establishing an extension of Theorem 3.11.6. For simplicity, we only consider functions of two variables and begin by outlining a proof of the following classical fact: *If u is a measurable function whose partial derivatives exist almost everywhere on a measurable set E, then u has an approximate total differential almost everywhere on E.* See Exercise 3.9 for the definition of approximate total differential.

STEP 1. By Lusin's theorem, we may assume that E is closed and that u is its partial derivatives are continuous on E.

STEP 2. For each $(x, y) \in E$ consider the differences

$$\Delta(x, y; h, k) = |u(x + h, y + k) - u(x, y) - hD_1u(x, y) - kD_2u(x, y)|$$

$$\Delta_1(x, y; h) = |u(x + h, y) - u(x, y) - hD_1u(x, y)|$$

$$\Delta_2(x, y; k) = |u(x, y + k) - u(x, y) - kD_2u(x, y)|$$

where $D_1 = \partial/\partial x$ and $D_2 = \partial/\partial y$. Choose positive numbers ε, τ. Using the information in Step 1, prove that there exists $\sigma > 0$ such that the set $A \subset E$ consisting of all points (x, y) with the property that

$$|\{x + h : \Delta_1(x, y; h) < \tau h,\ (x + h, y) \in E,$$
$$a \leq x \leq b,\ |b - a| < \sigma,\ |h| < |b - a|\}| \geq (1 - \varepsilon)|b - a|$$

satisfies $|E - A| < \varepsilon$. Perhaps the following informal description of A will be helpful. For fixed (x, y), let us agree to call a point $(x + h, y)$ "good" if $\Delta_1(x, y; h) \leq \tau h$ and $(x + h, y) \in E$. The set A consists of those points (x, y) with the property that if I_x is any interval parallel to the x-axis containing x whose length is less than σ, then the relative measure of the set of good points in I_x is large.

STEP 3. Now repeat the analysis of Step 2 with E replaced with A to obtain a positive number $\sigma_1 < \sigma$ and a closed set $B \subset A$, $|A - B| < \varepsilon$ which consists of all points (x, y) with the property that

$$|\{y + k : \Delta_2(x, y; k) \leq \tau k,\ (x, y + k) \in A,$$
$$a \leq y \leq b,\ |b - a| < \sigma_1,\ |k| < |b - a|\}| \geq (1 - \varepsilon)|b - a|.$$

STEP 4. Let $\sigma_2 < \sigma_1$ be such that

$$|D_1u(x+h_2, y+k_2) - D_1u(x+h_1, y+k_1)| < \tau$$

for any 2 points $(x+h_2, y+k_2)$, $(x+h_1, y+k_1)$ in E with $|h_2-h_1| < \sigma_2$, $|k_2 - k_1| < \sigma_1$.

STEP 5. Choose $(x_0, y_0) \in B$ and let $R = [a_1, b_1] \times [a_2, b_2]$ be any rectangle containing (x_0, y_0) whose diameter is less than $\sigma_2 < \sigma_1 < \sigma$. Let

$$E_2 = \{(y_0 + k) : \Delta_2(x_0, y_0; k) \leq \tau|k|, \\ (x_0, y_0 + k) \in A,\ a_2 \leq y_0 + k \leq b_2\},$$

and for each $(y_0 + k)$

$$E_1(y_0 + k) = \{(x_0 + h) : \Delta_1(x_0, y_0 + k; h) \leq \tau|h|, \\ (x_0 + h, y_0 + k) \in E\}.$$

Now for any (h, k) such that $y_0 + k \in E_2$ and $x_0 + h \in E_1(y_0 + k)$, we have $(x_0 + h, y_0 + k) \in E \cap R$ and therefore

$$\begin{aligned}\Delta(x_0, y_0; h, k) &\leq \Delta_1(x_0, y_0; h) + \Delta_2(x_0, y_0; k) \\ &\qquad + |h|\,|D_1u(x_0, y_0 + k) - D_1u(x_0, y_0)| \\ &\leq \tau(|h| + |k|).\end{aligned}$$

From this conclude that

$$\begin{aligned}&|B \cap R \cap \{(x_0 + h, y_0 + k) : \Delta(x_0, y_0; h, k) \leq 2\tau(|h| + |k|)\}| \\ &\qquad \geq\ (1-\varepsilon)^2(b_1 - a_1)(b_2 - a_2) = (1-\varepsilon)^2|R|.\end{aligned}$$

STEP 6. Take R to be a square with (x_0, y_0) as center and appeal to Exercise 3.8 to reach the desired conclusion.

3.14. We continue to outline the proof that a function whose partial derivatives exist almost everywhere agrees with a C^1 function on a set of large measure. Let u be a real valued function defined on a measurable set $E \subset R^n$, and for each positive number M and $x \in E$ let

$$A(x, M) = E \cap \{y : |u(y) - u(x)| \leq M|y - x|\}.$$

If $A(x, M)$ has density 1 at x, u is said to be of *approximate linear distortion at* x. Our objective is to show that if u is of approximate linear distortion at each point E, then there exists sets E_k such that $E = \cup_{k=1}^{\infty} E_k$ and u is Lipschitzian on each of the sets E_k.

STEP 1. If x_1 and x_2 are any two points of R^n, then

$$\alpha = \frac{|\overline{B}(x_1, |x_2 - x_1|) \cap \overline{B}(x_2, |x_2 - x_1|)|}{|\overline{B}(x_1, |x_2 - x_1|)| + |\overline{B}(x_2, |x_2 - x_1|)|}$$

is a positive number less than 1 which is independent of the choice of x_1 and x_2.

STEP 2. For each positive integer k, let E_k be the set of those points $x \in E$ such that $|u(x)| \leq k$ and that if r is any number such that $0 < r \leq 1/k$, then

$$\frac{|A(x,k) \cap \overline{B}(x,r)|}{|\overline{B}(x,r)|} > 1 - \alpha.$$

Prove that $E = \cup_{k=1}^{\infty} E_k$.

STEP 3. In order to show that u is Lipschitz on E_k, choose any two points $x_1, x_2 \in E_k$. If $|x_2 - x_1| > 1/k$, then

$$|u(x_2) - u(x_1)| \leq 2k^2 |x_2 - x_1|.$$

Thus, assume that

$$0 < |x_2 - x_1| \leq \frac{1}{k}.$$

Let

$$A_1 = A(x_1, k) \cap \overline{B}(x_1, |x_2 - x_1|),$$
$$A_2 = A(x_2, k) \cap \overline{B}(x_2, |x_2 - x_1|).$$

Prove that $|A_1 \cap A_2| > 0$. If $x^* \in A_1 \cap A_2$, show that

$$|u(x^*) - u(x_i)| \leq k|x^* - x_i|, \quad i = 1, 2$$
$$|x^* - x_i| \leq |x_2 - x_1|, \quad i = 1, 2.$$

Now conclude that

$$|u(x_2) - u(x_1)| \leq 2k|x_2 - x_1|.$$

STEP 4. If u has partial derivatives almost everywhere, appeal to the previous exercise to conclude that u is of approximate linear distortion at almost every point. Now refer to Theorem 3.11.6 to find a C^1 function that agrees with u on a set of arbitrarily large measure.

3.15. Suppose $u \in W^{1,p}(R^n)$. Prove that for $B_{1,p}$-q.e. $x \in R^n$, u is absolutely continuous on almost every ray λ_x whose endpoint is x.

3.16. Let f be a measurable function defined on $[0, 1]$ having the property that f' exists everywhere on $[0, 1]$ and that $|f'|$ is integrable. Prove that f is an absolutely continuous function.

Historical Notes

3.1. The idea that an integrable function has a representative that can be expressed as the limit of integral averages originates with Lebesgue [LE2]. The set of points for which the limit of integral averages does not exist (the exceptional set) is of measure zero. Several authors were aware that the exceptional sets associated with Sobolev functions or Riesz potentials were much smaller than sets of measure zero, cf. [DL], [ARS1], [FU], [FL], [GI]. However, optimal results for the exceptional sets in terms of capacity were obtained in [FZ], [BAZ], [ME2], [CFR]. The development in this section is taken from [MIZ].

3.2. The results in this section are merely a few of the many measure theoretic density theorems of a general nature; see [F, Section 2.10.9] for more.

3.3. Theorem 3.3.3 was first established in [FZ] for the case $k = 1$, and for general k in [BAZ], [ME2], and [CFR]. The concepts of thinness and fine continuity are found in classical potential theory although their development in the context of nonlinear potential theory was advanced significantly in [AM], [HE2], [HW], [ME3]. The proof of the theorem in Remark 3.3.5 was communicated to the author by Norman Meyers.

3.4. Derivatives of a function at a point in the L^p-sense were first studied in depth by Calderón and Zygmund [CZ]. They also proved Theorem 3.4.2 where the exceptional set was obtained as a set of Lebesgue measure zero. The proof of the theorem with the exceptional set expressed in terms of capacity appears in [BAZ], [ME2], and [CFR].

3.5. The spaces $T^k(E)$, $t^k(E)$, $T^{k,p}(x)$, and $t^{k,p}(x)$ were first introduced in [CZ] where also Theorem 3.5.7 was proved. These spaces introduce but one of many methods of dealing with the notion of "approximate differentiability." For other forms of approximate differentiability, see [F, Section 3.1.2], [RR].

3.6–3.8. The material in these sections is adopted from [CZ]. It should be noted that Theorem 8 in [CZ] is slightly in error. The error occurs in the following part of the statement of their theorem: "If in addition $f \in t_u^p(x_0)$ for all $x_0 \in Q$, then $f \in b_u(Q)$." The difficulty is that for this conclusion to hold, it is necessary that condition (1.2) in [CZ] holds uniformly. Indeed, the example in [WH] can be easily modified to show that this uniformity condition is necessary. Theorem 3.6.3 gives the correct version of their theorem. In order for this result to be applicable within the framework of Sobolev spaces, it is necessary to show that Sobolev functions are uniformly differentiable on the complement of sets of small capacity. This is established in Theorem 3.10.4.

In comparing Whitney's Extension theorem (Theorem 3.5.3) with the L^p-version (Theorem 3.6.3), observe that the latter is more general in the sense that the remainder term of the function u in question is required to approach zero only in L^p and not in L^∞. On the other hand, the function is required to be defined only on the set E in Whitney's theorem, while the condition $u \in t^{k,p}(x)$ in Theorem 3.5.3 implies that $u \in L^p[B(x,r)]$ (for all small $r > 0$) thus requiring u to be defined in a neighborhood of x.

3.9. Theorem 3.9.4 and the preceding lemmas are due to Calderón [CA4]; the remaining results are from [BAZ].

3.10. The main result of this section is Theorem 3.10.2 which easily implies that Sobolev functions are uniformly differentiable in L^p on the complement of sets of small capacity. The proof is due to Lars Hedberg. Observe that the proof of Theorem 3.10.2 becomes simpler if we are willing to accept a subsequence $\{r_{i_j}\}$ such that

$$\fint_{B(x,r_{i_j})} |u(y) - u(x)|^p dy \to 0$$

uniformly on $R^n - U$ where $B_{k,p}(U) < \varepsilon$. This can be proved by the methods of Lemma 2.6.4. However, this result would not be strong enough to apply Theorem 3.6.3, thus not making it possible to establish the approximation result in Theorem 3.10.5.

3.11. These results appear in [MIZ]. The main theorem (Theorem 3.11.6) is analogous to an interesting result proved by J.H. Michael [MI] in the setting of area theory. He proved that a measurable function f defined on a closed cube $Q \subset R^n$ can be approximated by a Lipschitz function g such that

$$|\{x : f(x) \neq g(x)\}| < \varepsilon$$

and $|A(f,Q) - A(g,Q)| < \varepsilon$ where $A(f,Q)$ denotes the Lebesgue area of f on Q. Theorem 3.11.6 was first proved by Liu [LI] in the case $m = \ell$.

4

Poincaré Inequalities—A Unified Approach

In Chapter 2, basic Sobolev inequalities were established for functions in the space $W_0^{k,p}(\Omega)$. We recall the following fundamental result which is a particular case of Theorem 2.4.2.

4.1.1. Theorem. *Let $\Omega \subset R^n$ be an open set and $1 \leq p < n$. There is a constant $C = C(p,n)$ such that if $u \in W_0^{1,p}(\Omega)$, then*

$$\|u\|_{p^*;\Omega} \leq C\|Du\|_{1,p;\Omega} \tag{4.1.1}$$

where $p^* = np/(n-p)$.

Clearly, inequality (4.1.1) is false in case u is the function that is identically equal to a non-zero constant, thereby ruling out the possibility that it may hold for all $u \in W^{1,p}(\Omega)$. One of the main objectives of this chapter is to determine the extent to which the hypothesis that u is "zero on the boundary of Ω" can be replaced by others. It is well known that there are a variety of hypotheses that imply (4.1.1). For example, if we assume that Ω is a bounded, connected, extension domain (see Remark 2.5.2) and that u is zero on a set S with $|S| = a > 0$, then it can be shown that (4.1.1) remains valid where the constant C now depends on a, n, and Ω. This inequality and others similar to it, are known as Poincaré-type inequalities. We will give a proof of this inequality which is based on an argument that is fundamental to the development of this chapter. A general and abstract version of this argument is given in Lemma 4.1.3.

There is no loss of generality in proving the inequality with p^* replaced by p. The proof proceeds as follows and is by contradiction. If (4.1.1) were false for the class of Sobolev functions that vanish on a set whose measure is greater than a, then for each integer i there is such a function u_i with the property that

$$\|u_i\|_{p;\Omega} \geq i\|Du_i\|_{p;\Omega}.$$

Clearly, we may assume that $\|u_i\|_{1,p;\Omega} = 1$. But then, there exist a subsequence (denoted by the full sequence) and $u \in W^{1,p}(\Omega)$ such that u_i tends weakly to u in $W^{1,p}(\Omega)$. By the Rellich–Kondrachov compactness theorem (Theorem 2.5.1, see also Remark 2.5.2) u_i tends strongly to u in $L^p(\Omega)$. Since $\|u_i\|_{1,p;\Omega} = 1$ it follows that $\|Du_i\|_{p;\Omega} \to 0$ and therefore that $\|Du\|_{p;\Omega} = 0$. Corollary 2.1.9 thus implies that u is constant on Ω. This constant is not 0 since $\|u\|_{p;\Omega} = 1$. Now each u_i is 0 on a set S_i whose measure is no less than a. The strong convergence of u_i to u in $L^p(\Omega)$ implies that (for a subsequence) $u_i \to u$ almost everywhere on $S = \bigcap_{j=1}^{\infty} \bigcup_{i=j}^{\infty} S_i$. Since $|S| \geq a > 0$, this contradicts the conclusion that u is equal to a non-zero constant on Ω.

A close inspection of the proof reveals that the result also remains valid if we assume $\int_\Omega u(x)dx = 0$ rather than $u = 0$ on a set of positive measure. In this chapter we will show that these two inequalities and many other related ones follow from a single, comprehensive inequality obtained in Theorem 4.2.1.

4.1 Inequalities in a General Setting

We now proceed to establish an abstract version of the argument given above which will lead to the general form of the Poincaré inequality, Theorem 4.2.1.

4.1.2. Definition. If X is a Banach space and $Y \subset X$ a *subspace,* then a bounded linear map $L : X \to Y$ onto Y is called a projection if $L \circ L = L$.

Note that

$$L(y) = y, \quad y \in Y, \tag{4.1.2}$$

for there exists $x \in X$ such that $L(x) = y$ and $y = L(x) = L[L(x)] = L(y)$.

4.1.3. Lemma. *Let X_0 be a normed linear space with norm $\| \ \|_0$ and let $X \subset X_0$ be a Banach space with norm $\| \ \|$. Suppose $\| \ \| = \| \ \|_0 + \| \ \|_1$ where $\| \ \|_1$ is a semi-norm and assume that bounded sets in X are precompact in X_0. Let $Y = X \cap \{x : \|x\|_1 = 0\}$. If $L : X \to Y$ is a projection, there is a constant C independent of L such that*

$$\|x - L(x)\|_0 \leq C\|L\| \, \|x\|_1 \tag{4.1.3}$$

for all $x \in X$.

Proof. First, select a particular projection $L' : X \to Y$. We will prove that there is a constant $C' = C'(\|L'\|)$ such that

$$\|x - L'(x)\|_0 \leq C'\|x\|_1, \tag{4.1.4}$$

for all $x \in X$. We emphasize that this part of the proof will produce a constant that depends on L'.

If (4.1.4) were false there would exist $x_i \in X$ such that

$$\|x_i - L'(x_i)\|_0 > i\|x_i\|_1, \quad i = 1, 2, \ldots .$$

Replacing x_i by $x_i/\|x_i - L'(x_i)\|_0$ it follows that

$$\|x_i - L'(x_i)\|_0 = 1 \quad \text{and} \quad \|x_i\|_1 \to 0.$$

Let $z_i = x_i - L'(x_i)$. Then

$$\begin{aligned}\|z_i\|_1 = \|x_i - L'(x_i)\|_1 &\leq \|x_i\|_1 + \|L'(x_i)\|_1 \\ &\leq \|x_i\|_1\end{aligned}$$

since $\|L'(x_i)\|_1 = 0$. Hence, z_i is a bounded sequence in X and therefore, by assumption, there exist a subsequence (which we still denote by $\{z_i\}$) and $z \in X_0$ such that $\|z_i - z\|_0 \to 0$. Since $\|z_i\|_1 \to 0$ it follows that z_i is a Cauchy sequence in X and therefore $\|z_i - z\| \to 0$. Note that $\|z\|_0 = 1$ and $\|z\|_1 = 0$. Thus $z \neq 0$, $z \in Y$, and $L'(z) = z$ by (4.1.2). But $L'(z_i) \to L'(z)$ and $L'(z_i) = 0$, a contradiction.

The next step is to prove (4.1.3) for any projection L where C does not depend on L. Let $L : X \to Y$ be a projection and observe that

$$x - L(x) = x - L'(x) - L(x - L'(x)).$$

Hence, by (4.1.4),

$$\begin{aligned}\|x - L(x)\|_0 &\leq \|x - L'(x)\|_0 + \|L(x - L'(x))\|_0 \\ &\leq C'\|x\|_1 + \|L(x - L'(x))\| \\ &\leq C'\|x\|_1 + \|L\| \, \|(x - L'(x))\| \\ &\leq C'\|x\|_1 + \|L\| \, [\|(x - L'(x))\|_0 + \|x\|_1]\end{aligned}$$

since $\|L'(x)\|_1 = 0$. Appealing again to (4.1.4) we obtain,

$$\begin{aligned}\|x - L(x)\|_0 &\leq C'\|x\|_1 + \|L\| \, [C'\|x\|_1 + \|x\|_1] \\ &= (C' + (C' + 1)\|L\|)\|x\|_1.\end{aligned}$$

Since L is a projection, $\|L\| \geq 1$ and the result now follows. □

We now will apply this result in the context of Sobolev spaces. In particular it will be convenient to take $X = W^{m,p}(\Omega)$.

For notational simplicity, in the following we will let the characteristic function of Ω be denoted by 1. That is, let $\chi_\Omega = 1$. Also, let $\mathcal{P}_k(R^n)$ denote the set of all polynomials in R^n of degree k.

4.1.4. Lemma. *Let k and m be integers with $0 \leq k < m$ and $p > 1$. Let $\Omega \subset R^n$ be an open set of finite Lebesgue measure and suppose $T \in (W^{m-k,p}(\Omega))^*$ has the property that $T(1) \neq 0$. Then there is a projection $L : W^{m,p}(\Omega) \to \mathcal{P}_k(R^n)$ such that for each $u \in W^{m,p}(\Omega)$ and all $|\alpha| \leq k$,*

$$T(D^\alpha u) = T(D^\alpha P) \tag{4.1.5}$$

where $P = L(u)$. Moreover, L has the form

$$L(u) = \sum_{|\alpha| \leq k} T(P_\alpha(Du))x^\alpha$$

where $P_\alpha \in \mathcal{P}_k(R^n)$, $Du = (D_1 u, D_2 u, \ldots, D_n u)$, and

$$\|L\| \geq C \cdot \left(\frac{\|T\|}{T(1)}\right)^{k+1},$$

$C = C(k, p, |\Omega|)$.

Proof. If $P \in \mathcal{P}_k(R^n)$ then P has the form $P(x) = \sum_{|\gamma|=0}^{k} a_\gamma x^\gamma$ and therefore

$$D^\alpha P(0) = \alpha! a_\alpha$$

for any multi-index α. Consequently, by Taylor's theorem for polynomials,

$$D^\alpha P(x) = \sum_{|\beta|=0}^{k-|\alpha|} \frac{D^{\alpha+\beta} P(0)}{\beta!} x^\beta$$

or

$$D^\alpha P(x) = \sum_{|\beta|=0}^{k-|\alpha|} a_{\alpha+\beta} \frac{(\alpha+\beta)!}{\beta!} x^\beta.$$

In particular,

$$D^\alpha P(x) = a_\alpha \alpha!$$

if $|\alpha| = k$. Thus, in order to satisfy (4.1.5), the coefficients a_α of the polynomial must satisfy

$$a_\alpha = \frac{T(D^\alpha u)}{\alpha! T(1)}, \tag{4.1.6}$$

if $|\alpha| = k$. Similarly, if $|\alpha| = k - 1$ then

$$D^\alpha P(x) = a_\alpha \alpha! + \sum_{|\beta|=1} a_{\alpha+\beta} \frac{(\alpha+\beta)!}{\beta!} x^\beta.$$

Consequently by using (4.1.6), (4.1.5) will hold if

$$a_\alpha = \frac{T(D^\alpha u)}{\alpha! T(1)} - \sum_{|\beta|=1} a_{\alpha+\beta} \frac{(\alpha+\beta)!}{\alpha! \beta!} \frac{T(x^\beta)}{T(1)},$$

where $|\alpha| = k - 1$. Proceeding recursively, for any $|\alpha| \leq k$ we have

$$a_\alpha = \frac{T(D^\alpha u)}{\alpha! T(1)} - \sum_{|\beta|=1}^{k-|\alpha|} a_{\alpha+\beta} \frac{(\alpha+\beta)!}{\alpha! \beta!} \frac{T(x^\beta)}{T(1)}. \tag{4.1.7}$$

It is easily verified that L is a projection since $L(u) = P$ implies $D^\alpha[L(u)] = D^\alpha P$ for any multi-index α. But then,

$$T(D^\alpha u) = T(D^\alpha P) = T[D^\alpha(Lu)]$$

and reference to (4.1.7) yields the desired conclusion.

In order to estimate the norm of L, let $u \in W^{m,p}(\Omega)$ with $\|u\|_{m,p;\Omega} \le 1$. Then $\|L\| \le \|L(u)\|_{m,p;\Omega} = \|P\|_{m,p;\Omega}$ where $P(x) = \sum_{|\gamma|=0}^{k} a_\gamma x^\gamma$. Now

$$\|P\|_{m,p;\Omega} \le C(|\Omega|) \sum_{|\gamma|=0}^{k} |a_\gamma|.$$

To estimate the series, first consider $|a_\alpha|$, $|\alpha| = k$. Note that for $|\alpha| = k$ and any non-negative integer ℓ,

$$\frac{\|T\|}{\alpha! T(1)} = \frac{\|T\| \cdot T(1)^\ell}{\alpha! T(1)^{1+\ell}} \le C(\ell, p, |\Omega|) \left(\frac{\|T\|}{T(1)}\right)^{\ell+1} \tag{4.1.8}$$

because $T(1) \le |\Omega|^{1/p} \|T\|$. In particular, this holds for $\ell = k$. Hence from (4.1.6) it follows that

$$|a_\alpha| \le \frac{\|T\|}{\alpha! T(1)} \le C(k, p, |\Omega|) \left(\frac{\|T\|}{T(1)}\right)^{k+1}. \tag{4.1.9}$$

If $|\alpha| = k-1$, $k \ge 1$, then from (4.1.7), (4.1.9) and the fact that $\|T\|/T(1) \ge |\Omega|^{-1/p}$,

$$\begin{aligned} |a_\alpha| &\le \frac{\|T\|}{\alpha! T(1)} + C(k, |\Omega|) \sum_{|\beta|=1} a_{\alpha+\beta} \frac{\|T\|}{T(1)} \\ &\le C(k, p, |\Omega|) \left(\frac{\|T\|}{T(1)}\right)^2 \\ &\le C'(k, p, |\Omega|) \left(\frac{\|T\|}{T(1)}\right)^{k+1}. \end{aligned}$$

In general, if $|\alpha| = k - i$, $k \ge i$, we have

$$\begin{aligned} |a_\alpha| &\le C(k, p, |\Omega|) \left(\frac{\|T\|}{T(1)}\right)^{i+1} \\ &\le C'(k, p, |\Omega|) \left(\frac{\|T\|}{T(1)}\right)^{k+1}. \end{aligned}$$

Proceeding in this way, we find that

$$\|L\| \le C(k, p, |\Omega|) \left(\frac{\|T\|}{T(1)}\right)^{k+1}. \qquad \square$$

In the preceding analysis, if we knew that the distribution T was a non-negative measure μ, then we would be able to improve the result. Indeed, suppose the measure satisfies the inequality

$$\int_{\Omega} x^{\alpha} d\mu(x) \leq M\mu(\Omega) \tag{4.1.10}$$

for every $|\alpha| \leq k$. Of course, such an M exists if either Ω or spt μ is bounded. Then the estimate of $\|L\|$ becomes sharper because, with $T = \mu$, the term $T(x^{\beta})/T(1)$ in (4.1.7) is bounded above by M, thus implying that

$$\|L\| \leq C(k,p,M)\frac{\|T\|}{T(1)}.$$

Hence, we have the following corollary.

4.1.5. Corollary. *Let k and m be integers with $0 \leq k < m$ and $p > 1$. Let $\Omega \subset R^n$ be an open set and suppose $\mu \in (W^{m-k,p}(\Omega))^*$ is a non-negative non-trivial measure satisfying* (4.1.10). *Then there is a projection $L : W^{m,p}(\Omega) \to \mathcal{P}_k(R^n)$ such that for each $u \in W^{m,p}(\Omega)$ and all $|\alpha| \leq k$,*

$$\mu(D^{\alpha}u) = \mu(D^{\alpha}P) \tag{4.1.11}$$

where $P = L(u)$. Moreover, L has the form

$$L(u) = \sum_{|\alpha| \leq k} \mu(P_{\alpha}(Du))x^{\alpha}$$

where $P_{\alpha} \in \mathcal{P}_k(R^n)$, $Du = (D_1u, D_2u, \ldots, D_nu)$, and

$$\|L\| \leq C \cdot \left(\frac{\|\mu\|}{\mu(\Omega)}\right),$$

$C = C(k,p,M)$.

4.2 Applications to Sobolev Spaces

We now consider some of the consequences of the previous two results when applied in the setting of Sobolev spaces. Thus, if $0 \leq k < m$ are integers, $p > 1$ and $\Omega \subset R^n$ is a bounded, connected, extension domain (see Remark 2.5.2), we employ Lemma 4.1.3 with $X = W^{m,p}(\Omega)$ and $X_0 = W^{k,p}(\Omega)$. It follows from the Rellich–Konrachov imbedding theorem (see Exercise 2.3) that bounded sets in $W^{m,p}(\Omega)$ are precompact in $W^{k,p}(\Omega)$. Set $\|u\|_0 = \|u\|_{k,p\Omega}$ and $\|u\|_1 = \|D^{k+1}u\|_{m-(k+1),p;\Omega}$ where $D^{k+1}u$ is considered as the vector $\{D^{\alpha}u\}$ $|\alpha| = k+1$. Clearly, $\|u\| = \|u\|_0 + \|u\|_1$ is an equivalent norm on $W^{m,p}(\Omega)$. Moreover, it follows from Exercise 2.7 that $\|u\|_1 = 0$ if and only if $u \in \mathcal{P}_k(R^n)$. If $T \in (W^{m-k,p}(\Omega))^*$ with $T(1) \neq 0$, then Lemma 4.1.4 asserts that there is a projection $L : W^{m,p}(\Omega) \to \mathcal{P}_k(R^n)$ such that

$$\|L\| \leq C\left(\frac{\|T\|}{T(1)}\right)^{k+1}.$$

Therefore, Lemma 4.1.3 implies

$$\begin{aligned}\|u - L(u)\|_{k,p;\Omega} &\leq C\|L\|\,\|D^{k+1}u\|_{m-(k+1),p;\Omega}\\ &\leq C\left(\frac{\|T\|}{T(1)}\right)^{k+1}\|D^{k+1}u\|_{m-(k+1),p;\Omega}.\end{aligned}$$

These observations are summarized in the following theorem.

4.2.1. Theorem. *Suppose $0 \leq k < m$ are integers and $p \geq 1$. Let $\Omega \subset R^n$ be a bounded, connected extension domain. Let $T \in (W^{m-k,p}(\Omega))^*$ be such that $T(1) \neq 0$. Then, if $L : W^{m,p}(\Omega) \to \mathcal{P}_k(R^n)$ is the projection associated with T,*

$$\|u - L(u)\|_{k,p;\Omega} \leq C\left(\frac{\|T\|}{T(1)}\right)^{k+1}\|D^{k+1}u\|_{m-(k+1),p;\Omega} \qquad (4.2.1)$$

where $C = C(k, p, \Omega)$.

It will now be shown that the norm on the left side of (4.2.1) can be replaced by the L^{p^*}-norm of $u - L(u)$, where $p^* = np/(n - mp)$. For this we need the following lemma.

4.2.2. Lemma. *Suppose $m > 1$ is an integer and $p \geq 1$. Let $\Omega \subset R^n$ be a bounded extension domain. Then for each integer k, $1 \leq k \leq m - 1$, and $\varepsilon > 0$ there is a constant $C = C(n, m, p, k, \varepsilon, \Omega)$ such that*

$$\|D^k u\|_{p;\Omega} \leq C\|u\|_{p;\Omega} + \varepsilon\|D^m u\|_{p;\Omega}, \quad u \in W^{m,p}(\Omega) \qquad (4.2.2)$$

whenever $u \in W^{m,p}(\Omega)$.

Proof. We proceed by contradiction. If the result were not true, then for each positive integer i there would exist $u_i \in W^{m,p}(\Omega)$ such that

$$\|D^k u_i\|_{p;\Omega} > i\|u_i\|_{p;\Omega} + \varepsilon\|D^m u_i\|_{p;\Omega}. \qquad (4.2.3)$$

By replacing u_i by $u_i/\|u_i\|_{m,p;\Omega}$ we may assume that $\|u_i\|_{m,p;\Omega} = 1$, $i = 1, 2, \ldots$. Hence, from Exercise 2.3 there is $u \in W^{m,p}(\Omega)$ and a subsequence (which we assume without loss of generality is the full sequence) such that $u_i \to u$ strongly in $W^{m-1,p}(\Omega)$. In particular $u_i \to u$ in $L^p(\Omega)$. Since

$$\|D^k u_i\|_{p;\Omega} \leq \|u_i\|_{m-1,p;\Omega} \leq \|u_i\|_{m,p;\Omega}, \qquad (4.2.4)$$

it follows from (4.2.3) that $u_i \to 0$ in $L^p(\Omega)$ and therefore $u = 0$. But then $u_i \to 0$ in $W^{m-1,p}(\Omega)$ and consequently $\|D^k u_i\|_{p;\Omega} \to 0$ by (4.2.4). This implies that $\|D^m u_i\|_{p;\Omega} \to 0$ or $u_i \to 0$ in $W^{m,p}(\Omega)$, a contradiction to the fact that $\|u_i\|_{m,p;\Omega} = 1$. □

If $v \in W^{m,p}(R^n)$ has compact support, then it follows from the fundamental Sobolev inequalities, namely Theorem 2.4.2, 2.9.1, and 2.4.4 that

$$\|v\|_{p^*} \le C\|v\|_{m,p}$$

where p^* is defined by

$$\frac{1}{p^*} = \frac{1}{p} - \frac{m}{n} \qquad \text{if } mp < n,$$

$$1 \le p^* < \infty \qquad \text{if } mp = n,$$

and

$$p^* = \infty \qquad \text{if } mp > n.$$

Since $\Omega \subset R^n$ is an extension domain, $u \in W^{m,p}(\Omega)$ has an extension to $v \in W^{m,p}(R^n)$ with compact support such that $\|v\|_{m,p} \le C\|u\|_{m,p;\Omega}$. Therefore,

$$\begin{aligned} \|u\|_{p^*;\Omega} &\le C\|v\|_{p^*} \\ &\le C\|v\|_{m,p} \\ &\le C\|u\|_{m,p;\Omega} \\ &\le C\left[\|u\|_{p;\Omega} + \|D^m u\|_{p;\Omega}\right] \end{aligned} \tag{4.2.5}$$

by Lemma 4.2.2. Now apply this to (4.2.1) while observing that $D^\alpha(L(u)) = 0$, $|\alpha| = m$, and obtain

$$\begin{aligned} \|u - L(u)\|_{p^*;\Omega} &\le C\left[\|u - L(u)\|_{p;\Omega} + \|D^m u\|_{p;\Omega}\right] \\ &\le C\left(\frac{\|T\|}{T(1)}\right)^{k+1} \|D^{k+1}u\|_{m-(k+1),p;\Omega}. \end{aligned}$$

We have thus established the following result.

4.2.3. Corollary. *With the hypotheses of Theorem* 4.2.1,

$$\|u - L(u)\|_{p^*;\Omega} \le C\left(\frac{\|T\|}{T(1)}\right)^{k+1} \|D^{k+1}u\|_{m-(k+1)p;\Omega}.$$

4.2.4. Remark. In many applications it is of interest to know when $L(u) = 0$. In this connection we remind the reader the coefficients of the polynomial $L(u)$ are given by (4.1.7) and will be zero if $T(D^\alpha u) = 0$ for $0 \le |\alpha| \le k$. The question of determining conditions under which $L(u) = 0$ will be pursued in Sections 4.4 and 4.5.

4.3 The Dual of $W^{m,p}(\Omega)$

In order to obtain more information from inequality (4.2.1) it will be helpful to have a representation of $(W^{m,p}(\Omega))^*$, the dual of $W^{m,p}(\Omega)$. This is easily accomplished by regarding $W^{m,p}(\Omega)$ as a closed subspace of the cartesian product of $L^p(\Omega)$.

To this end let

$$N = N(n,m) = \sum_{0 \le |\alpha| \le m} 1$$

be the number of multi-induces α with $0 \le |\alpha| \le m$. Let

$$L^p_N(\Omega) = \prod_{i=1}^{N} L^p(\Omega).$$

$L^p_N(\Omega)$ is endowed with the norm

$$\|v\|_{p,N;\Omega} = \begin{cases} \left(\sum_{|\alpha|=0}^{m} \|v_\alpha\|^p_{p;\Omega} \right)^{1/p} & \text{if } 1 \le p < \infty \\ \max_{0 \le |\alpha| \le m} \|v\|_{\infty;\Omega} & \text{if } p = \infty \end{cases}$$

where $v = \{v_\alpha\} \in L^p_N(\Omega)$.

4.3.1. Theorem. *Let $\Omega \subset R^n$ be an open set. Then each linear functional $T \in (W^{m,p}(\Omega))^*$, $1 \le p < \infty$, can be represented as*

$$T(u) = \sum_{|\alpha|=0}^{m} \int_\Omega v_\alpha(x) D^\alpha u(x) dx \quad \textit{for} \quad u \in W^{m,p}(\Omega), \tag{4.3.1}$$

where $v = \{v_\alpha\} \in L^{p'}_N(\Omega)$.

Proof. Clearly, the right side of (4.3.1) defines an element $T \in (W^{m,p}(\Omega))^*$ with

$$\|T\| \le C \|v\|_{p,N;\Omega},$$

see (2.1.5). In order to express $T(u)$ in the form of (4.3.1) first observe that $W^{m,p}(\Omega)$ can be identified as a subspace of $L^p_N(\Omega)$. The operator $D : W^{m,p}(\Omega) \to L^p_N(\Omega)$ defined by

$$D(u) = \{D^\alpha u\}, \quad 0 \le |\alpha| \le m$$

has a closed range since $W^{m,p}(\Omega)$ is complete. Define a linear functional T^* on the range of D by

$$T^*[D(u)] = T(u), \quad u \in W^{m,p}(\Omega).$$

By the Hahn–Banach theorem, there is a norm preserving extension T' of T^* to all of $L^p_N(\Omega)$. By the Riesz Representation theorem, there exists $v = \{v_\alpha\} \in L^{p'}_N(\Omega)$ such that

$$T'(w) = \sum_{|\alpha|=0}^{m} \int_\Omega v_\alpha(x) w_\alpha(x) dx$$

whenever $w = \{w_\alpha\} \in L^p_N(\Omega)$. Thus, if $u \in W^{m,p}(\Omega)$, we may regard

$Du = \{D_\alpha u\} \in L^p_N(\Omega)$ and therefore

$$\begin{aligned} T(u) &= T^*[D(u)] = T'(Du) \\ &= \sum_{|\alpha|=0}^{m} \int_\Omega v_\alpha(x) D^\alpha u(x) dx. \end{aligned}$$ □

In the event that $\Omega \subset R^n$ is a bounded extension domain, the representation of $(W^{m,p}(\Omega))^*$ is slightly simpler, as described in the following result.

4.3.2. Theorem. *If $\Omega \subset R^n$ is a bounded extension domain and $1 \le p < \infty$, then each element $T \in (W^{m,p}(\Omega))^*$ can be represented as*

$$T(u) = \int_\Omega \Big(vu + \sum_{|\alpha|=m} v_\alpha D^\alpha u \Big) dx \tag{4.3.2}$$

where v, $v_\alpha \in L^{p'}(\Omega)$, $|\alpha| = m$.

Proof. The proof is almost the same as in Theorem 4.3.1 except that now $W^{m,p}(\Omega)$ can be identified with a subspace of $L^p_N(\Omega)$ where $N = k(m)+1$, and $k(m) =$ the number of multi-indices α such that $|\alpha| = m$. Thus $u \in W^{m,p}(\Omega)$ is identified with $(u, \{D^\alpha u\}_{|\alpha|=m})$. In view of Lemma 4.2.2 this provides an isometric embedding of $W^{m,p}(\Omega)$ into $L^p_N(\Omega)$. □

It is useful to regard the restriction of the linear functional T in Theorems 4.3.1 and 4.3.2 to the space $\mathscr{D}(\Omega)$ as a distribution. Indeed, if $\varphi \in \mathscr{D}(\Omega)$ is a Schwartz test function (see Section 1.7), then from (4.3.1) we have

$$T(\varphi) = \sum_{|\alpha|=0}^{m} \int_\Omega v_\alpha D^\alpha \varphi \, dx \tag{4.3.3}$$

where $v_\alpha \in L^{p'}(\Omega)$. In the language of distributions, this states that T is a distribution in Ω with

$$T = \sum_{|\alpha|=0}^{m} (-1)^{|\alpha|} D^\alpha v_\alpha \tag{4.3.4}$$

where $v_\alpha \in L^{p'}(\Omega)$. Similarly, if T is the functional in Theorem 4.3.2, then

$$T = v + \sum_{|\alpha|=m} (-1)^{|\alpha|} D^\alpha v_\alpha \tag{4.3.5}$$

where $v, v_\alpha \in L^{p'}(\Omega)$. However, not every distribution T of the form (4.3.4) or (4.3.5) is necessarily in $(W^{m,p}(\Omega))^*$. In case one deals with $W_0^{m,p}(\Omega)$ instead of $W^{m,p}(\Omega)$, distributions of the form (4.3.4) or (4.3.5) completely describe the dual space, for if T is a distribution as in (4.3.4), for example, then it possesses a unique extension to $W_0^{m,p}(\Omega)$. To see this, consider $u \in W_0^{m,p}(\Omega)$ and let $\{\varphi_i\}$ be a sequence in $\mathscr{D}(\Omega)$ such that $\varphi_i \to u$ in $W_0^{m,p}(\Omega)$. Then

$$\begin{aligned}
|T(\varphi_i) - T(\varphi_j)| &= \left| \sum_{|\alpha|=0}^{m} \int_\Omega v_\alpha D^\alpha \varphi_i - v_\alpha D^\alpha \varphi_j dx \right| \\
&\le \sum_{|\alpha|=0}^{m} \|D^\alpha(\varphi_i - \varphi_j)\|_p \|v_\alpha\|_{p',\Omega} \\
&\le \|\varphi_i - \varphi_j\|_{m,p} \|v_\alpha\|_{p',\Omega} \to 0
\end{aligned}$$

as $i, j \to \infty$. Thus, $T(\varphi_i)$ converges to a limit, denoted by $\tilde{T}(u)$, which is well-defined. $\tilde{T}$ is clearly linear and bounded, for if $\varphi_i \to u$ in $W_0^{m,p}(\Omega)$, then

$$\begin{aligned}
|\tilde{T}(u)| &= \lim_{i\to\infty} |T(\varphi_i)| \le \lim_{i\to\infty} \|T\| \, \|\varphi_i\|_{m,p} \\
&= \|T\| \, \|u\|_{m,p}.
\end{aligned}$$

The norm $\|T\|$ in this context is defined relative to the space $W_0^{m,p}(\Omega)$.

These remarks are formalized in the following theorem.

4.3.3. Theorem. *Let $1 \le p < \infty$. If $\Omega \subset R^n$ is an open set, then the dual space $(W_0^{m,p}(\Omega))^*$ consists of all distributions T of the form*

$$T = \sum_{|\alpha|=0}^{m} (-1)^{|\alpha|} D^\alpha v_\alpha$$

where $v_\alpha \in L^{p'}(\Omega)$. If Ω is a bounded extension domain, then $(W_0^{m,p}(\Omega))^*$ consists of those T such that

$$T = v + \sum_{|\alpha|=m} (-1)^{|\alpha|} D^\alpha v_\alpha$$

$v, v_\alpha \in L^{p'}(\Omega)$.

The dual space $(W_0^{m,p}(\Omega))^*$ is denoted by $W^{-m,p'}(\Omega)$.

4.4 Some Measures in $(W_0^{m,p}(\Omega))^*$

We now exploit Theorem 4.2.1 and its Corollary 4.2.3 to derive some of the most basic and often used Poincaré-type inequalities. These inequalities are obtained below by considering Lebesgue measure and its variants as elements of $(W^{m,p}(\Omega))^*$.

In order to demonstrate the method that employs the results of Section 4.2, we begin by reproving the inequality

$$\|D^k u\|_p \le C\|D^m u\|_p \tag{4.4.1}$$

for $u \in C_0^\infty(R^n)$, where $0 \le k \le m$ are integers and $p \ge 1$. Suppose that the support of u is contained in some ball: $\operatorname{spt} u \subset B(0,r)$. Let $\Omega = B(0,2r)$. With this choice of Ω, we wish to apply Corollary 4.2.3 by selecting T so that the associated projection L will have the property that $L(u) = 0$. Then by appealing to (4.2.2), we will have established (4.4.1). Define $T \in (W^{m-k,p}(\Omega))^*$ by

$$T(w) = \int_\Omega vw\,dx$$

for $w \in W^{m-k,p}(\Omega)$, where $v = \chi_{B(0,2r)-B(0,r)}$. Since $\operatorname{spt} u \subset B(0,r)$,

$$T(D^\alpha u) = 0 \quad \text{for} \quad 0 \le |\alpha| \le k$$

and therefore $L(u) = 0$ by Remark 4.2.4. Hence, (4.4.1) is established.

In case Ω is a bounded open set and $u \in C_0^\infty(\Omega)$, a similar result can be established by defining u to be identically zero on the complement of Ω and by considering a ball $B(0,r)$ that contains Ω. Since $C_0^\infty(\Omega)$ is dense in $W_0^{k,p}(\Omega)$ the following is immediate. (Of course, this result also follows from the inequalities established in Chapter 2.)

4.4.1. Theorem. *Let $\Omega \subset R^n$ be a bounded set. Let $0 \le k \le m$ be integers and $p \ge 1$. Then, there is a constant $C = C(k, m, p, \operatorname{diam}\Omega)$ such that*

$$\|D^k u\|_{p;\Omega} \le C\|D^m u\|_{p;\Omega}.$$

for $u \in W_0^{m,p}(\Omega)$.

A slight variation of the preceding argument leads to the following results.

4.4.2. Theorem. *Suppose $0 \le k < m$ are integers and $p \ge 1$. Let Ω be a bounded extension domain. Suppose $u \in W^{m,p}(\Omega)$ has the property that*

$$\int_E D^\alpha u\,dx = 0 \quad \textit{for} \quad 0 \le |\alpha| \le k,$$

where $E \subset \Omega$ is a measurable set of positive Lebesgue measure. Then,

$$\|u\|_{k,p;\Omega} \le C\|D^{k+1}u\|_{m-(k+1),p;\Omega}$$

where $C = C(k, m, p, \Omega, |E|)$.

Proof. Define $T \in (W^{m-k,p}(\Omega))^*$ by

$$T(w) = \int_E w\,dx, \quad w \in W^{m-k,p}(\Omega).$$

Then $T(1) \neq 0$,

$$T(D^\alpha u) = 0 \quad \text{for} \quad 0 \le |\alpha| \le k,$$

and therefore by Remark 4.2.4 the associated functional L has the property that $L(u) = 0$. The result now follows from Theorem 4.2.1. □

4.4.3. Corollary. *If* $u \in W^{m,p}(\Omega)$ *has the property that* $D^\alpha u = 0$ *almost everywhere on* E *for* $0 \le |\alpha| \le k$, *then*

$$\|u\|_{k,p;\Omega} \le C\|D^{k+1}u\|_{m-(k+1),p;\Omega}.$$

Theorem 4.4.2 provides a Poincaré-type inequality provided the integral averages of the derivatives of u over a set E of positive measure are zero. In the next result, the integral average hypothesis is replaced by one involving the generalized notion of *median* of a function. If the sets A and B below are of equal measure, then we could think of 0 as being the median of u over $A \cup B$.

4.4.4. Theorem. *Let* $\Omega \in R^n$ *be a bounded extension domain and let* $u \in W^{1,p}(\Omega)$, $p \ge 1$. *Suppose* $u > 0$ *on* A *and* $u < 0$ *on* B, *where* A *and* B *are measurable subsets of* Ω *of positive Lebesgue measure. Then*

$$\|u\|_{p;\Omega} \le C\|Du\|_{p;\Omega}$$

where $C = C(p, n, |A|, |B|)$.

Proof. Let

$$\alpha = \int_A u\,dx \quad \text{and} \quad \beta = \int_B u\,dx$$

and define $T \in (W^{1,p}(\Omega))^*$ by

$$T(w) = \int_\Omega vw\,dx, \quad w \in W^{1,p}(\Omega)$$

where $v = (1/\alpha)\chi_A - (1/\beta)\chi_B$. Then $T(u) = 0$ and the result follows from Theorem 4.2.1 and Remark 4.2.4. □

4.4.5. Remark. In the remainder of this section, we will include a small development of the notion of trace of a Sobolev function on the boundary of a Lipschitz domain as well as some related Sobolev-type inequalities

(Theorem 4.4.6 and its corollary). This material will be subsumed in the development of BV functions in Chapter 5, but we include it here for the benefit of the reader who does not wish to pursue the BV theory.

If $\Omega \subset R^n$ is a bounded Lipschitz domain and $u \in W^{1,p}(\Omega)$, $1 < p \leq \infty$, it is possible to give a pointwise definition of u on $\partial\Omega$ in the following way. Since Ω is an extension domain, let $\tilde{u}$ denote an extension of u to all of R^n where $\tilde{u} \in W^{1,p}(R^n)$. Therefore, $\tilde{u}$ has a Lebesgue point everywhere on R^n except possibly for a set of $B_{1,p}$-capacity zero (Theorem 3.3.3). Since $p > 1$, we know from Theorem 2.6.16 that sets of $B_{1,p}$-capacity zero are of H^{n-1}-measure zero and therefore $\tilde{u}$ is defined H^{n-1}-almost everywhere on $\partial\Omega$. We define the trace of u on $\partial\Omega$ by setting $u = \tilde{u}$ on $\partial\Omega$.

We now show that this definition is independent of the extension $\tilde{u}$. For this purpose, we first show that at each Lebesgue point x_0 of $\tilde{u}$, there is a measurable set A such that the Lebesgue density of A at x_0 is 1 and that $\tilde{u}$ is continuous at x_0 relative to A. Since

$$\fint_{B(x_0,r)} |\tilde{u}(x) - \tilde{u}(x_0)|dx \to 0 \quad \text{as} \quad r \to 0,$$

for each positive integer i, there is a number r_i such that the set $E_i = R^n \cap \{x : |u(x) - u(x_0)| > 1/i\}$ has the property that

$$\frac{|B(x_0,r) \cap E_i|}{|B(x_0,r)|} < 2^{-i}, \quad \text{for } r \leq r_i. \tag{4.4.2}$$

We may assume that the sequence $\{r_i\}$ is strictly decreasing. Let

$$E = \bigcup_{i=1}^{\infty} [B(x_0,r_i) - B(x_0,r_{i-1})] \cap E_i.$$

We now will show that the Lebesgue density of E at x_0 is zero, that is

$$\lim_{r\to 0} \frac{|B(x_0,r) \cap E|}{|B(x,r)|} = 0. \tag{4.4.3}$$

Choose a small $r > 0$ and let k be that unique index such that $r_{k+1} < r < r_k$. For notational simplicity, let $B(r) = B(x_0,r)$. Then from (4.4.2) it follows that

$$\begin{aligned}
|B(r) \cap E| &\leq \left| \bigcup_{i=k}^{\infty} [(B(r) \cap E_i) \cap (B(r_i) - B(r_{i+1}))] \right| \\
&\leq 2^{-k}|B(r)| + \sum_{i=k+1}^{\infty} 2^{-i}|B(r_i)| \\
&\leq 2^{-k}|B(r)| + 2^{-k}|B(r)|
\end{aligned}$$

which establishes (4.4.3). Clearly, if we set $A = R^n - E$ we have that u is continuous at x_0 relative to A and that the Lebesgue density of A at x_0 is one.

Because Ω is a Lipschitz domain, the boundary of Ω is locally representable as the graph of a Lipschitz function. Thus, the boundary can be expressed locally as $\{(x, f(x)) : x \in U\}$, where U is an open ball in R^{n-1} and f is a Lipschitz function. Recall from Theorem 2.2.1 that a Lipschitz function is differentiable almost everywhere. Moreover, the function $\overline{f} : R^{n-1} \to R^n$ defined by

$$\overline{f}(x) = (x, f(x))$$

is Lipschitz and carries sets of Lebesgue measure zero in R^{n-1} into sets of H^{n-1}-measure zero in R^n. Consequently, $\partial\Omega$ possesses a tangent plane at all H^{n-1}-almost all points of $\partial\Omega$. From this it is not difficult to show that

$$\lim_{r\to 0} \frac{|B(x_0, r) \cap \Omega|}{|B(x_0, r)|} = \frac{1}{2},$$

for H^{n-1}-almost all $x_0 \in \partial\Omega$. Since the Lebesgue density of A at x_0 is equal to one, it follows that

$$\lim_{r\to 0} \frac{|B(x_0, r) \cap \Omega \cap A|}{|B(x_0, r)|} = \frac{1}{2}.$$

Also, because u is continuous at x_0 relative to A, it is clear that

$$\lim_{\substack{x\to x_0 \\ x\in\Omega\cap A}} u(x) = \tilde{u}(x_0).$$

This shows that the value of $\tilde{u}(x_0)$ is determined by u in Ω, thus proving that the trace of u on the boundary of Ω is independent of the extension $\tilde{u}$.

In the statement of the next theorem, we will let μ denote the restriction of $(n-1)$-dimensional Hausdorff measure to $\partial\Omega$. That is $\mu(A) = H^{n-1}(A \cap \partial\Omega)$ whenever $A \subset R^n$.

4.4.6. Theorem. *Let $\Omega \subset R^n$ be a bounded Lipschitz domain and suppose $u \in W^{1,p}(\Omega)$, $1 < p < \infty$. Let*

$$c(u) = \int_{\partial\Omega} u \, dH^{n-1} = \int_{\partial\Omega} u \, d\mu.$$

Then $\mu \in (W^{1,p}(\Omega))^*$ and

$$\left(\int_\Omega |u - c(u)|^{p^*} dx\right)^{1/p^*} \le C \left(\int_\Omega |Du|^p dx\right)^{1/p},$$

where $p^* = np/(n-p)$ *and* $C = C(n, p, \Omega)$.

Proof. Because Ω is a bounded Lipschitz domain, u has an extension $\tilde{u}$ to all of R^n such that $\|\tilde{u}\|_{1,p} \leq C\|u\|_{1,p;\Omega}$. By multiplying $\tilde{u}$ by a function $\varphi \in C_0^\infty(R^n)$ with $\varphi \equiv 1$ on Ω, we may assume that $\tilde{u}$ has compact support.

In order to show that $\mu \in (W^{1,p}(\Omega))^*$, we will first prove that

$$\int v \, d\mu \leq C\|v\|_{1,p} \tag{4.4.4}$$

whenever v is a non-negative function in $C_0^\infty(R^n)$. From Lemma 1.5.1, we have

$$\int v \, d\mu = \int_0^\infty \mu(E_t)dt \tag{4.4.5}$$

where $E_t = \{x: v(x) > t\}$. By the Morse–Sard theorem, for almost all t, E_t is bounded by a smooth manifold. We now borrow an essentially self-contained result of Chapter 5. That is, we employ Lemma 5.9.3 and Remark 5.4.2 to conclude that for all such t, E_t can be covered by balls $B(x_i, r_i)$ such that

$$\sum_{i=1}^\infty r_i^{n-1} \leq CH^{n-1}[v^{-1}(t)], \tag{4.4.6}$$

where C is a constant depending only on n. Because $\partial\Omega$ is locally the graph of a Lipschitz function, it follows from (1.4.6) that there is a constant C such that $\mu(B(x,r)) \leq Cr^{n-1}$. Thus, from (4.4.6) it follows that

$$\begin{aligned}\mu(E_t) &\leq \sum_{i=1}^\infty \mu(B(x_i, r_i)) \\ &\leq C\sum_{i=1}^\infty r_i^{n-1} \leq CH^{n-1}(v^{-1}(t)).\end{aligned}$$

Appealing to (4.4.5) and co-area formula (Theorem 2.7.1), we have

$$\begin{aligned}\int v \, d\mu &= \int_0^\infty \mu(E_t)dt \\ &\leq C\int_0^\infty H^{n-1}(v^{-1}(t))dt \\ &= C\|Dv\|_1 \\ &\leq C(\Omega)\|Dv\|_p \\ &\leq C\|v\|_{1,p},\end{aligned}$$

thus establishing (4.4.4).

If v is now assumed to be a bounded, non-negative function in $W^{1,p}(R^n)$, we may apply (4.4.4) to the mollified function v_ε. From Theorem 1.6.1 we have that $\|v_\varepsilon - v\|_{1,p} \to 0$ and that $v_\varepsilon(x) \to v(x)$ whenever x is a Lebesgue point for v. From Theorem 3.3.3 we know that v has a Lebesgue point at all x except possibly for a set of $B_{1,p}$ capacity zero, therefore of H^{n-1}-measure 0, and therefore of μ-measure 0. Consequently, by Lebesgue's dominated convergence theorem,

$$\int v_\varepsilon d\mu \to \int v\, d\mu.$$

It now follows that (4.4.4) is established whenever v is a non-negative, bounded function in $W^{1,p}(R^n)$.

If we drop the assumption that v is bounded, then we may apply (4.4.4) to the functions

$$v_k(x) = \begin{cases} k & \text{if } \ v(x) \geq k \\ v(x) & \text{if } \ v(x) < k. \end{cases}$$

It follows from Corollary 2.1.8 that $v_k \in W^{1,p}(R^n)$ for $k = 1, 2, \ldots$. Thus, an application of the Monotone Convergence theorem yields (4.4.4) for non-negative functions in $W^{1,p}(R^n)$, in particular, for $\tilde{u}^+$ and $\tilde{u}^-$. Hence (4.4.4) is established for $\tilde{u}$.

From Remark 4.4.5 we have that $u = \tilde{u}$ H^{n-1}-almost everywhere on $\partial\Omega$, and therefore

$$\begin{aligned} \int u\, d\mu &= \int \tilde{u}\, d\mu \\ &\leq C\|\tilde{u}\|_{1,p} \\ &\leq C\|u\|_{1,p;\Omega}. \end{aligned}$$

Thus, we have shown that $\mu \in (W^{1,p}(\Omega))^*$, and reference to Corollary 4.2.3 completes the proof. □

The following is an immediate consequence of Theorem 4.4.6.

4.4.7. Corollary. *If Ω is a bounded Lipschitz domain and $u \in W^{1,p}(\Omega)$, $p > 1$, then*

$$\int_{\partial\Omega} u\, dH^{n-1} \leq C\left[\|u\|_{p^*;\Omega} + \|Du\|_{p;\Omega}\right]$$

and

$$\|u\|_{p^*;\Omega} \leq C\left[\|Du\|_{p;\Omega} + \int_{\partial\Omega} u\, dH^{n-1}\right].$$

As mentioned in the beginning of Remark 2.4.5, these inequalities will be extended to the situation when $u \in BV$, thus including the case $p = 1$.

4.5 Poincaré Inequalities

Here we further develop the results in Section 4.2 to obtain Poincaré-type inequalities for which the term $L(u)$ in inequality (4.2.1) is zero. We will show that this term vanishes provided the set $\{x : u(x) = 0\}$ is sufficiently large when measured by an appropriate capacity.

First, recall from Corollary 2.6.9 that if $A \subset R^n$ is a Suslin set, then

$$B_{\ell,p}(A) = \sup\{B_{\ell,p}(K) : K \subset A, K \text{ compact}\}.$$

Moreover, if $K \subset R^n$ is compact Theorem 2.6.12 implies that there is a non-negative measure μ such that $\operatorname{spt}\mu \subset K$,

$$\|g_\ell * \mu\|_{p'} \leq 1,$$

and

$$\mu(R^n) = [B_{\ell,p}(K)]^{1/p}.$$

Now consider $u \in W^{m-k,p}(\Omega)$ where $\Omega \subset R^n$ is a bounded extension domain. Then u has an extension $\tilde{u}$ defined on R^n such that $\|\tilde{u}\|_{m-k,p} \leq C\|u\|_{m-k,p;\Omega}$. Without loss of generality, we may assume that $\tilde{u}$ has compact support. From Theorem 2.6.1 it follows that $\tilde{u}$ has the representation

$$\tilde{u} = g_{m-k} * f$$

where $f \in L^p(R^n)$ and $\|f\|_p \sim \|\tilde{u}\|_{m-k,p}$.

Now suppose that μ is a non-negative measure with the properties that $\operatorname{spt}\mu \subset \overline{\Omega}$ and

$$g_{m-k} * \mu \in L^{p'}(R^n)$$

where k is an integer, $0 \leq k < m$. Observe that μ can be considered as an element of $(W^{m-k}(\Omega))^*$ for if we define $T : W^{m-k}(\Omega) \to R^1$ by

$$T(u) = \int u \, d\mu,$$

then,

$$\begin{aligned}
\int \mu \, d\mu &= \int \tilde{u} \, d\mu \\
&= \int g_{m-k} * f \, d\mu \\
&= \int g_{m-k} * \mu \cdot f \, dx, \quad \text{by Fubini's theorem,} \\
&\leq \|g_{m-k} * \mu\|_{p'} \|f\|_p, \quad \text{by Hölder's inequality,} \\
&\leq C\|g_{m-k} * \mu\|_{p'} \|\tilde{u}\|_{m-k,p'} \\
&\leq C\|g_{m-k} * \mu\|_{p'} \|u\|_{m-k,p;\Omega}. \qquad (4.5.1)
\end{aligned}$$

Thus, $\mu \in (W^{m-k,p}(\Omega))^*$.

This leads to another application of Theorem 4.2.1 which allows the main constant in the inequality to be estimated by the capacity of the set on which u vanishes.

4.5.1. Theorem. *Let $\Omega \in R^n$ be a bounded extension domain and let $A \subset R^n$ be a Suslin set with $B_{m-k,p}(A) > 0$ where $0 \le k < m$ are integers and $p \ge 1$. Then, there exists a projection $L\colon W^{m,p}(\Omega) \to \mathcal{P}_k(R^n)$ such that*

$$\|u - L(u)\|_{k,p;\Omega} \le C\,(B_{m-k,p}(A))^{-1/p}\,\|D^{k+1}u\|_{m-(k+1),p;\Omega}$$

where $C = C(k, m, p, \Omega)$.

Proof. From the above discussion, there exists a non-negative measure μ such that μ is supported in A,

$$\|g_{m-k} * \mu\|_{p'} \le 1$$

and

$$\mu(R^n) \ge \frac{1}{2}\,(B_{m-k,p}(A))^{1/p}.$$

If we set $T = \mu$ in Theorem 4.2.1, we have $T(1) = \mu(R^n) > 0$ and from (4.5.1) that

$$\|T\| \le C\|g_{m-k} * \mu\|_{p'} \le C.$$

The result now follows from Theorem 4.2.1 and Corollary 4.1.5. □

4.5.2. Corollary. *Let $u \in W^{m,p}(\Omega)$ and let*

$$N = \overline{\Omega} \cap \{x : D^\alpha u(x) = 0 \text{ for all } 0 \le |\alpha| \le k\}.$$

If $B_{m-k,p}(N) > 0$ then

$$\|u\|_{k,p;\Omega} \le C\,(B_{m-k,p}(N))^{-1/p}\,\|D^{k+1}u\|_{m-(k+1),p;\Omega}$$

and

$$\|u\|_{p^*;\Omega} \le C\,(B_{m-k,p}(N))^{-1/p}\,\|D^{k+1}u\|_{m-(k+1),p;\Omega}.$$

Proof. The coefficients of the polynomial $L(u)$ in Theorem 4.5.1 depend upon

$$T(D^\alpha u) = \int D^\alpha u\, d\mu$$

for $0 \le |\alpha| \le k$, and thus are all zero, (see Remark 4.2.4). The second inequality follows from Corollary 4.2.3. □

Because of the importance of the case $m = 1$, $k = 0$, we state the Poincaré inequality separately in this situation.

4.5.3. Corollary. *If $u \in W^{1,p}(\Omega)$, then*

$$\|u\|_{p^*;\Omega} \le C\,(B_{1,p}(N))^{-1/p}\,\|Du\|_{p;\Omega} \tag{4.5.2}$$

where $N = \{x : u(x) = 0\}$.

4.6 Another Version of Poincaré's Inequality

We can improve the inequalities of Corollary 4.5.2 if we allow dependence on the set N and not merely on its capacity. In particular, if j, k, and m are integers such that $0 < j \le k < m$, then the assumption $B_{m-k,p}(N) > 0$ will be replaced by the weaker one, $B_{m-(k-j),p}(N) > 0$, provided an additional condition is added which requires dependence on the set N in the resulting inequality.

To make this precise let Ω be a bounded extension domain and let $N \subset \overline{\Omega}$ be a Suslin set with the property that

$$B_{m-(k-j),p}(N - Z) > 0 \tag{4.6.1}$$

for every set Z of the form

$$Z = \bigcap_{|\alpha| \le k-j} \{x : D^\alpha P(x) = 0,\ 0 \ne P \in \mathcal{P}_k(R^n)\}. \tag{4.6.2}$$

These sets comprise a subclass of the class of *algebraic varieties.* Thus, for any algebraic variety of the form (4.6.2), we require some subset of N of positive capacity to lie in the complement of Z.

Let $\mathcal{M}(N)$ denote the set of all non-negative Radon measures μ compactly supported in N such that

$$g_{m-(k-j)} * \mu \in L^{p'}(R^n).$$

Consider all functionals of the form

$$T(u) = \int D^\alpha u \, d\mu, \quad \mu \in \mathcal{M}(N), \tag{4.6.3}$$

where $|\alpha| \le k-j$. We will verify that all such T are elements of $(W^{m,p}(\Omega))^*$. Let $u \in W^{m,p}(\Omega)$. Since Ω is an extension domain, there is an extension $\tilde{u}$ of u to R^n with $\|\tilde{u}\|_{m,p} \le C\|u\|_{m,p;\Omega}$. From Theorem 2.6.1, we know that $\tilde{u}$ has the representation $\tilde{u} = g_m * f$, where $f \in L^p(R^n)$ and $\|f\|_p \sim \|\tilde{u}\|_{m,p}$. Since μ is supported in $N \subset \overline{\Omega}$, it follows that

$$\int D^\alpha u \, d\mu = \int D^\alpha \tilde{u} \, d\mu.$$

But $D^\alpha \tilde{u} \in W^{m-|\alpha|,p}(R^n)$ and therefore

$$D^\alpha \tilde{u} = g_{m-|\alpha|} * g$$

where $g \in L^p(R^n)$. By Fubini's theorem we have

$$\int D^\alpha \tilde{u}\, d\mu = \int g_{m-|\alpha|} * g\, d\mu$$
$$= \int g_{m-|\alpha|} * \mu \cdot g\, dx.$$

It follows from (2.6.2) that

$$g_{m-|\alpha|} * \mu = g_\ell * (g_{m-(k-j)} * \mu),$$

where $\ell = (k-j) - |\alpha|$. Since $g_{m-(k-j)} * \mu \in L^{p'}(R^n)$ by assumption and $g_\ell \in L^1(R^n)$, it follows from Young's inequality that $g_{m-|\alpha|} * \mu \in L^{p'}(R^n)$. Therefore an application of Hölder's inequality yields

$$\int D^\alpha u\, d\mu = \int g_{m-|\alpha|} * \mu \cdot g\, dx$$
$$\leq \|g_{m-|\alpha|} * \mu\|_{p'} \|g\|_p$$
$$\leq \|g_{m-|\alpha|} * \mu\|_{p'} \|D^\alpha \tilde{u}\|_{m-|\alpha|,p}.$$

However,

$$\|D^\alpha \tilde{u}\|_{m-|\alpha|,p} \leq \|\tilde{u}\|_{m,p} \leq C\|u\|_{m,p},$$

thus proving that $T \in (W^{m,p}(\Omega))^*$.

Let $V \subset (W^{m,p}(\Omega))^*$ be the space spanned by all such functionals T as defined in (4.6.3). Let

$$V_0 = \{T \mid \mathcal{P}_k : T \in V\},$$

so that $V_0 \subset \mathcal{P}_k^*$. Observe that

$$\dim V_0 = \dim \mathcal{P}_k(R^n)$$

or

$$V_0 = [\mathcal{P}_k(R^n)]^*,$$

for if this were not true, there would exist $0 \neq P \in \mathcal{P}_k(R^n)$ such that $T(P) = 0$ for every $T \in V$. This would imply

$$\int D^\alpha P\, d\mu = 0, \quad |\alpha| \leq k - j$$

for all $\mu \in \mathcal{M}(N)$. That is, from Theorem 2.6.12, this would imply

$$D^\alpha P(x) = 0,$$

for $B_{m-(k-j)}$-q.e. $x \in N$ and $|\alpha| \leq k-j$, a contradiction to (4.6.1) and (4.6.2). Therefore $\dim V_0 = \dim \mathcal{P}_k(R^n)$ or $V_0 = [\mathcal{P}_k(R^n)]^*$. This implies the existence of $T_\alpha \in V$ such that

$$T_\alpha(x^\beta) = \delta_{\alpha,\beta}. \tag{4.6.4}$$

Hence, if we define $L : W^{m,p}(\Omega) \to \mathcal{P}_k(R^n)$ by

$$L(u) = \sum_{|\alpha|=0}^{m} T_\alpha(u)x^\alpha, \tag{4.6.5}$$

then (4.6.4) shows that L is a projection. An appeal to (4.1.3) results in the following theorem.

4.6.1. Theorem. *Let $\Omega \subset R^n$ be a bounded extension domain. Suppose $N \subset \overline{\Omega}$ is a Suslin set such that* (4.6.1) *is satisfied. Then, with L given by* (4.6.5), *there is a constant $C = C(j, k, m, N, \Omega)$ such that*

$$\|u - L(u)\|_{k,p;\Omega} \leq C\|D^{k+1}u\|_{m-(k+1),p;\Omega}.$$

The special nature of the projection L is what makes this result interesting. For example if we assume that $D^\alpha u = 0$ on N except possibly for a $B_{m-(k-j),p}$-null set, then all T of the form (4.6.3) are zero and therefore $L(u) = 0$. The following is a consequence.

4.6.2. Corollary. *Let $\Omega \subset R^n$ be a bounded extension domain. Suppose $N \subset \overline{\Omega}$ is a Suslin set such that* (4.6.1) *is satisfied. If $u \in W^{m,p}(\Omega)$ is such that*

$$D^\alpha u(x) = 0 \quad \textit{for} \quad B_{m-(k-j),p}\text{-q.e. } x \in N$$

and all $0 \leq |\alpha| \leq k - j$, then

$$\|u\|_{k,p;\Omega} \leq C\|D^{k+1}u\|_{m-(k+1),p;\Omega}$$

where $C = C(j, k, m, N, \Omega)$.

4.7 More Measures in $(W^{m,p}(\Omega))^*$

The general inequality (4.2.1) involves a projection operator $L : W^{m,p}(\Omega) \to \mathcal{P}_k(R^n)$ which is determined by an element $T \in (W^{m-k,p}(\Omega))^*$. It is therefore of importance to have an ample supply of elements in the dual of $W^{m-k,p}(\Omega)$ that are useful in applications. In Section 4.4 we have already seen that Lebesgue measure (more precisely, suitably normalized measures which are absolutely continuous with respect to Lebesgue measure) and normalized $(n-1)$-dimensional Hausdorff measure belong to $(W^{m-k,p}(\Omega))^*$. The fact that these measures are elements of $(W^{m-k,p}(\Omega))^*$ allowed us to deduce interesting Poincaré-type inequalities. In this section we will perform a finer analysis to establish that a large class of measures belong to $(W^{m-k,p}(\Omega))^*$, including those that are obtained as the restriction of Hausdorff measure to sub-manifolds of appropriate dimension in R^n.

We begin with a result that provides a generalization of the Sobolev inequality for Riesz potentials and also gives us a method of exhibiting a large class of measures that are elements of $(W^{m-k,p}(\Omega))^*$. It will depend on the *Marcinkiewicz Interpolation Theorem* which we state here without proof.

Let (p_0, q_0) and (p_1, q_1) be pairs of numbers such that $1 \le p_i \le q_i < \infty$, $i = 0, 1$, $p_0 < p_1$, and $q_0 \neq q_1$. Let μ be a Radon measure defined on R^n and suppose T is an additive operator defined on $C_0^\infty(R^n)$ whose values are μ-measurable functions. The operator T is said to be of *weak-type* (p_i, q_i) if there is a constant C_i such that for any $f \in C_0^\infty(R^n)$, and $\alpha > 0$,

$$\mu(\{x : |(Tf)(x)| > \alpha\}) \le (\alpha^{-1} C_i \|f\|_{p_i})^{q_i}.$$

4.7.1. Theorem (Marcinkiewicz Interpolation Theorem). *Suppose T is simultaneously of weak-types (p_0, q_0) and (p_1, q_1). If $0 < \theta < 1$, and*

$$1/p = \frac{1-\theta}{p_0} + \frac{\theta}{p_1}$$

$$1/q = \frac{1-\theta}{q_0} + \frac{\theta}{p_1},$$

then T is of strong type (p, q); that is,

$$\|Tf\|_{q;\mu} \le C\, C_0^{1-\theta} C_1^{\theta} \|f\|_p, \quad f \in C_0^\infty(R^n),$$

where $C = C(p_i, q_i, \theta)$, $i = 0, 1$.

We are now in a position to prove the basic estimate of this section which is expressed in terms of the Riesz kernel, I_k, that was introduced in Section 2.6.

4.7.2. Theorem. *Let μ be a Radon measure on R^n such that for all $x \in R^n$ and $0 < r < \infty$, there is a constant M with the property that*

$$\mu[B(x, r)] \le M r^a$$

where $a = q/p(n - kp)$, $k > 0$, $1 < p < q < \infty$, and $kp < n$. If $f \in L^p(R^n)$, then

$$\left(\int |I_k * f|^q d\mu \right)^{1/q} \le C M^{1/q} \|f\|_p$$

where $C = C(k, p, q, n)$.

This inequality is obviously an extension of the Sobolev inequality for Riesz potentials that was established in Theorem 2.8.4. In that situation, the measure μ is taken as Lebesgue measure. In Remark 4.7.3, we will discuss further what other measures play an important role in the inequality of Theorem 4.7.2.

Proof of Theorem 4.7.2. For $t > 0$ let

$$A_t = \{y : I_k * |f|(y) > t\}.$$

Our objective is to estimate $\mu(A_t)$ in terms of $\|f\|_p$. Let $\mu_t = \mu \,|\, A_t$. Then

$$\begin{aligned} t\mu(A_t) &\leq \int_{A_t} I_k * |f| d\mu = \int I_k * |f| d\mu_t \\ &= \int_{R^n} I_k * \mu_t(x) |f(x)| dx \end{aligned} \tag{4.7.1}$$

where the last equality is a consequence of Fubini's theorem. Referring to Lemma 1.5.1 it follows that

$$\begin{aligned} I_k * \mu_t(x) &= \frac{1}{\gamma(k)} \int_0^\infty \mu_t \left[B\left(x, r^{1/(k-n)}\right)\right] dr \\ &= \frac{(n-k)}{\gamma(k)} \int_0^\infty \mu_t[B(x,r)] r^{k-n-1} dr. \end{aligned}$$

For $R > 0$ which will be specified later, (4.7.1) becomes

$$\begin{aligned} t\mu(A_t) \leq & \frac{(n-k)}{\gamma(k)} \int_0^R \int_{R^n} |f(x)| \mu_t[B(x,r)] r^{k-n-1} dx\, dr \\ & + \frac{(n-k)}{\gamma(k)} \int_R^\infty \int_{R^n} |f(x)| \mu_t[B(x,r)] r^{k-n-1} dx\, dr \\ = & I_1 + I_2. \end{aligned} \tag{4.7.2}$$

Since $\mu[B(x,r)] \leq Mr^a$ by hypothesis, the first integral, I_1, is estimated by observing that

$$\mu_t[B(x,r)] \leq \mu_t[B(x,r)]^{1/p'} (Mr^a)^{1/p}$$

and then applying Hölder's inequality to obtain

$$I_1 \leq \frac{(n-k)}{\gamma(k)} \|f\|_p M^{1/p} \int_0^R \left(\int_{R^n} \mu_t[B(x,r)] dx \right)^{1/p'} r^{k-n-1+(a/p)} dr. \tag{4.7.3}$$

We now will evaluate

$$\int_{R^n} \mu_t[B(x,r)] dx.$$

For this purpose, consider the diagonal

$$D = (R^n \times R^n) \cap \{(x,y) : x = y\}$$

and define for $r > 0$,

$$D_r = (R^n \times R^n) \cap \{(x,y) : |x-y| < r\}.$$

Finally, let $F = \chi_{D_r}$. Then, by Fubini's theorem,

$$\begin{aligned}\int_{R^n} \mu_t[B(x,r)]dx &= \int_{R^n}\int_{B(x,r)} d\mu_t(y)dx \\ &= \int_{R^n}\int_{R^n} F(x,y)d\mu_t(y)dx \\ &= \int_{R^n}\int_{R^n} F(x,y)dx\, d\mu_t(y) \\ &= \int_{R^n} |B(y,r)|\, d\mu_t(y) \\ &= \alpha(n)r^n\mu(A_t).\end{aligned}$$

Therefore (4.7.3) yields

$$I_1 \le \frac{p(n-k)}{\gamma(k)[kp-(n-a)]}\|f\|_p M^{1/p}\alpha(n)^{1/p'}\mu(A_t)^{1/p'}R^{k-(n-a)p}.$$

Similarly, by employing the elementary estimate

$$\mu_t[B(x,r)] \le \mu_t[B(x,r)]^{1/p'}\mu(A_t)^{1/p},$$

we have

$$\begin{aligned}I_2 &\le \frac{(n-k)}{\gamma(k)}\|f\|_p\mu(A_t)^{1/p}\int_R^\infty\left(\int_{R^n}\mu_t[B(x,r)]dx\right)^{1/p'} r^{k-n-1}dr \\ &\le \frac{p(n-k)}{[n-kp]\gamma(k)}\|f\|_p\mu(A_t)\alpha(n)^{1/p'}R^{k-n/p}.\end{aligned}$$

Hence

$$I_1 + I_2 \le \frac{p(n-k)}{\gamma(k)}\alpha(n)^{1/p'}\|f\|_p\left[\frac{M^{1/p}\mu(A_t)^{1/p'}R^{k-(n-a)/p}}{kp-(n-a)} + \frac{\mu(A_t)R^{k-n/p}}{n-kp}\right].$$

In order for this inequality to achieve its maximum effectiveness, we seek that value of R for which the right-hand side attains a minimum. An elementary calculation shows that

$$R = \left(\frac{\mu(A_t)}{M}\right)^{1/a},$$

and the value of the right-hand side for this value of R is

$$\frac{p(n-k)a}{\gamma(k)(n-kp)(kp-n+a)}\alpha(n)^{1/p'}M^{1/q}\mu(A_t)^{1-1/q}\|f\|_p.$$

Consequently, from (4.7.2)

$$\begin{aligned} t\mu(A_t)^{1/q} &= t\mu(A_t)\mu(A_t)^{1/q-1} \\ &\le \left(\frac{p(n-k)a}{\gamma(k)(n-kp)(kp-n+a)}\alpha(n)^{1/p'}M^{1/q}\right) \\ &\quad \cdot \left(\mu(A_t)^{1-1/q}\mu(A_t)^{1/q-1}\|f\|_p\right) \\ &\le \frac{pq}{\gamma(k)(n-kp)(q-p)}\alpha(n)^{1/p'}M^{1/q}\|f\|_p. \end{aligned} \tag{4.7.4}$$

Expression (4.7.4) states that the Riesz potential operator I_k is of weak type (p,q) whenever p and q are numbers such that

$$1 < p < q < \infty, \quad kp < n. \tag{4.7.5}$$

Hence, if (p_0, q_0), (p, q) and (p_1, q_1) are pairs of numbers such that (p_0, q_0), (p_1, q_1) satisfy (4.7.5) and for $0 \le \theta < 1$,

$$1/p = \frac{1-\theta}{p_0} + \frac{\theta}{p_1}$$

$$1/q = \frac{1-\theta}{q_0} + \frac{\theta}{p_1},$$

then the Marcinkiewicz Interpolation Theorem states that I_k is of type (p,q), with

$$\|I_k * f\|_{q;\mu} \le CM^{1/q}\|f\|_p,$$

thus establishing our result. □

4.7.3. Remark. The number a that appears in the statement of Theorem 4.7.2 is equal to n when $q = np/(n-kp) = p^*$. In this case the conditions of the theorem are satisfied by any measure μ that is absolutely continuous with respect to Lebesgue and that has bounded density. In particular, if we take μ as Lebesgue measure, we can recover Theorem 2.3.6, which is Sobolev's inequality for Riesz potentials.

Theorem 4.7.2 also provides an inequality for Riesz potentials restricted to a lower dimensional submanifold M^λ of R^n. For example, if M^λ is a compact, smooth λ-dimensional submanifold of R^n, then it is easy to verify that λ-dimensional Hausdorff measure restricted to M satisfies the condition of Theorem 4.7.2. That is, if we define μ by

$$\mu(E) = H(E \cap M^\lambda),$$

for every Borel set $E \subset R^n$, then there is a constant M such that

$$\mu[B(x,r)] \leq Mr^\lambda \tag{4.7.6}$$

for every ball $B(x,r) \subset R^n$. Now let $f \in L^p(R^n)$ and consider the potential

$$u = I_k * f.$$

By Theorem 4.7.2 we have

$$\left(\int u^{\lambda^*} d\mu\right)^{1/\lambda^*} \leq C\|f\|_p$$

where $\lambda^* = \lambda p/(n-kp)$, $n - \lambda < kp < n$. In other words,

$$\left(\int_{M^\lambda} |u(x)|^{\lambda^*} dH^\lambda(x)\right)^{1/\lambda^*} \leq C\|f\|_p \tag{4.7.7}$$

where $C = C(k,p,\lambda,n,M)$. Note that the constant C depends on M which, in turn, theory, it is sometimes possible to obtain an equality similar to (4.7.7) where the constant is independent of the manifold.

Inequality (4.7.7) is valid for Riesz potentials $u = I_k * f$ and thus does not automatically include Sobolev functions. However, it is immediate that Theorem 4.7.2 and (4.7.7) apply as well to Sobolev functions $u \in W^{k,p}(R^n)$ because Theorem 2.6.2 states that u can be represented as

$$u = g_k * f$$

where g_k is the Bessel kernel, $f \in L^p(R^n)$ and $\|f\|_p \sim \|u\|_{k,p}$. Moreover, we know from (2.6.3) that there is a constant C such that $g_k(x) \leq CI_k(x)$, $x \neq 0$.

To reassure ourselves that the integral on the left-side of (4.7.7) is meaningful, recall from Theorem 3.3.3 that u is defined pointwise everywhere on R^n except possibly for a set A with $R_{k,p}(A) = 0$. Therefore, by Theorem 2.6.16, $H^{n-kp+\varepsilon}(A) = 0$ for every $\varepsilon > 0$. By assumption, $\lambda > n - kp$ and consequently $H^\lambda(A) = 0$. Thus, u is defined H^λ almost everywhere on M^λ which is in accord with inequality (4.7.7).

Also, we observe that if μ is a non-negative measure on R^n with compact support, and otherwise satisfies the conditions of Theorem 4.7.2, then $\mu \in (W^{k,p}(\Omega))^*$ whenever Ω is a bounded extension domain. To see this, let $u \in W^{k,p}(\Omega)$ and let $\tilde{u}$ be an extension of u to R^n such that $\|\tilde{u}\|_{k,p} \leq C\|u\|_{k,p;\Omega}$. Because $\tilde{u} \in W^{k,p}(R^n)$, we have

$$\tilde{u} = g_k * f$$

where $\|f\|_p \sim \|\tilde{u}\|_{k,p}$. Hence, by Theorem 4.7.2 and the fact that $\operatorname{spt}\mu$ is compact,

$$\int |u| d\mu \leq C\left(\int |u|^q d\mu\right)^{1/q} \leq C\|f\|_p$$

$$= C\|\tilde{u}\|_{k,p}$$
$$\leq C\|u\|_{k,p;\Omega}.$$

This establishes the following result.

4.7.4. Theorem. *Let $\Omega \subset R^n$ be a bounded extension domain and suppose μ is a compactly supported Radon measure on R^n with the property that if $\varepsilon > 0$, there is a constant M such that*

$$\mu[B(x,r)] \leq Mr^{n-kp+\varepsilon},$$

for all $x \in R^n$ and all $r > 0$, where $kp < n$, $p > 1$. Then $\mu \in (W^{k,p}(\Omega))^*$.

This result obviously is not sharp and thereby invites the question of determining an optimal condition for μ to be an element of $(W^{k,p}(R^n))^*$. By using a different approach, it is possible to find a condition, related to the one in the theorem above, that provides a characterization of those Radon measures that are elements of $(W^{k,p}(R^n))^*$.

For this purpose, we need a few preliminaries. If μ is a Radon measure, we will use the fractional maximal operator

$$\mathcal{M}_k\mu(x) = \sup\{r^{k-n}\mu[B(x,r)] : r > 0\}.$$

There is an obvious relationship between the Riesz potential of μ and the fractional maximal operator: $\mathcal{M}_k\mu(x) \leq CI_k * \mu(x)$ for every $x \in R^n$, where $C = C(k,n)$. The opposite inequality in integrated form is not so obvious and is implied by a result due to [MW]. It states that for every $1 < p < \infty$ and $0 < k < n$, there exists $C = C(k,p,n)$ such that

$$\|I_k * \mu\|_p \leq C\|\mathcal{M}_k\mu\|_p. \tag{4.7.8}$$

The (k,p)-energy of μ is defined as

$$\mathcal{E}_{k,p}(\mu) = \int_{R^n} (g_k * \mu)^{p'} dx.$$

Since the Bessel kernel is dominated by the Riesz kernel, we have

$$\begin{aligned}
\mathcal{E}_{k,p}(\mu) &\leq C\int_{R^n} (I_k * \mu)^{p'} dx \\
&= \int_{R^n} (I_k * \mu) \cdot (I_k * \mu)^{1/(p-1)} dx \\
&= \int_{R^n} I_k * (I_k * \mu)^{1/(p-1)} d\mu, \quad \text{by Fubini's theorem.}
\end{aligned} \tag{4.7.9}$$

The expression

$$I_k * (I_k * \mu)^{1/(p-1)}$$

is called the *non-linear potential* of μ.

4.7.5. Theorem. *Let $p > 1$ and $kp \le n$. If μ is a Radon measure, then $\mu \in (W^{k,p}(R^n))^*$ if and only if*

$$\int_{R^n} \int_0^1 \left(\frac{\mu[B(y,r)]}{r^{n-kp}} \right)^{1/(p-1)} \frac{dr}{r} d\mu(y) < \infty. \tag{4.7.10}$$

Proof. In order to avoid technical details involving the behavior at infinity, we will give a proof for measures μ with compact support.

If μ is such a measure with $g_k * \mu \in L^{p'}$, then by Fubini's theorem and with $u = g_k * f$ we can write

$$\begin{aligned} \int u\, d\mu &= \int g_k * f\, d\mu \\ &= \int g_k * \mu \cdot f\, dx \\ &\le \|g_k * \mu\|_{p'} \|f\|_p \\ &\le C \|g_k * \mu\|_{p'} \|u\|_{k,p}, \end{aligned}$$

which implies that $\mu \in (W^{k,p}(R^n))^*$. Conversely, if $\mu \in (W^{k,p}(R^n))^*$, then the reflexivity of L^p implies that $g_k * \mu \in L^{p'}$. Therefore μ is an element of $(W^{k,p}(R^n))^*$ if and only if $\|g_k * \mu\|_{p'} < \infty$, i.e., if and only if the (k,p)-energy of μ is finite.

We proceed to find a (sharp) condition on μ that will ensure the finiteness of its (k,p)-energy. For each $r > 0$,

$$\begin{aligned} \frac{\mu[B(x,r)]}{r^{n-k}} &\le C \left(\int_r^{2r} \left[\frac{\mu[B(x,t)]}{t^{n-k}} \right]^{p'} \frac{dt}{t} \right)^{1/p'} \\ &\le C \left(\int_0^{\infty} \left[\frac{\mu[B(x,t)]}{t^{n-k}} \right]^{p'} \frac{dt}{t} \right)^{1/p'}. \end{aligned}$$

Thus,

$$\mathcal{M}_k \mu(x) \le C \left(\int_0^{\infty} \left[\frac{\mu[B(x,t)]}{t^{n-k}} \right]^{p'} \frac{dt}{t} \right)^{1/p'}.$$

Therefore,

$$\begin{aligned} \mathcal{E}_{k,p}(\mu) &\le C \int_{R^n} (I_k * \mu)^{p'} dx \\ &\le C \int_{R^n} (\mathcal{M}_k \mu)^{p'} dx, \text{ by (4.7.8)}, \\ &\le C \int_{R^n} \int_0^{\infty} \left[\frac{\mu[B(x,t)]}{t^{n-k}} \right]^{p'} \frac{dt}{t} dx. \end{aligned}$$

Now to evaluate the last term, we have

$$\begin{aligned}\int_{R^n} \mu[B(x,t)]^{p'}dx &= \int_{R^n} \mu[B(x,t)]^{1/(p-1)}\mu[B(x,t)]dx \\ &\leq \int_{R^n}\left(\mu[B(x,t)]^{1/(p-1)}\int_{B(x,t)} d\mu(y)\right)dx \\ &\leq \int_{R^n}\left(\int_{B(x,t)} \mu[B(y,2t)]^{1/(p-1)}d\mu(y)\right)dx \\ &\leq \int_{R^n}\left(\int_{R^n} F(x,y)\mu[B(y,2t)]^{1/(p-1)}d\mu(y)\right)dx \\ &\leq \int_{R^n} \mu[B(y,2t)]^{1/(p-1)}|B(x,t)|d\mu(y)\end{aligned}$$

where $F(x,y) = \chi_{D_t}$ and $D_t = R^n \times R^n \cap \{(x,y) : |x-y| < t\}$. Therefore,

$$\begin{aligned}\mathcal{E}_{k,p}(\mu) &\leq C\int_0^\infty (t^{k-n})^{p'}\int_{R^n} \mu[B(y,2t)]^{1/(p-1)}|B(x,t)|d\mu(y)\frac{dt}{t} \\ &\leq C\int_{R^n}\int_0^\infty \left(\frac{\mu[B(y,t)]}{t^{n-kp}}\right)^{1/(p-1)}\frac{dt}{t}d\mu(y).\end{aligned}$$

Since μ has compact support and finite total mass, it is evident that the expression on the right side of the above inequality is finite if and only if

$$\int_{R^n}\int_0^1 \left(\frac{\mu[B(y,t)]}{t^{n-kp}}\right)^{1/(p-1)}\frac{dt}{t}d\mu(y) < \infty.$$

This establishes the sufficiency of condition (4.7.10).

For the proof of necessity, we employ the estimate

$$g_k(x) \geq C|x|^{k-n}e^{-2|x|} \quad \text{for} \quad x \in R^n,\ x \neq 0 \tag{4.7.11}$$

(see Section 2.6). As in (4.7.9), we have

$$\mathcal{E}_{k,p}(\mu) = \int_{R^n} (g_k * \mu)^{p'}dx = \int_{R^n} g_k * (g_k * \mu)^{1/(p-1)}d\mu.$$

To estimate the last integral, let $f = (g_k * \mu)^{1/(p-1)}$ and use Lemma 1.5.1 and (4.7.11) to obtain

$$g_k * f(x) \geq C\int_0^\infty \left(\int_{B(x,r)} f(y)dy\right) r^{k-n}e^{-2r}\frac{dr}{r}.$$

Clearly, for $r > 0$,

$$f(y) \geq \left(\int_{B(y,r)} g_k(y-z)d\mu(z)\right)^{1/(p-1)}$$

and therefore,

$$\begin{aligned} g_k * f(x) &\geq C \int_0^\infty \left(\int_{B(x,r)} f(y)dy \right) r^{k-n} e^{-2r} \frac{dr}{r} \\ &\geq C \int_0^\infty \left(r^n \left(\int_{B(y,r)} g_k(y-z)d\mu(z) \right)^{1/(p-1)} dy \right) r^{k-n} e^{-2r} \frac{dr}{r} \\ &\geq C \int_0^\infty \left(r^n \left(\int_{B(y,r)} r^{k-n} e^{-2r} d\mu(z) \right)^{1/(p-1)} dy \right) r^{k-n} e^{-2r} \frac{dr}{r} \\ &\geq C \int_0^\infty \left(\frac{\mu[B(y,r)]}{r^{n-kp}} \right)^{1/(p-1)} e^{-2p'r} \frac{dr}{r}. \end{aligned}$$

This implies

$$\begin{aligned} \mathcal{E}_{k,p}(\mu) &\geq C \int_{R^n} \int_0^\infty \left(\frac{\mu[B(y,r)]}{r^{n-kp}} \right)^{1/(p-1)} e^{-2p'r} \frac{dr}{r} d\mu(y) \\ &\geq C \int_{R^n} \int_0^1 \left(\frac{\mu[B(y,r)]}{r^{n-kp}} \right)^{1/(p-1)} e^{-2p'r} \frac{dr}{r} d\mu(y) \\ &\geq Ce^{-2p'} \int_{R^n} \int_0^1 \left(\frac{\mu[B(y,r)]}{r^{n-kp}} \right)^{1/(p-1)} \frac{dr}{r} d\mu(y). \end{aligned}$$ □

4.8 Other Inequalities Involving Measures in $(W^{k,p})^*$

We now return to the inequality (4.2.1) for another application. It states that

$$\|u - L(u)\|_{k,p;\Omega} \leq C \left(\frac{\|T\|}{T(1)} \right)^{k+1} \|D^{k+1}u\|_{m-(k+1),p;\Omega},$$

where $T \in (W^{m-k,p(\Omega)})^*$ and $L : W^{m,p}(\Omega) \to \mathcal{P}_k(R^n)$ is the associated projection. $L(u)$ has the form

$$L(u) = \sum_{|\alpha| \leq k} T(P_\alpha(D)u)x^\alpha$$

where P_α is a polynomial of degree $|\alpha|$ whose argument is $D = (D_1, \ldots, D_n)$. In Corollary 4.5.2 we found that $L(u) = 0$ if $B_{m-k,p}(N) > 0$ where

$$N = \overline{\Omega} \cap \{x : D^\alpha u(x) = 0 \text{ for all } 0 \leq |\alpha| \leq k\}.$$

This was proved by the establishing the existence of a measure $\mu \geq 0$ supported in N with $\mu \in (W^{m-k,p}(\Omega))^*$. By taking $T = \mu$ it clearly follows that

$$L(u) = \sum_{|\alpha| \le k} \int P_\alpha(D) u \, d\mu = 0. \tag{4.8.1}$$

Now, if μ is taken as any non-negative measure in $(W^{m-k,p}(\Omega))^*$ with the property that

$$\int D^\alpha u \, d\mu = 0 \quad \text{for all} \quad 0 \le |\alpha| \le k,$$

then (4.8.1) holds. This observation along with Theorem 4.2.1 and Corollary 4.2.3 yield the following result.

4.8.1. Theorem. *Let $p > 1$ and suppose $0 \le k \le m$ are integers. Let $\Omega \subset R^n$ be a bounded extension domain. If μ is a non-negative measure on R^n such that $\mu \in (W^{m-k,p}(\Omega))^*$, $\mu(R^n) \ne 0$ and*

$$\int D^\alpha u \, d\mu = 0 \quad \textit{for all} \quad 0 \le |\alpha| \le k,$$

then

$$\|u\|_{k,p;\Omega} \le C \|D^{k+1} u\|_{m-(k+1),p;\Omega}$$

and

$$\|u\|_{p^*;\Omega} \le C \|D^{k+1} u\|_{m-(k+1),p;\Omega}$$

where $C = (k, p, m, \mu, \Omega)$.

In particular, with $k = 0$ and $m = 1$, we have

$$\|u\|_{p^*;\Omega} \le C \|Du\|_{p;\Omega}$$

if $\mu \in (W^{1,p}(\Omega))^*$ and

$$\int u \, d\mu = 0.$$

From the preceding section we have found that a non-negative measure μ with compact support belongs to $(W^{m-k,p}(\Omega))^*$ if, for some $\varepsilon > 0$,

$$\sup \left\{ \frac{\mu[B(x,r)]}{r^{n-(m-k)p+\varepsilon}} : x \in R^n, \; r > 0 \right\} < \infty. \tag{4.8.2}$$

Consequently, if λ is an integer such that $\lambda \ge n - (m-k)p + \varepsilon$ and M^λ is a smooth compact manifold of dimension λ, then $H^\lambda \mid M^\lambda$ is a measure in $(W^{m-k,p}(\Omega))^*$. As an immediate consequence of Theorem 4.8.1, we have

4.8.2. Corollary. *Let λ be an integer such that $\lambda \ge n - (m-k)p + \varepsilon$ where $p > 1$ and $\varepsilon > 0$. Suppose M^λ is a smooth compact submanifold*

of dimension λ of R^n, $\Omega \subset R^n$ is an extension domain and suppose $u \in W^{m,p}(\Omega)$ has the property that

$$\int_{M^\lambda \cap \Omega} D^\alpha u \, dH^\lambda = 0 \quad \text{for all} \quad 0 \leq |\alpha| \leq k$$

where $H^\lambda(M^\lambda \cap \Omega) \neq 0$. Then

$$\|u\|_{k,p;\Omega} \leq C\|D^{k+1}u\|_{m-(k+1),p;\Omega}$$

and

$$\|u\|_{p^*;\Omega} \leq C\|D^{k+1}u\|_{m-(k+1),p;\Omega}$$

where $C = C(k, p, m, M^\lambda, \Omega)$.

4.9 The Case $p = 1$

The development thus far in this chapter has excluded the case $p = 1$, a situation which almost always requires special treatment in L^p-theory. Our objective here is to extend Theorem 4.7.2 to include the case $p = 1$. Since the analysis will depend upon estimates involving $\|Du\|_1$, it is not surprising that the co-area formula (Theorem 2.7.1) will play a critical role. We begin with the following lemma that serves as a first approximation to Theorem 4.7.2 in the case $p = 1$. We will return to this later (in Chapter 5) for a complete development in the setting of BV functions.

4.9.1. Lemma. *Let $\mu \geq 0$ be a Radon measure on R^n and q a number such that $1 \leq q \leq n/n-1$. If*

$$\sup\left\{\frac{\mu[B(x,r)]}{r^{q(n-1)}} : x \in R^n, \ r > 0\right\} \leq M$$

for some $M \geq 0$, then there exists $C = C(q, n)$ such that

$$\left(\int_{R^n} u^q d\mu\right)^{1/q} \leq CM^{1/q}\|Du\|_1 \tag{4.9.1}$$

whenever $u \in C_0^\infty(R^n)$.

Proof. First consider $q = 1$ and refer to Lemma 1.5.1 to conclude that

$$\int u \, d\mu = \int_0^\infty \mu(E_t)dt \tag{4.9.2}$$

whenever $u \in C_0^\infty(R^n)$ is non-negative. Here $E_t = \{x : u(x) > t\}$. Because u is continuous, $\partial E_t \subset u^{-1}(t)$ for each $t > 0$; moreover the smoothness of u and the Morse–Sard theorem states that $u^{-1}(t)$ is a smooth $(n-1)$-manifold for almost all $t > 0$. Consequently, Lemma 5.9.3 and Remark 5.4.2 imply that for all such t there exists a covering of E_t by a sequence of balls $B(x_i, r_i)$ such that

$$\sum_{i=1}^{\infty} r_i^{n-1} \leq CH^{n-1}(\partial E_t) \leq CH^{n-1}(\{u^{-1}(t)\})$$

where $C = C(n)$. Hence,

$$\begin{aligned} \mu(E_t) &\leq \sum_{i=1}^{\infty} \mu[B(x_i, r_i)] \\ &\leq M \sum_{i=1}^{\infty} r_i^{n-1} \\ &\leq CMH^{n-1}[\{u^{-1}(t)\}] \end{aligned}$$

where $C = C(n)$. Referring to (4.9.2) and the co-area formula (Theorem 2.7.1) we have

$$\begin{aligned} \int u \, d\mu &= \int_0^{\infty} \mu(E_t) dt \\ &\leq CM \int_0^{\infty} H^{n-1}(\{u^{-1}(t)\}) dt \\ &= CM \int_{R^n} |Du| \, dx. \end{aligned}$$

If u is not non-negative, write $|u| = u^+ - u^-$, and apply the preceding argument to u^+ and u^- to establish our result for $q = 1$.

Now consider $q > 1$ and let $g \in L^{q'}(\mu)$, $g \geq 0$. Then, Hölder's inequality implies

$$\begin{aligned} \int_{B(x,r)} g \, d\mu &\leq \left(\int_{B(x,r)} g^{q'} d\mu \right)^{1/q'} \mu[B(x,r)]^{1/q} \\ &\leq M^{1/q} \|g\|_{q';\mu} r^{n-1}. \end{aligned}$$

Thus, $g\mu$ is a Radon measure which satisfies the conditions of the lemma for $q = 1$. Consequently, if $u \in C_0^{\infty}(R^n)$ we have

$$\int_R |u| g \, d\mu \leq CM^{1/q} \|g\|_{q';\mu} \int_{R^n} |Du| dx$$

for all $g \in L^{q'}(\mu)$, $g \geq 0$. However, by the Riesz Representation theorem,

$$\|u\|_{q;\mu} = \sup \left\{ \int_{R^n} |u| g \, d\mu : \|g\|_{q';\mu} \leq 1, \ g \geq 0 \right\}$$

and our result is established. □

4.9.2. Remark. The restriction $q \le n/n-1$ in the lemma is not essential. If $q > n/n-1$, the lemma would require a Radon measure μ to satisfy

$$\mu[B(x,r)] \le Mr^m$$

for all $x \in R^n$, all $r > 0$, and some $m > n$. However, there is no non-trivial Radon measure with this property. In order to see this, let $U \subset R^n$ be a bounded open set. Choose $\varepsilon > 0$ and for each $x \in U$, consider the family $\mathcal{G}_x$ of closed balls $\overline{B}(x,r)$ such that $0 < r < \varepsilon$ and $\overline{B}(x,r) \subset U$. Defining

$$\mathcal{G} = \{B : B \in \mathcal{G}_x, x \in U\}$$

we see that $\mathcal{G}$ covers U and thus, by Theorem 1.3.1 there is a disjointed subfamily $\mathcal{F} \subset \mathcal{G}$ such that

$$U \subset \cup\{B : B \in \mathcal{G}\} \subset \cup\{\hat{B} : B \in \mathcal{F}\}.$$

Hence, denoting the radius of B_i by r_i, we have

$$\begin{aligned} \mu(U) &\le \sum_{B \in \mathcal{F}} \mu(\hat{B}_i) \le M \sum_{i=1}^{\infty} (5r_i)^m \\ &\le M5^m \varepsilon^{m-n} \sum_{i=1}^{\infty} r_i^n \\ &\le M5^m \varepsilon^{m-n} |U|. \end{aligned}$$

Since U is bounded and ε is arbitrary, it follows that $\mu(U) = 0$. □

Our next objective in this section is to extend inequality (4.9.1) by replacing $\|Du\|_1$ on the right side by $\|D^\ell u\|_1$. For this purpose it will be necessary to first establish the following lemma.

4.9.3. Lemma. *Let $\mu \ge 0$ be a Radon measure on R^n, $\ell < n$, $1 \le q < (n-\ell+1)(n-\ell)$ and $\tau^{-1} = 1-(q-1)(n-\ell)/n$. Then there is a constant C such that for all $x \in R^n$ and $r > 0$,*

$$r^{\ell-(n+1)} \|I_1 * \mu\|_{\tau;B(x,r)} \le C \sup\{r^{(\ell-n)q} \mu[B(x,r)] : x \in R^n, r > 0\}.$$

Proof. It will suffice to prove the lemma for $x = 0$. An application of Minkowski's inequality for integrals yields

$$\left(\int_{B(0,r)} \left(\int_{B(0,2r)} \frac{d\mu(y)}{|x-y|^{n-1}} \right)^\tau dx \right)^{1/\tau}$$

$$\leq \int_{B(0,2r)} \left(\int_{B(0,r)} \frac{dx}{|x-y|^{(n-1)\tau}} \right)^{1/\tau} d\mu(y). \tag{4.9.3}$$

Observe that

$$\int_{B(0,r)} \frac{dx}{|x-y|^{(n-1)\tau}} = \int_{B(0,r)\cap B(y,r)} \frac{dx}{|x-y|^{(n-1)\tau}} + \int_{B(0,r)-B(y,r)} \frac{dx}{|x-y|^{(n-1)\tau}}.$$

The first integral can be estimated by

$$\int_{B(y,r)} \frac{dx}{|x-y|^{(n-1)\tau}} \leq Cr^{n-\tau(n-1)}$$

and the second integral is dominated by $r^{n-\tau(n-1)}$. Here we have used the fact that $(n-1)\tau < n$. Thus,

$$\int_{B(0,r)} \frac{dx}{|x-y|^{(n-1)\tau}} \leq Cr^{n-\tau(n-1)},$$

and therefore from (4.9.3)

$$r^{\ell-(n+1)} \left(\int_{B(0,r)} \left(\int_{B(0,2r)} \frac{d\mu(y)}{|x-y|^{(n-1)}} \right)^{\tau} dx \right)^{1/\tau} \leq Cr^{(\ell-n)q} \mu[B(0,2r)]. \tag{4.9.4}$$

If $|x| < r$ and $|y| \geq 2r$, then $|y| \leq 2|y-x|$. Consequently,

$$r^{\ell-(n+1)} \left(\int_{B(0,r)} \left(\int_{|y|\geq 2r} \frac{d\mu(y)}{|x-y|^{(n-1)}} \right)^{\tau} dx \right)^{1/\tau} \leq Cr^{(\ell-1)-(q-1)(n-\ell)} \int_{|y|\geq 2r} \frac{d\mu(y)}{|y|^{n-1}}. \tag{4.9.5}$$

Appealing to Lemma 1.5.1, we have

$$\int_{|y|\geq 2r} \frac{d\mu(y)}{|y|^{n-1}} \leq (n-1) \int_{2r}^{\infty} \mu[B(0,t)] t^{-n} dt.$$

Now define a measure ν on R^1 by $\nu = t^{(n-\ell)q-n} dt$ and write

$$\begin{aligned} \int_{2r}^{\infty} \mu[B(0,t)] t^{-n} dt &= \int_{2r}^{\infty} \mu[B(0,t)] t^{(\ell-n)q} d\nu \\ &\leq \sup_{r>0} \left\{ r^{(\ell-n)q} \mu[B(0,r)] \right\} \int_{2r}^{\infty} d\nu \\ &\leq Cr^{(n-\ell)q-n+1} \sup_{r>0} r^{(\ell-n)q} \mu[B(0,r)]. \end{aligned}$$

This combined with (4.9.5) and (4.9.4) yield the desired result. □

We are now prepared for the main result of this section.

4.9.4. Theorem. *Let $\mu \geq 0$ be a Radon measure on R^n and let $\ell \leq n$, $q \geq 1$. Then if*

$$\sup\left\{\frac{\mu[B(x,r)]}{r^{q(n-\ell)}} : x \in R^n,\ r > 0\right\} = M < \infty,$$

there exists $C = C(q,n)$ such that

$$\left(\int_{R^n} u^q d\mu\right)^{1/q} \leq CM^{1/q}\|D^\ell u\|_1$$

whenever $u \in C_0^\infty(R^n)$.

Proof. If $\ell = n$, then for $u \in C_0^\infty(R^n)$ and $x \in R^n$,

$$u(x) \leq \int_{R^n} |D^n u| dy,$$

from which the result follows.

If $1 = \ell < n$, the result follows from Lemma 4.9.1.

Next, consider the case $\ell < n$, $\ell > 1$ and $q > n/(n-1)$. Since $u \in C_0^\infty(R^n)$, it follows that $u \in W_0^{\ell-1,n/(n-1)}(R^n)$ and therefore $u = g_{\ell-1} * f$, $f \in L^{n/(n-1)}(R^n)$ with $\|f\|_{n/(n-1)} \sim \|u\|_{\ell-1,n/(n-1)} \sim \|D^{\ell-1}u\|_{n/(n-1)}$. Thus, Theorem 4.7.2 implies

$$\|u\|_{q;\mu} \leq CM^{1/q}\|D^{\ell-1}u\|_{n/(n-1)}.$$

Since $\|D^{\ell-1}\|_{n/n-1} \leq C\|D^\ell u\|_1$ by Theorem 2.4.1, our result is established in this case.

Finally, consider $\ell < n$, $\ell > 1$, and $q \leq n/(n-1)$. We proceed by induction on ℓ, assuming that the result holds for derivatives of orders up to and including $\ell - 1$. As in (2.4.5),

$$\begin{aligned}
\int_{R^n} |u(x)|^q d\mu(x) &\leq C\int_{R^n}\left|\int_{R^n}\frac{|D(|u(y)|^q)}{|x-y|^{n-1}}dy\right| d\mu(x) \\
&\leq Cq\int_{R^n} |Du|\,|u|^{q-1}I_1 * \mu\, dy \text{ (by Fubini's theorem)} \\
&\leq Cq\|u\|_{n/(n-\ell)}^{q-1}\|\,|Du|I_1 * \mu\|_\tau \text{ (by Hölder's inequality)}
\end{aligned}$$

where $\tau^{-1} = 1 - (q-1)(n-\ell)n^{-1}$. By Sobolev's inequality,

$$\|u\|_{n/(n-\ell)}^{q-1} \leq C\|D^\ell u\|_1^{q-1}. \tag{4.9.6}$$

To estimate $\| |Du| I_1 * \mu \|_\tau$ let m be a measure on R^n defined by $m = (I_1 * \mu)^\tau dx$ and apply the induction hypothesis to obtain

$$\begin{aligned}\| |Du| I_1 * \mu \|_\tau &= \| |Du| \|_{\tau;m} \\ &\le C \sup\{r^{(\ell-1)-n} m[B(x,r)]^{1/\tau} : r > 0, x \in R^n\} \|D^\ell u\|_1 \\ &= C \sup\{r^{(\ell-1)-n} \|I_1 * \mu\|_{\tau;B(x,r)} : r > 0, x \in R^n\} \|D^\ell u\|_1.\end{aligned}$$

This combined with (4.9.6) and Lemma 4.9.3 establishes the proof. □

Exercises

4.1. Give a proof of Corollary 4.5.3 based on the argument that immediately precedes Section 4.1. You will need the material in Section 2.6.

4.2. The following provides another method that can be used to define the trace of a Sobolev function $u \in W^{1,p}(\Omega)$ on the boundary of a Lipschitz domain Ω.

STEP 1. Assume first that $u \in C^1(\overline{\Omega})$, $u \ge 0$. For each $x_0 \in \partial\Omega$ and with the $(n+1)$-cube centered at x_0 with side length $2r$ denoted by $C(x_0, r)$, we may assume (after a suitable rotation and relabeling of coordinate axes) that there exists $r > 0$ such that $C(x_0, r) \cap \partial\Omega$ can be represented as the graph of Lipschitz function f where the unit exterior normal ν can be expressed as

$$\frac{(D_1 f, \ldots, D_n, f, 1)}{\sqrt{1 + |Df|^2}}$$

H^n-a.e. on $C(x_0, r) \cap \partial\Omega$. With $e_{n+1} = (0, \ldots, 1)$ and under the assumption that $\operatorname{spt} u \subset C(x_0, r)$, appeal to the Gauss–Green theorem (see Theorem 5.8.7) to conclude that there exists a non-negative constant C, depending only on the Lipschitz constant of f, such that

$$\begin{aligned}\int_{\partial\Omega} u \, dH^n &\le C \int_{\partial\Omega} (u e_{n+1}) \cdot \nu \, dH^n \\ &= C \int_\Omega \operatorname{div}(u e_{n+1}) dx \\ &\le C \int_\Omega |Du| dx.\end{aligned}$$

If u assumes both positive and negative values, write $|u| = u^+ + u^-$ to obtain

$$\int_{\partial\Omega} |u| \, dH^n \le C \int_\Omega |Du| \, dx.$$

STEP 2. With no restriction on $\operatorname{spt} u$, use a partition of unity to obtain

$$\int_{\partial\Omega} |u|\, dH^n \leq C \int_{\Omega} (|u| + |Du|) dx.$$

STEP 3. Prove that

$$\int_{\partial\Omega} |u|^p dH^n \leq C \int_{\Omega} (|u|^p + |Du|^p) dx$$

by replacing $|u|$ by $|u|^p$ in the preceding step.

STEP 4. Now under the assumption that $u \in W^{1,p}(\Omega)$, refer to Exercise 2.17 to obtain a sequence of smooth functions u_k such that $\|u_k - u\|_{1,p;\Omega} \to 0$ and

$$\int_{\partial\Omega} |u_k - u_\ell|^p dH^n \to 0 \quad \text{as} \quad k, \ell \to \infty.$$

The limiting function $u^* \in L^p(\partial\Omega)$ is called the *trace* of u.

4.3. Prove that u^* obtained in the preceding exercise is equivalent to the trace obtained in Remark 4.4.5.

4.4. Prove the following Poincaré-type inequality which provides an estimate of the measure of $\{|u| \geq k\}$ in terms of $\|Du\|_1$. Let $u \in W^{1,p}(B)$ where B is an open ball of radius r and suppose μ is a measure of total mass 1 supported in $B \cap \{x : u(x) = 0\}$. Then, if $k \geq 0$,

$$k|\{x : |u(x)| \geq k\}| \leq Cr \int_B |Du| + Cr^n \int_B (I_1 * \mu) \cdot |Du|,$$

where R_1 is the Riesz kernel (see Section 2.6). Hint: Choose $x, y \in B$ with $u(y) = 0$ and obtain

$$|u(x)| \leq |u(x) - u(z)| + |u(z) - u(y)|$$

whenever $z \in B$. An application of polar coordinates yields

$$u(x) \leq C[I_1 * (\chi_B \cdot |Du|)(x) + I_1 * (\chi_B \cdot |Du|)(y)].$$

4.5. The technique in the preceding exercise yields yet another proof of Corollary 4.5.3 which is outlined as follows. Let $u \in W^{1,p}(B(r))$ where $B(r)$ is a ball of radius r and let $N = \{x : u(x) = 0\}$. Let φ be a non-negative smooth function with $\operatorname{spt} \varphi$ contained in the ball of radius $2r$ and such that φ is identically one on $B(r)$. Select $x \in B(r)$ in accordance with the result of Exercise 3.15 and define $h = \varphi[u(x) - u]$. Then, for each $y \in R^n$,

$$\varphi(y)u(x) = \varphi(y)u(y) + h(y).$$

Recall from Theorem 2.6.1 that the operator $J : L^p(R^n) \to W^{1,p}(R^n)$ defined in terms of the Bessel kernel g_1 by $J(f) = g_1 * f$ is an isometry. Therefore, with μ as any non-negative Radon measure μ, an application of Fubini's theorem yields

$$\int h\,d\mu = \int g_1 * (J^{-1}h)d\mu = \int (J^{-1}h)\cdot g_1 * \mu\,dx.$$

Thus, if μ is concentrated on $B(r)\cap N$ and satisfies $\|g_1 * \mu\|_{p'} \le 1$, it follows that

$$\mu[B(r)\cap N]|u(x)| \le \|J^{-1}[\varphi(u-u(x))]\|_p \le C\|\varphi(u-u(x)\|_{1,p}.$$

Taking the supremum over all such μ leads to

$$B_{1,p}[B(r)\cap N]|u(x)|^p \le C\left[r^{-p}\int_{B(2r)}|u(x)-u(y)|^p dy + \int_{B(2r)}|Du|^p dy\right].$$

Use Exercise 3.15 to estimate

$$\int_{B(2r)}|u(x)-u(y)|^p dy$$

in terms of the norm of Du.

4.6. Poincaré's inequality states that if $u \in W^{1,p}(\Omega)$ and u vanishes on a set N of positive $B_{1,p}$-capacity, then $\|u\|_{p^*;\Omega} \le C\|Du\|_{p;\Omega}$, where C depends on Ω and the capacity of N. In the event that more is known about u, this result can be improved. Using the indirect proof of Section 4.1, prove that if $u \in W^{1,p}(\Omega)$ is a harmonic function that vanishes at some point $x_0 \in \Omega$, then there exists $C = C(x_0,\Omega)$ such that

$$\|u\|_{p^*;\Omega} \le C\|Du\|_{p;\Omega}.$$

4.7. Lemma 4.2.2 is one of many interpolation results involving different orders of derivatives of a given function. In this and the next exercise, we will establish another one that has many useful applications. Prove the following: Let g be a measurable function on R^n, and let $0 < \alpha < n$, $0 < \varepsilon < 1$. Then

$$|I_{\alpha\varepsilon}(g)(x)| \le C(Mg(x))^{1-\varepsilon}\cdot(I_\alpha(|g|)(x))^\varepsilon,$$

where Mg is the maximal function of g. Refer to the proof of Theorem 2.8.4 and choose δ in(2.8.4) as

$$\delta^\alpha = \frac{I_\alpha(|g|)(x)}{Mg(x)}.$$

4.8. Let $f = I_\alpha(g)$, $g \geq 0$. Prove the interpolation inequality

$$\|D^k f\|_r \leq C\|f\|_{\alpha,p}^{|k|/\alpha}\|f\|_s^{(\alpha-|k|)/s}$$

where k is any multi-index with $0 < |k| < \alpha$, $1/r = |k|/\alpha p + (1-|k|)/s$, and $p < s \leq \infty$. Use the previous exercise to prove

$$\|I_{\alpha\varepsilon}(g)\|_r \leq C\|g\|_p^{1-\varepsilon}\|I_\alpha(|g|)\|_s^\varepsilon$$

where $1 < p < \infty$, $1/r = (1-\varepsilon)/p + \epsilon/s$, $p < s \leq \infty$. Then let $f = I_\alpha(g)$ and observe that

$$|D^k f(x)| \leq I_{\alpha-k}(g)(x).$$

4.9. Prove the following as a consequence of Theorem 4.2.1. Let $\Omega \subset R^n$ be a bounded, connected, extension domain. Suppose $v \in L^{p'}(\Omega)$, $n \neq 0$, $p > 1$. Prove that there exists $C = C(p, v, \Omega)$ such that

$$\|u\|_{p;\Omega} \leq C\|Du\|_{p;\Omega}$$

whenever $u \in W^{1,p}(\Omega)$, $\int_\Omega uv\,dx = 0$.

4.10. When $\Omega \subset R^{n+1}$ is a Lipschitz domain, Exercise 4.2 shows one way of defining the trace, $u^* \in L^p(\partial\Omega)$, when $u \in W^{1,p}(\Omega)$, $p > 1$. Note that

$$\int_{\partial\Omega} |u^*|^p dH^n \leq C \int_\Omega (|u|^p + |Du|^p)dx.$$

Let $v \in L^{p'}(\partial\Omega)$, $v \neq 0$. Prove that there exists $C = C(p, v, \Omega)$ such that

$$\|u\|_{p;\Omega} \leq C\|Du\|_{p;\Omega}$$

whenever $u \in W^{1,p}(\Omega)$, $\int_{\partial\Omega} u^* v^* dH^n = 0$.

4.11. At the beginning of this chapter an indirect proof of the following Poincaré inequality is given: If $u \in W^{1,p}(\Omega)$ and $u = 0$ on a set of positive measure S, then $\|u\|_{p;\Omega} \leq C\|Du\|_{p;\Omega}$. Show that essentially the same argument will establish the same conclusion if it is only assumed that $u = 0$ on a set of positive $B_{1,p}$-capacity.

Historical Notes

4.1. Lemmas 4.1.3 and 4.1.4 provide the main idea that serves as the keystone for the developments in this chapter. They are due to Norman Meyers [ME4] and many other results in this chapter, such as those in Sections 2, 5, and 6 are taken from this paper. It should be emphasized that Lemma 4.1.3 is an abstract version of the usual indirect proof of the basic Poincaré inequalities.

4.4. In Remark 4.4.5 an approach to the subject of trace theory on the boundary is indicated which is based on the material in Chapter 3 concerning the property of Sobolev functions being defined everywhere in the complement of small exceptional sets. Another approach to this subject is presented in [LM].

In the proof of Theorem 4.4.6 it is not necessary to use the Morse–Sard theorem if we are willing to use the full strength of the "Boxing Inequality" [GU] and not the version reflected in Lemma 5.9.3. The inequality in [GU] states that there is a constant $C = C(n)$ such that any compact set $K \subset R^n$ can be covered by a sequence of balls $\{B(r_i)\}$ such that

$$\sum_{i=1}^{\infty} r_i^{n-1} \leq CH^{n-1}(\partial K).$$

This inequality could be used to establish (4.4.6) if K is taken as $\overline{E}_t$ and by observing that $\partial \overline{E}_t \subset v^{-1}(t)$ since v is continuous.

4.5. The proof of the Poincaré inequality here is, of course, based on the material in the previous sections, particularly Theorem 4.2.1. This proof is contained in [ME4]. There are several other proofs of the Poincaré inequality including the one in [P] which is especially interesting.

4.7. All of the Sobolev-type inequalities discussed thus far are in terms of inequalities defined on R^n. There also are similar inequalities that hold for functions defined on submanifolds of R^n. For example, in minimal surface theory, Sobolev inequalities are known to hold for functions defined on submanifolds where the inequality includes a term involving the mean curvature of the submanifold, cf. [MS]. In case of a minimal surface, the mean curvature is 0. Theorem 4.7.2 is a result of the same ilk in that the left side involves integration with respect to a measure μ which can be taken as a suitable Hausdorff measure restricted to some submanifold. However, it is different in the respect that the right side of the inequality involves the L^p-norm of the gradient relative to Lebesgue measure on R^n and not the norm relative to Hausdorff measure restricted to the submanifold. This interesting result was proved by David Adams [AD2]. Theorem 4.7.4 states that measures with suitable growth over all balls are elements of the dual of $W^{k,p}(R^n)$. Thus, Theorem 4.7.2 is closely related to (4.2.1).

Theorem 4.7.5 which yields a characterization of those measures in the dual of $W^{k,p}(R^n)$ is due to Hedberg and Wolff [HW] although the proof we give is adapted from [AD7].

4.9. Inequality (4.9.1) is due to Meyers–Ziemer [MZ] in case $q = 1$. The proof for the case $1 < q < n/(n-1)$ is taken from [MA3]. This inequality is also established in Chapter 5 in the setting of BV functions, cf. Theorem 5.12.5. Corollary 4.1.5 is an observation that was communicated to the author by David Adams. This result when applied to Theorem 4.5.1 yields more information if Lemma 4.1.4 were used. This is an interesting example of the critical role played by the sharpness of a constant, in this instance, the exponent of $B_{m-k,p}(A)$ in Theorem 4.5.1. Indeed, in the work of Hedberg [HE2], it was essential that the best exponent appear. He gave a different proof of Theorem 4.5.1.

5

Functions of Bounded Variation

A function of bounded variation of one variable can be characterized as an integrable function whose derivative in the sense of distributions is a signed measure with finite total variation. This chapter is directed to the multivariate analog of these functions, namely the class of L^1 functions whose partial derivatives are measures in the sense of distributions. Just as absolutely continuous functions form a subclass of BV functions, so it is that Sobolev functions are contained within the class of BV functions of several variables. While functions of bounded variation of one variable have a relatively simple structure that is easy to expose, the multivariate theory produces a rich and beautiful structure that draws heavily from geometric measure theory. An interesting and important aspect of the theory is the analysis of sets whose characteristic functions are BV (called sets of finite perimeter). These sets have applications in a variety of settings because of their generality and utility. For example, they include the class of Lipschitz domains and the fact that the Gauss–Green theorem is valid for them underscores their usefulness. One of our main objectives is to establish Poincaré-type inequalities for functions of bounded variation in a context similar to that developed in Chapter 4 for Sobolev functions. This will require an analysis of the structure of BV functions including the notion of trace on the boundary of an open set.

5.1 Definitions

5.1.1. Definition. A function $u \in L^1(\Omega)$ whose partial derivatives in the sense of distributions are measures with finite total variation in Ω is called a *function of bounded variation.* The class of all such functions will be denoted by $BV(\Omega)$. Thus $u \in BV(\Omega)$ if and only if there are Radon (signed measures) measures $\mu_1, \mu_2, \ldots, \mu_n$ defined in Ω such that for $i = 1, 2, \ldots, n$, $|D\mu_i|(\Omega) < \infty$ and

$$\int u D_i \varphi \, dx = -\int \varphi \, d\mu_i \tag{5.1.1}$$

for all $\varphi \in C_0^\infty(\Omega)$.

The gradient of u will therefore be a vector valued measure with finite total variation:

$$\|Du\| = \sup\{\int_\Omega u \operatorname{div} v\, dx : v = (v_1, \ldots, v_n) \in C_0^\infty(\Omega; R^n),$$

$$|v(x)| \le 1 \text{ for } x \in \Omega\} < \infty. \tag{5.1.2}$$

The divergence of a vector field v is denoted by $\operatorname{div} v$ and is defined by $\operatorname{div} v = \sum_{i=1}^n D_i v_i$. Observe that in (5.1.1) and (5.1.2), the space $C_0^\infty(\Omega)$ may be replaced by $C_0^1(\Omega)$. The space $BV(\Omega)$ is endowed with the norm

$$\|u\|_{BV} = \|u\|_{1;\Omega} + \|Du\|. \tag{5.1.3}$$

If $u \in BV(\Omega)$ the total variation $\|Du\|$ may be regarded as a measure, for if f is a non-negative real-valued continuous function with compact support Ω, define

$$\|Du\|(f) = \sup\{\int_\Omega u \operatorname{div} v\, dx : v = (v_1, \ldots, v_n) \in C_0^\infty(\Omega; R^n),$$

$$|v(x)| \le f(x) \text{ for } x \in \Omega\}. \tag{5.1.4}$$

5.1.2. Remark. In order to see that $\|Du\|$ as defined by (5.1.4) is in fact a measure, an appeal to the Riesz Representation Theorem shows it is sufficient to prove that $\|Du\|$ is a positive linear functional on $C_0(\Omega)$ which is continuous under monotone convergence. That is, if $\{f_i\}$ is a sequence of non-negative functions in $C_0(\Omega)$ such that $f_i \uparrow g$ for $g \in C_0(\Omega)$, then $\|Du\|(f_i) \to \|Du\|(g)$, cf. [F4, Theorem 2.5.5]. In order to prove that $\|Du\|$ has these properties, let $\mu = Du$ and refer to (5.1.1) to see that μ satisfies

$$\int u \operatorname{div} \varphi\, dx = -\int \varphi \cdot d\mu$$

where $\varphi \in C_0^\infty(\Omega; R^n)$. Therefore, we may write (5.1.4) as

$$\|Du\|(f) = \sup\{\int_\Omega v \cdot d\mu : v = (v_1, \ldots, v_n) \in C_0(\Omega; R^n)$$

$$|v(x)| \le f(x) \text{ for } x \in \Omega\}. \tag{5.1.5}$$

To show that $\|Du\|$ is additive, let $f, g \in C_0(\Omega)$ be non-negative functions and suppose $v \in C_0(\Omega, R^n)$ is such that $|v| \le f + g$. Let $h = \inf\{f, |v|\}$ and define

$$w(x) = \begin{cases} h(x)\dfrac{v(x)}{|v(x)|} & |v(x)| \neq 0 \\ 0 & |v(x)| = 0. \end{cases}$$

It is easy to verify that $w \in C_0(\Omega)$ and $|v - w| = |v| - h \leq g$. Therefore, since $|w| = h \leq f$,

$$\int_\Omega v \cdot d\mu = \int_\Omega w \cdot d\mu + \int_\Omega (v - w) \cdot d\mu$$
$$\leq \|Du\|(f) + \|Du\|(g).$$

This implies that $\|Du\|(f + g) \leq \|Du\|(f) + \|Du\|(g)$. The opposite inequality is obvious and consequently it follows that $\|Du\|$ is additive. It is clearly positively homogeneous. It remains to show that it is continuous under monotone convergence. For this purpose, let $f_i \uparrow g$ and let $v \in C_0(\Omega, R^n)$ be such that $|v| \leq g$. Also, define $h_i = \inf\{f_i, |v|\}$ and

$$w_i(x) = \begin{cases} h_i(x)\dfrac{v(x)}{|v(x)|} & |v(x)| \neq 0 \\ 0 & |v(x)| = 0. \end{cases}$$

Note that $w_i \in C_0(\Omega)$, $|w_i| = h_i \leq f_i$, and that $|v - w_i| = |v| - h_i \downarrow 0$ as $i \to \infty$. Since $|v - w_i| = |v| - h_i \leq 2|v|$, Lebesgue's Dominated Convergence Theorem implies

$$\int v \, d\mu = Du \cdot v = \lim_{i \to \infty} Du \cdot w_i \leq \lim_{i \to \infty} \|Du\|(h_i).$$

By taking the supremum of the left side over all such v it follows that $\|Du\|(g) \leq \lim_{i \to \infty} \|Du\|(h_i)$. Since $h_i \leq g$ for all $i = 1, 2, \ldots$, we have $\|Du\|(g) = \lim_{i \to \infty} \|Du\|(h_i)$. This establishes that $\|Du\|$ is a non-negative Radon measure on Ω.

We know that the space of absolutely continuous u with $u' \in L^1(R^1)$ is contained within $BV(R^1)$. Analogously, in R^n we have that a Sobolev function is also BV. That is, $W^{1,1}(\Omega) \subset BV(\Omega)$, for if $u \in W^{1,1}(\Omega)$, then

$$\int_\Omega u \operatorname{div} v \, dx = -\int_\Omega \sum_{i=1}^n D_i u v \, dx$$

and the gradient of u has finite total variation with

$$\|Du\|(\Omega) = \int_\Omega |Du| dx.$$

5.2 Elementary Properties of BV Functions

In this section we establish a few results concerning convergence properties of BV functions. We begin with the following which is almost immediate from definitions, but yet extremely useful.

5.2.1. Theorem. *Let $\Omega \subset R^n$ be an open set and $u_i \in BV(\Omega)$ a sequence* *of functions that converge to a function u in* $L^1_{\text{loc}}(\Omega)$. *Then*

$$\liminf_{i\to\infty} \|Du_i\|(U) \geq \|Du\|(U)$$

for every open set $U \subset \Omega$.

Proof. Let v be a vector field such that $v \in C_0^\infty(U; R^n)$ and $|v(x)| \leq 1$ for $x \in U$. Then

$$\int_U u \operatorname{div} v \, dx = \lim_{i\to\infty} \int_U u_i \operatorname{div} v \, dx \leq \liminf_{i\to\infty} \|Du_i\|(U).$$

The result follows by taking the supremum over all such v. □

5.2.2. Remark. Note that the above result does not assert that the limit function u is an element of $BV(\Omega)$. However, if $u \in L^1(\Omega)$ and we assume that

$$\sup\{\|Du_i\|(\Omega) : i = 1, 2, \ldots\} < \infty$$

then $u \in BV(\Omega)$. Indeed, if $\varphi \in C_0^\infty(\Omega)$, and Du_i is any partial derivative of u_i, then

$$\lim_{i\to\infty} \int_\Omega \varphi \, Du_i dx = -\lim_{i\to\infty} \int_\Omega u_i D\varphi \, dx = \int_\Omega u \, D\varphi \, dx$$

and therefore

$$\left| \int_\Omega u \, D\varphi \, dx \right| \leq \sup |\varphi| \liminf_{i\to\infty} \|Du_i\|(\Omega) < \infty.$$

Since $C_0^\infty(\Omega)$ is dense in the space of continuous functions with compact support, we have that

$$Du(\varphi) = -\int_\Omega u \, D\varphi \, dx$$

is a bounded functional on $C_0(\Omega)$. That is, Du is a measure on Ω.

Theorem 5.2.1 established the lower semicontinuity of the total variation of the gradient measure relative to convergence in L^1_{loc}. We now will prove an elementary result that provides upper semicontinuity.

5.2.3. Theorem. *Let* $\{u_i\} \in BV(\Omega)$ *be a sequence such that* $u_i \to u$ *in* $L^1_{\text{loc}}(\Omega)$ *and*

$$\lim_{i\to\infty} \|Du_i\|(\Omega) = \|Du\|(\Omega).$$

Then,

$$\limsup_{i\to\infty} \|Du\|(\overline{U} \cap \Omega) \leq \|Du\|(\overline{U} \cap \Omega)$$

whenever U is an open subset of Ω.

Proof. Since $V = \Omega - \overline{U}$ is open, it follows from Theorem 5.2.1 that

$$\|Du\|(U) \leq \liminf_{i\to\infty} \|Du_i\|(U)$$
$$\|Du\|(V) \leq \liminf_{i\to\infty} \|Du_i\|(V).$$

But,

$$\begin{aligned}\|Du\|(\overline{U}\cap\Omega) + \|Du\|(V) = \|Du\|(\Omega) &= \lim_{i\to\infty} \|Du_i\|(\Omega) \\ &\geq \limsup_{i\to\infty} \|Du_i\|(\overline{U}\cap\Omega) + \liminf_{i\to\infty} \|Du_i\|(V) \\ &\geq \limsup_{i\to\infty} \|Du_i\|(\overline{U}\cap\Omega) + \|Du\|(V). \qquad \square\end{aligned}$$

In view of the last result and Theorem 5.2.1, the following is immediate.

5.2.4. Corollary. *If $\{u_i\} \in BV(\Omega)$ is a sequence such that $u_i \to u$ in $L^1_{\text{loc}}(\Omega)$, $\lim_{i\to\infty} \|Du_i\|(\Omega) = \|Du\|(\Omega)$, and $\|Du\|(\partial\overline{U}) = 0$, where U is an open subset of Ω, then*

$$\lim_{i\to\infty} \|Du_i\|(U) = \|Du\|(U).$$

5.3 Regularization of BV Functions

Here we collect some results that employ the technique of regularization introduced in Section 1.6. Thus, for each $\varepsilon > 0$, φ_ε is the regularizing kernel and $u_\varepsilon = u*\varphi_\varepsilon$. From the proof of Theorem 1.6.1, it follows that if $U \subset\subset \Omega$, and $u \in L^1_{\text{loc}}(\Omega)$, then $\|u_\varepsilon\|_{1;U} \leq \|u\|_{1;\Omega}$ for all sufficiently small $\varepsilon > 0$. In this sense, regularization does not increase the norm. We begin by showing that a similar statement is valid when the BV norm is considered.

5.3.1. Theorem. *Suppose U is an open set with $\overline{U} \subset \Omega$ and let $u \in BV(\Omega)$. Then, for all sufficiently small $\varepsilon > 0$,*

$$\|u_\varepsilon\|_{BV(U)} \leq \|u\|_{BV(\Omega)}.$$

Proof. In view of Theorem 5.2.1, it suffices to show that $\|Du_\varepsilon\|(U) \leq \|Du\|(\Omega)$ for all sufficiently small $\varepsilon > 0$. Select $v \in C^1_0(U, R^n)$ with $|v| \leq 1$. Choose $\eta > 0$ such that $\{x : d(x,U) < \eta\} \subset \Omega$. Note that $|v_\varepsilon| \leq 1$ and $\operatorname{spt} v_\varepsilon \subset \{x : d(x,U) < \eta\}$ for all small $\varepsilon > 0$. For all such $\varepsilon > 0$, Fubini's Theorem yields

$$\begin{aligned}\int_U u_\varepsilon \operatorname{div} v\, dx = \int_\Omega u_\varepsilon \operatorname{div} v\, dx &= \int_\Omega u(\operatorname{div} v)_\varepsilon dx \\ &= \int_\Omega u \operatorname{div} v_\varepsilon dx \leq \|Du\|(\Omega).\end{aligned}$$

The result follows by taking the supremum over all such v. □

5.3.2. Proposition. *Let $u \in BV(\Omega)$ and $f \in C_0^\infty(\Omega)$. Then $fu \in BV(\Omega)$ and $D(fu) = Dfu + fDu$ in the sense of distributions.*

Proof. Let U be an open set such that $\operatorname{spt} f \subset U \subset \overline{U} \subset \Omega$. Then, $u_\varepsilon f \in C_0^\infty(U)$ with $D(fu_\varepsilon) = (Df)u_\varepsilon + fDu_\varepsilon$ at all points in U. However, $\|u_\varepsilon - u\|_{1;U} \to 0$ as $\varepsilon \to 0$. (Of course, we consider only those $\varepsilon > 0$ for which $u_\varepsilon(x)$ is defined for $x \in U$.) In particular, when considered as distributions, $u_\varepsilon \to u$. That is, $u_\varepsilon \to u$ in $\mathscr{D}'(U)$ and therefore $Du_\varepsilon \to Du$ in $\mathscr{D}'(U)$, (see Section 1.7). Since $f \in C_0^\infty(U)$, it follows that $fDu_\varepsilon \to fDu$ in $\mathscr{D}'(U)$. Clearly, $(Df)u_\varepsilon \to (Df)u$ in $\mathscr{D}'(U)$. Finally, with the observation that $fu_\varepsilon \to fu$ in $\mathscr{D}'(U)$ and therefore that $D(fu_\varepsilon) \to D(fu)$ in $\mathscr{D}'(U)$, the conclusion readily follows. □

We now proceed to use the technique of regularization to show that BV functions can be approximated by smooth functions and thus obtain a result somewhat analogous to Theorem 2.3.2 which states that $C^\infty(\Omega) \cap \{u : \|u\|_{k,p;\Omega} < \infty\}$ is dense in $W^{k,p}(\Omega)$. Of course, it is not possible to obtain a strict analog of this result for BV functions because a sequence $\{u_i\} \in C^\infty(\Omega)$ that is fundamental in the BV norm will converge to a function in $W^{1,1}(\Omega)$. However, we obtain the following approximation result.

5.3.3. Theorem. *Let $u \in BV(\Omega)$. Then there exists a sequence $\{u_i\} \in C^\infty(\Omega)$ such that*

$$\lim_{i \to \infty} \int_\Omega |u_i - u| dx = 0$$

and

$$\lim_{i \to \infty} \|Du_i\|(\Omega) = \|Du\|(\Omega).$$

Proof. In view of Theorem 5.2.1, it suffices to show that for every $\varepsilon > 0$, there exists a function $v_\varepsilon \in C^\infty(\Omega)$ such that

$$\int_\Omega |u - v_\varepsilon| dx < \varepsilon \quad \text{and} \quad \|Dv_\varepsilon\|(\Omega) < \|Du\|(\Omega) + \varepsilon. \tag{5.3.1}$$

Proceeding as in Theorem 2.3.2, let Ω_i be subdomains of Ω such that $\Omega_i \subset\subset \Omega_{i+1}$ and $\cup_{i=0}^\infty \Omega_i = \Omega$. Since $\|Du\|$ is a measure we may assume, by renumbering if necessary, that $\|Du\|(\Omega - \Omega_0) < \varepsilon$. Let $U_0 = \Omega_1$ and $U_i = \Omega_{i+1} - \overline{\Omega}_{i-1}$ for $i = 1, 2, \ldots$. By Lemma 2.3.1, there is a partition of unity subordinate to the covering $U_i = \Omega_{i+1} - \overline{\Omega}_{i-1}$, $i = 0, 1, \ldots$. Thus, there exist functions f_i such that $f_i \in C_0^\infty(U_i)$, $0 \le f_i \le 1$, and $\sum_{i=0}^\infty f_i \equiv 1$ on Ω. Let φ_ε be a regularizer as discussed at the beginning of this section. Then, for each i there exists $\varepsilon_i > 0$ such that

$$\operatorname{spt}((f_i u)_{\varepsilon_i}) \subset U_i, \tag{5.3.2}$$

$$\int_\Omega |(f_i u)_{\varepsilon_i} - f_i u| dx < \varepsilon 2^{-(i+1)} \tag{5.3.3}$$

$$\int_\Omega |(uDf_i)_{\varepsilon_i} - uDf_i| dx < \varepsilon 2^{-(i+1)}. \tag{5.3.4}$$

Define

$$v_\varepsilon = \sum_{i=0}^{\infty} (uf_i)_{\varepsilon_i}.$$

Clearly, $v_\varepsilon \in C^\infty(\Omega)$ and $u = \sum_{i=0}^{\infty} uf_i$. Therefore, from (5.3.3)

$$\int_\Omega |v_\varepsilon - u| dx \le \sum_{i=0}^{\infty} \int_\Omega |(uf_i)_{\varepsilon_i} - uf_i| dx < \varepsilon. \tag{5.3.5}$$

Reference to Proposition 5.3.2 leads to

$$\begin{aligned} Dv_\varepsilon &= \sum_{i=0}^{\infty} (f_i Du)_{\varepsilon_i} + \sum_{i=0}^{\infty} (uDf_i)_{\varepsilon_i} \\ &= \sum_{i=0}^{\infty} (f_i Du)_{\varepsilon_i} + \sum_{i=0}^{\infty} [(uDf_i)_{\varepsilon_i} - uDf_i]. \end{aligned}$$

Here we have used the fact that $\Sigma\, Df_i = 0$ on Ω. Therefore,

$$\int_\Omega |Dv_\varepsilon| dx \le \sum_{i=1}^{\infty} \int_\Omega |(f_i Du)_{\varepsilon_i}| dx + \sum_{i=0}^{\infty} \int_\Omega |(uDf_i)_{\varepsilon_i} - uDf_i| dx.$$

The last term is less than ε by (5.3.4). In order to estimate the first term, let $\psi \in C_0^\infty(\Omega; R^n)$ with $\sup|\psi| \le 1$. Then, with $\varphi_\varepsilon * \psi = \psi_\varepsilon$,

$$\begin{aligned} \left| \int \varphi_\varepsilon * (f_i Du) \cdot \psi \, dx \right| &= \left| \int \psi_\varepsilon f_i d(Du) \right| \quad \text{by Fubini's theorem,} \\ &= \left| \int u \operatorname{div}(\psi_\varepsilon f_i) dx \right| \\ &\le \|Du\|(U_i) \end{aligned}$$

since $\operatorname{spt} \psi_\varepsilon f_i \subset U_i$ and $|\psi_\varepsilon f_i| \le 1$. Taking the supremum over all such ψ yields

$$\int_\Omega |(f_i Du)_{\varepsilon_i}| dx \le \|Du\|(U_i), \quad i = 0, 1, \ldots .$$

Therefore, since each $x \in \Omega$ belongs to at most two of the sets U_i,

$$\begin{aligned} \int_\Omega |Dv_\varepsilon| dx &\le \sum_{i=0}^{\infty} \|Du\|(U_i) + \varepsilon \\ &\le \|Du\|(\Omega_1) + \sum_{i=1}^{\infty} \|Du\|(U_i) + \varepsilon \\ &\le \|Du\|(\Omega_1) + 2\|Du\|(\Omega - \Omega_0) + \varepsilon \\ &\le \|Du\|(\Omega) + 3\varepsilon. \end{aligned}$$

Since $\varepsilon > 0$ is arbitrary, this along with (5.3.5) establishes (5.3.1). □

5.3.4. Corollary. *Let $\Omega \subset R^n$ be a bounded extension domain for $W^{1,1}(\Omega)$. Then $BV(\Omega) \cap \{u : \|u\|_{BV} \leq 1\}$ is compact in $L^1(\Omega)$.*

Proof. Let $u_i \in BV(\Omega)$ be a sequence of functions with the property that $\|u_i\|_{BV} \leq 1$. By Theorem 5.3.3 there exist functions $v_i \in C^\infty(\Omega)$ such that

$$\int_\Omega |v_i - u_i| dx < i^{-1} \quad \text{and} \quad \int_\Omega |Dv_i| dx \leq 2.$$

Thus, the sequence $\{\|v_i\|_{1,1;\Omega}\}$ is bounded. Then, by the Rellich–Kondrachov compactness theorem (Theorem 2.5.1), there is a subsequence of $\{v_i\}$ that converges to a function v in $L^1(\Omega)$. Referring to Remark 5.2.2, we obtain that $v \in BV(\Omega)$. □

In Theorem 2.1.4 we found that $u \in W^{1,p}$ if and only if $u \in L^p$ and u has a representative that is absolutely continuous on almost all line segments parallel to the coordinate axes and whose partial derivatives belong to L^p. We will show that a similar result holds for BV functions.

Since we are concerned with functions for which changes on sets of measure zero have no effect, it will be necessary to replace the usual notion of variation of a function by *essential variation.* If u is defined on the interval $[a, b]$, the essential variation of u on $[a, b]$ is defined as

$$\text{ess } V_a^b(u) = \sup \left\{ \sum_{i=1}^k |u(t_i) - u(t_{i-1})| \right\}$$

where the supremum is taken over all finite partitions $a < t_0 < t_1 \ldots t_k < b$ such that each t_i is a point of approximate continuity of u. (See Remarks 3.3.5 and 4.4.5 for discussions relating to approximate continuity.) From Exercise 5.1, we see that $u \in BV(a, b)$ if and only if $\text{ess } V_a^b(u) < \infty$. Moreover, $\text{ess } V_a^b(u) = \|Du\|[(a, b)]$. We will use this fact in the following theorem. As in Theorem 2.1.4, if $1 \leq i \leq n$, we write $x = (\tilde{x}, x_i)$ where $\tilde{x} \in R^{n-1}$ and we define $u_i(x_i) = u(\tilde{x}, x_i)$. Note that u_i depends on the choice of $\tilde{x}$ but for simplicity, this dependence will not be exhibited in the notation. Also, we consider rectangular cells R of the form $R = (a_1, b_1) \times (a_2, b_2) \times \cdots \times (a_n, b_n)$.

5.3.5. Theorem. *Let $u \in L^1_{\text{loc}}(R^n)$. Then $u \in BV_{\text{loc}}(R^n)$ if and only if*

$$\int_{\tilde{R}} \text{ess } V_{a_i}^{b_i}(u_i) d\tilde{x} < \infty$$

for each rectangular cell $\tilde{R} \subset R^{n-1}$, each $i = 1, 2, \ldots, n$, and $a_i < b_i$.

Proof. Assume first that $u \in BV_{\text{loc}}(R^n)$. For $1 \leq i \leq n$ it will be shown that

$$\int_{\tilde{R}} \operatorname{ess} V_{a_i}^{b_i}(u_i) d\tilde{x} < \infty$$

for each rectangular cell $\tilde{R} \subset R^{n-1}$ and $a_i < b_i$. For notational simplicity, we will drop the dependence on i and take R of the form $R = \tilde{R} \times [a, b]$. Now consider the mollified function $u_\varepsilon = \varphi_\varepsilon * u$ and note that

$$\int_R |u_\varepsilon - u| dx \to 0 \quad \text{as} \quad \varepsilon \to 0$$

and

$$\limsup_{\varepsilon \to 0} \int_R |Du_\varepsilon| dx < \infty \quad \text{(Theorem 5.3.1)}.$$

Consequently, with $u_{\varepsilon,i}(x_i) = u_\varepsilon(\tilde{x}, x_i)$, it follows that $u_{\varepsilon,i} \to u_i$ in $L^1(a, b)$ for H^{n-1}-a.e. $\tilde{x} \in \tilde{R}$. Theorem 5.2.1 implies that $\liminf_{\varepsilon \to 0} \|Du_{\varepsilon,i}\|[(a, b)] \geq \|Du_i\|[(a, b)]$ and therefore, from Exercise 5.1,

$$\operatorname{ess} V_a^b(u_i) \leq \liminf_{\varepsilon \to 0} \operatorname{ess} V_a^b(u_{\varepsilon,i})$$

for H^{n-1}-a.e. $\tilde{x} \in \tilde{R}$. Fatou's lemma yields

$$\begin{aligned}
\int_{\tilde{R}} \operatorname{ess} V_a^b(u_i) dH^{n-1}(\tilde{x}) &\leq \liminf_{\varepsilon \to 0} \int_{\tilde{R}} \operatorname{ess} V_a^b(u_{\varepsilon,i}) dH^{n-1}(\tilde{x}) \\
&= \liminf_{\varepsilon \to 0} \int_R |D_i u_\varepsilon| dx \\
&\leq \limsup_{\varepsilon \to 0} \int_R |Du_\varepsilon| dx < \infty.
\end{aligned}$$

For the other half of the theorem, let $u \in L^1_{\text{loc}}(R^n)$ and assume

$$\int_{\tilde{R}} \operatorname{ess} V_a^b(u_i) dH^{n-1}(\tilde{x}) < \infty$$

for each $1 \leq i \leq n$, $a < b$, and each rectangular cell $\tilde{R} \subset R^{n-1}$. Choose $\varphi \in C_0^\infty(R)$, $|\varphi| \leq 1$, where $R = \tilde{R} \times (a, b)$ and employ Exercise 5.1 to obtain

$$\int_{R^n} u \, D_i \varphi \, dx \leq \int_{\tilde{R}} \operatorname{ess} V_a^b(u_i) dH^{n-1}(\tilde{x}) < \infty.$$

This shows that the partial derivatives of u are totally finite measures over R and therefore that $u \in BV_{\text{loc}}(R^n)$. □

5.4 Sets of Finite Perimeter

The Gauss–Green theorem is one of the fundamental results of analysis and although its proof is well understood for smoothly bounded domains or even domains with piece-wise smooth boundary, the formulation of the result in its ultimate generality requires the notion of an exterior normal to a set with no smoothness properties in the classical sense. In this section, we introduce a large class of subsets of R^n for which the Gauss–Green theorem holds. These sets are called sets of finite perimeter and it will be shown that they possess an exterior normal which is defined in the same spirit as Lebesgue points of L^p-derivatives. The Gauss–Green theorem in the setting of sets of finite perimeter will be proved in Section 5.8.

5.4.1. Definition. A Borel set $E \subset R^n$ is said to have *finite perimeter in an open set* Ω provided that the characteristic function of E, χ_E, is a function of bounded variation in Ω. Thus, the partial derivatives of χ_E are Radon measures in Ω and the perimeter of E in Ω is defined as

$$P(E,\Omega) = \|D\chi_E\|(\Omega).$$

A set E is said to be of *locally finite perimeter* if $P(E,\Omega) < \infty$ for every bounded open set Ω. If E is of finite perimeter in R^n, it is simply called a set of *finite perimeter.* From (5.1.4), it follows that

$$P(E,\Omega) = \sup\left\{\int_E \operatorname{div} v\, dx : v = (v_1,\ldots,v_n) \in C_0^\infty(\Omega, R^n), |v(x)| \le 1\right\}. \tag{5.4.1}$$

5.4.2. Remark. We will see later that sets with minimally smooth boundaries, say Lipschitz domains, are of finite perimeter. In case E is a bounded open set with C^2 boundary, by a simple application of the Gauss–Green theorem it is easy to see that E is of finite perimeter. For if $v \in C_0^\infty(\Omega; R^n)$ with $\|v\|_\infty \le 1$, then

$$\int_E \operatorname{div} v\, dx = \int_{\partial E} v \cdot \nu\, dH^{n-1} \le H^{n-1}(\Omega \cap \partial E) < \infty$$

where $\nu(x)$ is the unit exterior normal to E at x. Therefore, by (5.4.1), $P(E,\Omega) < \infty$ whenever Ω is an open set.

Moreover, it is clear that $P(E,\Omega) = H^{n-1}(\Omega \cap \partial E)$. Indeed, since E is a C^2-domain, there is an open set, U, containing ∂E such that $d(x) = d(x,E)$ is C^1 on $U - \partial E$ and $Dd(x) = (x - \xi(x))/d(x)$ where $\xi(x)$ is the unique point in ∂E that is nearest to x. Therefore, the unit exterior normal ν to E has an extension $\tilde\nu \in C_0^1(R^n)$ such that $|\tilde\nu| \le 1$. Hence, if $v = \eta\tilde\nu$ with $\eta \in C_0^\infty(\Omega)$, we have,

$$\int_E \operatorname{div} v\, dx = \int_E \operatorname{div} \eta\tilde\nu\, dx = \int_{\partial E} \eta\, dH^{n-1}.$$

This implies

$$\begin{aligned} P(E,\Omega) &\ge \sup\left\{\int_{\partial E} \eta dH^{n-1} : \eta \in C_0^\infty(\Omega), |\eta| \le 1\right\} \\ &= H^{n-1}(\Omega \cap \partial E). \end{aligned}$$

Intuitively, the measure $D\chi_E$ is nothing more than surface measure (H^{n-1}-measure) restricted to the boundary of E, at least if E is a smoothly bounded set. One of the main results of this chapter is to show that this idea still remains valid if E is a set of finite perimeter. Of course, since we are in the setting of measure theory, the topological boundary of E is no longer the appropriate object of study. Rather, it will be seen that a subset of the topological boundary, defined in terms of metric density, will carry the measure $D\chi_E$.

In Theorem 2.7.4 we observed that the isoperimetric inequality lead to the Sobolev inequality via the co-area formula. Conversely, in Remark 2.7.5 we indicated that the Sobolev inequality can be used to establish the isoperimetric inequality. We now return to this idea and place it in the appropriate context of sets of finite perimeter. We will establish the classical isoperimetric inequality for sets of finite perimeter and also a local version, called the *relative isoperimetric inequality.*

5.4.3. Theorem. *Let $E \subset R^n$ be a bounded set of finite perimeter. Then there is a constant $C = C(n)$ such that*

$$|E|^{(n-1)/n} \leq C\|D\chi_E\|(R^n) = CP(E). \tag{5.4.2}$$

Moreover, for each ball $B(r) \subset R^n$,

$$\min\{|B(r) \cap E|, |B(r) - E)|\}^{(n-1)/n} \leq C\|D\chi_E\|(B(r)) = CP(E, B(r)). \tag{5.4.3}$$

Proof. The inequality (5.4.2) is a special case of the Sobolev inequality for BV functions since χ_E is BV. We will give a general treatment of Sobolev-type inequalities in Section 11. If $u \in BV(R^n)$, refer to Theorem 5.3.3 to find functions $u_i \in C_0^\infty(R^n)$ such that

$$\lim_{i\to\infty} \int |u_i - u| dx = 0,$$

$$\lim_{i\to\infty} \|Du_i\|(R^n) = \|Du\|(R^n).$$

By passing to a subsequence, we may assume that $u_i \to u$ a.e. Then, by Fatou's lemma and Sobolev's inequality (Theorem 2.4.1),

$$\begin{aligned} \|u\|_{n/(n-1)} &\leq \liminf_{i\to\infty} \|u_i\|_{n/(n-1)} \\ &\leq \lim_{i\to\infty} C\|Du_i\|(R^n) \\ &\leq C\|Du\|(R^n). \end{aligned}$$

To prove the relative isoperimetric inequality (5.4.3), a similar argument along with Poincaré's inequality for smooth functions (Theorem 4.4.2), yields

$$\|u - \overline{u}(r)\|_{n/(n-1);B(r)} \leq C\|Du\|(B(r))$$

where $\overline{u}(r) = \fint_{B(r)} u(x)dx$ and $B(r)$ is any ball in R^n. Now let $u = \chi_E$ and obtain

$$\int_{B(r)} |u(x) - \overline{u}(r)|^{n/(n-1)} dx = \left(\frac{|B(r) - E|}{|B(r)|}\right)^{n/(n-1)} |B(r) \cap E| + \left(\frac{|B(r) \cap E|}{|B(r)|}\right)^{n/(n-1)} |B(r) - E|.$$

If $|B(r) - E| \geq |B(r) \cap E|$, then $(|B(r) - E|)/(|B(r)|) \geq \frac{1}{2}$ and

$$\begin{aligned} C\|D\chi_E\|(B(r)) = C\|Du\|(B(r)) &\geq \|u - \overline{u}(r)\|_{n/(n-1);B(r)} \\ &\geq \left(\frac{|B(r) - E|}{|B(r)|}\right) |B(r) \cap E|^{(n-1)/n} \\ &\geq \frac{1}{2} \min\left(\frac{|B(r) \cap E|}{|B(r)|}, \frac{|B(r) - E|}{|B(r)|}\right)^{(n-1)/n} \end{aligned}$$

A similar argument treats the case $|B(r) \cap E| \geq |B(r) - E|$. □

We now return to the topic of the co-area formula which was proved in Theorem 2.7.1 for smooth functions. Simple examples show that (2.7.1) cannot hold for BV functions (consider a step function). However, a version is valid if the perimeters of level sets are employed. In the following, we let

$$E_t = \Omega \cap \{x : u(x) > t\}.$$

5.4.4. Theorem. *Let $\Omega \subset R^n$ be open and $u \in BV(\Omega)$. Then*

$$\|Du\|(\Omega) = \int_{R^1} \|D\chi_{E_t}\|(\Omega)dt.$$

Moreover, if $u \in L^1(\Omega)$ and E_t has finite perimeter in Ω for almost all t with

$$\int_{R^1} \|D\chi_{E_t}\|(\Omega)dt < \infty,$$

then $u \in BV(\Omega)$.

Proof. We will first proof the second assertion of the theorem. For each $t \in R^1$, define a function $f_t : R^n \to R^1$ by

$$f_t = \begin{cases} \chi_{E_t} & \text{if } t \geq 0 \\ -\chi_{R^n - E_t} & \text{if } t < 0. \end{cases}$$

Thus,

$$u(x) = \int_{R^1} f_t(x)dt, \quad x \in R^n.$$

Now consider a test function $\varphi \in C_0^\infty(\Omega)$, such that $\sup|\varphi| \leq 1$. Then

$$\begin{aligned}\int_{R^n} u(x)\varphi(x)dx &= \int_{R^n}\int_{R^1} f_t(x)\varphi(x)dtdx \\ &= \int_{R^1}\int_{R^n} f_t(x)\varphi(x)dxdt. \qquad (5.4.4)\end{aligned}$$

Now (5.4.4) remains valid if φ is replaced by any one of its first partial derivatives. Also, it is not difficult to see that the mapping $t \to \|D\chi_{E_t}\|(\Omega)$ is measurable. Therefore, if φ is taken as $\varphi \in C_0^\infty(\Omega; R^n)$ with $\sup|\varphi| \leq 1$, we have

$$\begin{aligned}Du(\varphi) = -\int_{R^n} u \cdot \operatorname{div}\varphi\, dx &= -\int_{R^1}\int_{R^n} f_t(x)\operatorname{div}\varphi(x)dxdt \\ &\leq \int_{R^1} Df_t(\varphi)dt \leq \int_{R^1} \|D\chi_{E_t}\|(\Omega)dt < \infty. \qquad (5.4.5)\end{aligned}$$

However, the sup of (5.4.5) over all such φ equals $\|Du\|(\Omega)$, which establishes the second assertion.

In order to prove the opposite inequality under the assumption that $u \in BV(\Omega)$, let $\{P_k\}$ be a sequence of polyhedral regions invading Ω and $L_k : P_k \to R^1$ piecewise linear maps such that

$$\lim_{k\to\infty}\int_{P_k} |L_k - u|dx = 0 \qquad (5.4.6)$$

and

$$\lim_{k\to\infty}\int_{P_k} |DL_k|dx = \|Du\|(\Omega), \qquad (5.4.7)$$

(see Exercise 5.2). Let

$$\begin{aligned}E_t^k &= P_k \cap \{x : L_k(x) > t\}, \\ \chi_t^k &= \chi_{E_t^k}.\end{aligned}$$

From (5.4.6) it follows that there is a countable set $S \subset R^1$ such that for each $j = 1, 2, \ldots$

$$\lim_{k\to\infty}\int_{F_j} |\chi_t(x) - \chi_t^k(x)|dx = 0 \qquad (5.4.8)$$

whenever $t \notin S$. Thus, for $t \notin S$, and $\varepsilon > 0$, refer to (5.4.1) to find $\varphi \in C_0^\infty(\Omega; R^n)$ such that $|\varphi| \leq 1$ and

$$\|D\chi_{E_t}\|(\Omega) - \int_{E_t} \operatorname{div}\varphi\, dx < \frac{\varepsilon}{2}. \qquad (5.4.9)$$

Let $M = \int_{R^n} |\operatorname{div}\varphi|dx$ and choose j such that $\operatorname{spt}\varphi \subset P_j$. Choose $k_0 \geq j$ such that for $k \geq k_0$,

$$\int_{P_j} |\chi_t - \chi_t^k|dx < \frac{\varepsilon}{2M}.$$

For $k \geq k_0$,

$$\left| \int_{E_t} \operatorname{div} \varphi \, dx - \int_{E_t^k} \operatorname{div} \varphi \, dx \right| \leq M \int_{P_j} |\chi_t - \chi_t^k| dx < \frac{\varepsilon}{2}. \tag{5.4.10}$$

Therefore, from (5.4.9) and (5.4.10)

$$\begin{aligned} \|D\chi_{E_t}\|(\Omega) &\leq \int_{E_t^k} \operatorname{div} \varphi \, dx + \varepsilon \\ &\leq \|D\chi_{E_t^k}\|(\Omega) + \varepsilon. \end{aligned}$$

Thus, for $t \notin S$,

$$\|D\chi_{E_t}\|(\Omega) \leq \liminf_{k \to \infty} \|D\chi_{E_t^k}\|(\Omega).$$

Therefore, Fatou's lemma implies

$$\begin{aligned} \int_{R^1} \|D\chi_{E_t}\|(\Omega) dt &\leq \liminf_{k \to \infty} \int_{R^1} \|D\chi_{E_t^k}\|(\Omega) dt \\ &\leq \liminf_{k \to \infty} \int_{R^1} H^{n-1}[L_k^{-1}(t) \cap \Omega] dt \quad \text{(by Remark 5.4.2)} \\ &\leq \liminf_{k \to \infty} \int_{P_k} |DL_k| dx \quad \text{(by (2.7.1))} \\ &= \|Du\|(\Omega) \quad \text{(by (5.4.7))}. \end{aligned}$$

□

5.5 The Generalized Exterior Normal

In Remark 5.4.2 we observed that a smoothly bounded set has finite perimeter. We now begin the investigation of the converse by determining the regularity properties possessed by the boundary of a set of finite perimeter.

5.5.1. Definition. Let E be of locally finite perimeter. The *reduced boundary* of E, $\partial^- E$, consists of all points $x \in R^n$ for which the following hold:

(i) $\|D\chi_E\|[B(x,r)] > 0$ for all $r > 0$,

(ii) If $\nu_r(x) = -D\chi_E[B(x,r)]/\|D\chi_E\|[B(x,r)]$, then the limit $\nu(x) = \lim_{r \to 0} \nu_r(x)$ exists with $|\nu(x)| = 1$.

$\nu(x)$ is called the *generalized exterior normal* to E at x. We will employ the notation $\nu(x) = \nu(x, E)$ in case there is a possibility of ambiguity. The notation ∂^- is used in $\partial^- E$ to indicate that the normal to E is pointing in the direction opposite to the gradient.

Observe that $\nu(x, E)$ is essentially the Radon–Nikodym derivative of $D\chi_E$ with respect to $\|D\chi_E\|$. To see this, let $\rho(x)$ be the vector-valued function defined by

$$\rho(x) = -\lim_{r\to 0} \frac{D\chi_E[B(x,r)]}{\|D\chi_E\|[B(x,r)]}.$$

From the theory of differentiation of measures in Chapter 1 (see Remark 1.3.9) this implies that ρ is the Radon–Nikodym derivative of $D\chi_E$ with respect to $\|D\chi_E\|$ and that

$$D\chi_E(B) = -\int_B \rho(x)d\|D\chi_E\|(x)$$

for all Borel sets $B \subset R^n$. Moreover,

$$\int_E \operatorname{div} v\, dx = -\int v(x)\cdot \rho(x) d\|D\chi_E\|$$

whenever $v \in C_0^1(R^n; R^n)$. Consequently, by (5.1.2), $|\rho(x)| = 1$ for $\|D\chi_E\|$-a.e. $x \in R^n$ and therefore, $\rho(x) = \nu(x, E)$ for $\|D\chi_E\|$-a.e. $x \in R^n$. Thus, we have

$$D\chi_E(B) = -\int_{B\cap\partial^- E} \nu(x,E)d\|D\chi_E\|(x),$$
$$\|D\chi_E\|(R^n - \partial^- E) = 0.$$

The next lemma is a preliminary version of the Gauss–Green theorem.

5.5.2. Lemma. *Suppose E is of locally finite perimeter and let $f \in C_0^\infty(R^n)$. Then, for almost all $r > 0$,*

$$\int_{E\cap B(r)} D_i f\, dx = -\int_{B(r)} f d(D_i\chi_E) + \int_{E\cap\partial(B(r))} f(y)\nu_i(y, B(r))dH^{n-1}(y)$$

where $B(r) = B(x,r)$ and $\nu_i(y, B(r))$ is the ith component of the unit exterior normal.

Proof. To simplify notation, we will take $x = 0$. From Proposition 5.3.2, we have that $f\chi_E \in BV(\Omega)$. Let S be the countable set of r such that $\|D_i(f\chi_E)\|[\partial(B(r))] \neq 0$. Select $r \notin S$ and let η_ε be a piecewise linear function on $(0,\infty)$ such that $\eta_\varepsilon \equiv 1$ on $(0, r]$ and $\eta_\varepsilon \equiv 0$ on $(r+\varepsilon, \infty)$. Since $D_i[f\chi_E]$ is a measure, we have

$$\begin{aligned}\int_{R^n} f(x)\chi_E(x)D_i[\eta_\varepsilon(|x|)]dx &= -\int_{R^n} \eta_\varepsilon(|x|)d(D_i[f\chi_E])(x)\\ &= -D_i(f\chi_E)[B(r)]\\ &\quad - \int_{B(r+\varepsilon)-B(r)} \eta_\varepsilon(|x|)d(D_i[f\chi_E])(x).\end{aligned}$$

Therefore

$$-\frac{1}{\varepsilon}\int_{B(r+\varepsilon)-B(r)} f(x)\chi_E(x)\frac{x_i}{|x|}dx$$

$$= -D_i(f\chi_E)[B(r)] - \int_{B(r+\varepsilon)-B(r)} \eta_\varepsilon(|x|)d(D_i[f\chi_E])(x).$$

Since $r \notin S$ and $|\eta| \leq 1$, the integral on the right converges to 0 as $\varepsilon \downarrow 0$. By the co-area formula (Theorem 2.7.3), the integral on the left can be expressed as

$$-\frac{1}{\varepsilon}\int_{B(r+\varepsilon)-B(r)} f(x)\chi_E(x)\frac{x_i}{|x|}dx$$

$$= -\frac{1}{\varepsilon}\int_r^{r+\varepsilon}\int_{E\cap\partial(B(r))} f(x)\frac{x_i}{|x|}dH^{n-1}(x)dt.$$

Therefore

$$-\frac{1}{\varepsilon}\int_{B(r+\varepsilon)-B(r)} f(x)\chi_E(x)\frac{x_i}{|x|}dx \to -\int_{E\cap\partial(B(r))} f(x)\frac{x_i}{|x|}dH^{n-1}(x)$$

which implies

$$\int_{E\cap\partial(B(r))} f(x)\frac{x_i}{|x|}dH^{n-1}(x) = D_i(f\chi_E)[B(r)]$$

for almost all $r > 0$. Moreover, from Proposition 5.3.2,

$$\begin{aligned} D_i[f\chi_E](B(r)) &= (D_i f)\chi_E(B(r)) + fD_i\chi_E(B(r)) \\ &= \int_{E\cap B(r)} D_i f(x)dx + \int_{B(r)} fd(D_i\chi_E). \end{aligned}$$ □

5.5.3. Corollary. *If E has finite perimeter in Ω, then for almost all $r > 0$ with $\overline{B}(r) \subset \Omega$,*

$$P(E\cap B(r),\Omega) \leq P(E,B(r)) + H^{n-1}[E\cap\partial(B(r))].$$

Proof. Choose $v \in C_0^\infty(\Omega, R^n)$ with $|v| \leq 1$ and let $r > 0$ be a number for which the preceding lemma holds. Then

$$\begin{aligned} \int_{E\cap B(r)} \operatorname{div} v\, dx &= -\int_{B(r)} v\cdot d(D\chi_E) \\ &\quad + \int_{E\cap\partial(B(r))} v(x)\cdot\nu(x,B(r))dH^{n-1}(x) \\ &\leq \|D\chi_E\|(B(r)) + H^{n-1}[E\cap\partial(B(r))]. \end{aligned}$$

Taking the supremum over all such v establishes the result. □

Remark. Equality actually holds in the above corollary, but this is not needed in the immediate sequel.

The next lemma will be needed later when we begin to investigate boundary regularity of sets of finite perimeter.

5.5.4. Lemma. *Let E be a set with locally finite perimeter. Then, for each $x \in \partial^- E$, there is a positive constant $C = C(n)$ such that for all sufficiently small $r > 0$,*

$$r^{-n}|B(x,r) \cap E| \geq C, \tag{5.5.1}$$

$$r^{-n}|B(x,r) - E| \geq C, \tag{5.5.2}$$

$$C \leq r^{1-n}\|D\chi_E\|(B(x,r)) \leq C^{-1}. \tag{5.5.3}$$

Proof. To simplify the notation, we may assume that $x = 0$. Since $0 \in \partial^- E$, there is a positive constant $C = C(n)$ such that

$$|\nu_r(0)| = |D\chi_E(B(r))|/\|D\chi_E\|[B(r)] \geq C \tag{5.5.4}$$

for all small $r > 0$. For almost all $r > 0$, it follows from Lemma 5.5.2 that

$$D\chi_E(B(r)) = \int_{E\cap\partial(B(R))} \frac{x}{|x|} dH^{n-1}(x)$$

and therefore

$$|D\chi_E(B(r))| \leq H^{n-1}[E \cap \partial(B(r))].$$

Consequently, (5.5.4) implies

$$\|D\chi_E\|(B(r)) \leq C^{-1}H^{n-1}[E \cap \partial(B(r))] < C^{-1}r^{n-1}. \tag{5.5.5}$$

Note that (5.5.5) holds for all small values of r since the left side is a left-continuous function of r. This establishes the upper bound in (5.5.3).

To establish (5.5.1), recall from Corollary 5.5.3 and (5.5.5) that for almost all $r > 0$,

$$P(E \cap B(r)) \leq P(E, B(r)) + H^{n-1}[E \cap \partial(B(r))]$$

and

$$P(E, B(r)) \leq C^{-1}H^{n-1}[E \cap \partial(B(r))].$$

Thus, an application of the isoperimetric inequality (Theorem 5.4.3) and the previous two inequalities lead to

$$|E \cap B(r)|^{(n-1)/n} \leq CP(E \cap B(r)) \leq CH^{n-1}[E \cap \partial(B(r))],$$

for some constant $C = C(n)$. Let $h(r) = |E \cap B(r)|$ and observe that the co-area formula (Theorem 2.7.3) yields

$$h(r) = \int_{E \cap B(r)} |D(|x|)| dx = \int_0^r H^{n-1}[E \cap \partial(B(t))] dt.$$

Hence, $h'(r) \geq Ch(r)^{(n-1)/n}$ and therefore that $h(r)^{(1/n)-1}h'(r) = n(h^{1/n}(r))' \geq C$. This implies $h(r)^{1/n} \geq Cr$, thus establishing (5.5.1).

Note that (5.5.1) implies (5.5.2) since $P(E) = P(R^n - E)$ and $\nu_E = -\nu_{R^n - E}$.

The lower bound in (5.5.3) follows immediately from (5.5.1), (5.5.2), and the relative isoperimetric inequality (Theorem 5.4.3)

$$\frac{\|D\chi_E\|(B(r))}{r^{n-1}} \geq C \min\left(\frac{|B(r) \cap E|}{r^n}, \frac{|B(r) - E|}{r^n}\right)^{(n-1)/n} \qquad \square$$

5.6 Tangential Properties of the Reduced Boundary and the Measure-Theoretic Normal

Now that we have introduced the definition of the unit exterior normal to a set of finite perimeter, we ask whether the existence of the exterior normal implies some type of regularity of the boundary. In order for the theory to run parallel to the classical development, the hyperplane orthogonal to the generalized normal in some sense should be tangent to the reduced boundary (see Definition 5.5.1). Although it cannot be expected that this plane is tangent in the usual sense, it will be shown that it is so in the measure-theoretic sense.

For this purpose, we will employ a "blow-up" technique which views the local behavior of a set at a point by examining a sequence of dilations of the set at the point. Specifically, let E be a set of locally finite perimeter and suppose for notational simplicity that $0 \in \partial^- E$. For each $\varepsilon > 0$, consider the dilation $T_\varepsilon(x) = x/\varepsilon$ and let $E_\varepsilon = T_\varepsilon(E)$. Note that $\chi_{E_\varepsilon} = \chi_E \circ T_\varepsilon^{-1}$ and that the scaling of $D\chi_{E_\varepsilon}$ is of order $n-1$. That is,

$$\begin{aligned} D\chi_{E_\varepsilon}[B(r/\varepsilon)] &= \varepsilon^{1-n} D\chi_E[B(r)] \quad \text{for } r > 0 \\ \|D\chi_{E_\varepsilon}\|[B(r/\varepsilon)] &= \varepsilon^{1-n} \|D\chi_E\|[B(r)] \quad \text{for } r > 0. \end{aligned} \tag{5.6.1}$$

The proof of the second equation, for example, can be obtained by choosing a sequence $\{u_i\} \in C^\infty[B(r)]$ such that $u_i \to \chi_E$ in $L^1[B(r)]$ and $\int_{B(r)} |Du_i| dx \to \|D\chi_E\|[B(r)]$ (Theorem 5.3.3). However,

$$\int_{B(r/\varepsilon)} |Du_{i,\varepsilon}| dx = \varepsilon^{1-n} \int_{B(r)} |Du_i| dx$$

where $u_{i,\varepsilon} = u_i \circ T_\varepsilon^{-1}$. Then, $u_{i,\varepsilon} \to \chi_E \circ T_\varepsilon^{-1} = \chi_{E_\varepsilon}$ in $L^1[B(\varepsilon/r)]$ and by Theorem 5.2.1,

$$\liminf_{i\to\infty} \int_{B(r/\varepsilon)} |Du_{i,\varepsilon}|dx \geq \|D\chi_{E_\varepsilon}\|[B(r/\varepsilon)].$$

Hence,

$$\|D\chi_{E_\varepsilon}\|[B(r/\varepsilon)] \leq \varepsilon^{1-n}\|D\chi_E\|[B(r)].$$

The reverse inequality is obtained by a similar argument involving a sequence of smooth function approximating $\chi_E \circ T_\varepsilon^{-1}$.

5.6.1. Definition. For $x \in \partial^- E$, let $\pi(x)$ denote the $(n-1)$-plane orthogonal to $\nu(x, E)$, the generalized exterior normal to E at x. Also, define the half-spaces

$$H^+(x) = \{y : \nu(x) \cdot (y - x) > 0\}$$
$$H^-(x) = \{y : \nu(x) \cdot (y - x) < 0\}.$$

5.6.2. Theorem. *If E is of locally finite perimeter and $0 \in \partial^- E$, then*

$$\chi_{E_\varepsilon} \to \chi_{H^-} \quad \textit{in} \quad L^1_{\text{loc}}(R^n) \quad \textit{as} \quad \varepsilon \downarrow 0$$

and

$$\|D\chi_{E_\varepsilon}\|(U) \to \|D\chi_{H^-}\|(U)$$

whenever U is a bounded open set with $H^{n-1}[(\partial U) \cap \pi(0)] = 0$.

Proof. Without loss of generality, we may assume that the exterior normal to E at 0 is directed along the x_n-axis so that $\nu_n(0) = 1$ and $\nu_1(0) = \ldots = \nu_{n-1}(0) = 0$. It is sufficient to show that for each sequence $\{\varepsilon_i\} \to 0$, there is a subsequence (which we denote by the full sequence) such that

$$\int |\chi_{E_{\varepsilon_i}} - \chi_{H^-}|dx \to 0 \quad \text{and} \quad \|D\chi_{E_{\varepsilon_i}}\|(U) \to \|D\chi_{H^-}\|(U) \tag{5.6.2}$$

as $\varepsilon_i \downarrow 0$.

From (5.6.1) and (5.5.3) we obtain for each $r > 0$,

$$\|D\chi_{E_\varepsilon}\|[B(r)] = \varepsilon^{1-n}\|D\chi_E\|[B(\varepsilon r)] \leq C^{-1}\varepsilon^{1-n}(\varepsilon r)^{n-1} = C^{-1}r^{n-1}, \tag{5.6.3}$$

and

$$\|D\chi_{E_\varepsilon}\|[B(r)] = \varepsilon^{1-n}\|D\chi_E\|[B(\varepsilon r)] \geq C\varepsilon^{1-n}(\varepsilon r)^{n-1} = Cr^{n-1} \tag{5.6.4}$$

for all sufficiently small $\varepsilon > 0$. Thus, for each $B(r)$, $Cr^{n-1} \leq \|\chi_{E_\varepsilon}\|_{BV(B(r))} \leq C^{-1}r^{n-1}$ for all sufficiently small $\varepsilon > 0$. Therefore we may invoke the compactness of BV functions (Corollary 5.3.4) and a diagonalization process to conclude that $\chi_{E_{\varepsilon_i}} \to \chi_A$ in $L^1_{\text{loc}}(R^n)$. For each bounded open set Ω, $\chi_{E_{\varepsilon_i}} \to \chi_A$ in $\mathscr{D}'(\Omega)$ (in the sense of distributions) and therefore $D\chi_{E_{\varepsilon_i}} \to D\chi_A$ in $\mathscr{D}'(\Omega)$. Note that $D\chi_A \neq 0$ from (5.6.4). Moreover $D\chi_{E_{\varepsilon_i}} \to D\chi_A$ weakly in the sense of Radon measures and therefore, for all but countably many $r > 0$,

$$D\chi_{E_{\varepsilon_i}}[B(r)] \to D\chi_A[B(r)]. \tag{5.6.5}$$

From (5.6.1), and the definition of the generalized exterior normal,

$$\begin{aligned} &\lim_{\varepsilon \to 0} D_i\chi_{E_\varepsilon}[B(r)]/\|D\chi_{E_\varepsilon}\|[B(r)] \\ &= \lim_{\varepsilon \to 0} D_i\chi_E[B(\varepsilon r)]/\|D\chi_E\|[B(\varepsilon r)] = 0, \quad i = 1, 2, \ldots, n-1, \end{aligned} \tag{5.6.6}$$

whereas

$$\lim_{\varepsilon \to 0} D_n\chi_{E_\varepsilon}[B(r)]/\|D\chi_{E_\varepsilon}\|[B(r)] = -1. \tag{5.6.7}$$

Thus, from (5.6.7) and (5.6.3),

$$\lim_{i \to \infty} \|D\chi_{E_{\varepsilon_i}}\|[B(r)] = -\lim_{i \to \infty} D_n\chi_{E_{\varepsilon_i}}[B(r)] = -D_n\chi_A[B(r)]. \tag{5.6.8}$$

From the lower semicontinuity of the total variation measure (Theorem 5.2.1) we obtain

$$\liminf_{i \to \infty} \|D\chi_{E_{\varepsilon_i}}\|[B(r)] \geq \|D\chi_A\|[B(r)]$$

and therefore $\|D\chi_A\|[B(r)] \leq -D_n\chi_A[B(r)]$ from (5.6.8). Since the opposite inequality is always true, we conclude that

$$\|D\chi_A\|[B(r)] = -D_n\chi_A[B(r)] \tag{5.6.9}$$

for all $r > 0$. Therefore, by Theorem 1.3.8 and Remark 1.3.9,

$$\|D\chi_A\|(B(r)) = -D_n\chi_A(B(r)) = \int_{B(r)} \nu_n(x, A)d\|D\chi_A\|(x).$$

This implies that $\nu_n(x, A) = 1$ for $\|D\chi_A\|$-a.e. x and thus that $\nu_i(x, A) = 0$ for $\|D\chi_A\|$-a.e. x, $i = 1, 2, \ldots, n-1$. Consequently, we conclude that the measures $D_i\chi_A$ are identically zero, $i = 1, 2, \ldots, n-1$. Hence, χ_A depends only on x_n and is a non-increasing function of that variable. Let

$$\lambda = \sup\{x_n : \chi_A(x) = 1\}.$$

Since $D\chi_A \neq 0$, we know that $\lambda \neq \infty$. The proof will be completed by showing that $\lambda = 0$. If $\lambda < 0$, we would have $B(r) \subset R^n - A$ for $r < |\lambda|$ and since $\chi_{E_{\varepsilon_i}} \to \chi_A$ in $L^1_{\text{loc}}(R^n)$,

$$\begin{aligned} 0 = |B(r) \cap A| &= \lim_{i \to \infty} |E_{\varepsilon_i} \cap B(r)| = \lim_{i \to \infty} \varepsilon_i^{-n}|B(r\varepsilon_i) \cap E| \\ &= \lim_{i \to \infty} r^n(r\varepsilon_i)^{-n}|B(r\varepsilon_i) \cap E| \end{aligned}$$

which contradicts (5.5.1). A similar contradiction is reached if $\lambda > 0$. Therefore, $A = H^-$ and by (5.6.8) and (5.6.9),

$$\lim_{i\to\infty} \|D\chi_{E_{\varepsilon_i}}\|[B(r)] = \|D_n\chi_{H^-}\|[B(r)]$$

for all but countably many $r > 0$. If U is an open set with $U \subset B(r)$ for such an $r > 0$ and

$$\|D\chi_{H^-}\|(\partial U) = H^{n-1}[\pi(0) \cap \partial U] = 0,$$

then Corollary 5.2.4 implies $\|D\chi_{E_{\varepsilon_i}}\|(U) \to \|D\chi_{H^-}\|(U)$. □

We now will explore the sense in which the hyperplane $\pi(x)$ introduced in Definition 5.6.1 is tangent to $\partial^- E$ at x. For this, we introduce the following.

5.6.3. Definition. Let $\nu \in R^n$ with $|\nu| = 1$. For $x \in R^n$ and $\varepsilon > 0$, let

$$C(x, \varepsilon, \nu) = R^n \cap \{y : |(y - x) \cdot \nu| > \varepsilon|y - x|\}.$$

In what follows, it will be clear from the context that both x and ν are fixed and therefore, we will simply write $C(\varepsilon) = C(x, \varepsilon, \nu)$.

$C(\varepsilon)$ is a cone with vertex at x whose major axis is parallel to the vector ν. If M were a smooth hypersurface with ν normal to M at x, then for each $\varepsilon > 0$

$$C(\varepsilon) \cap M \cap B(x, r) = \emptyset \qquad (5.6.10)$$

for all $r > 0$ sufficiently small. When M is replaced by $\partial^- E$, Theorem 5.6.5 below yields an approximation to (5.6.10).

Before we begin the proof of Theorem 5.6.5, we introduce another concept for the exterior normal to a set. This one states, roughly, that a unit vector n is normal to a set E at a point x if E lies completely on one (the appropriate) side of the hyperplane orthogonal to n, in the sense of metric density. The precise definition is as follows.

5.6.4. Definition. Let $E \subset R^n$ be a Lebesgue measurable set. A unit vector n is called the *measure-theoretic normal* to E at x if

$$\lim_{r\to 0} r^{-n}|B(x, r) \cap \{y : (y - x) \cdot n < 0, y \notin E\}| = 0$$

and

$$\lim_{r\to 0} r^{-n}|B(x, r) \cap \{y : (y - x) \cdot n > 0, y \in E\}| = 0.$$

The measure-theoretic normal to E at x will be denoted by $n(x, E)$ and we define

$$\partial^* E = \{x : n(x, E) \text{ exists}\}.$$

The following result proves that the measure theoretic normal exists whenever the generalized exterior normal does. Thus, $\partial^- E \subset \partial^* E$.

5.6.5. Theorem. *Let E be a set with locally finite perimeter. Suppose $0 \in \partial^- E$. Let ν be the generalized exterior normal to E at 0 and $\pi(0)$ the hyperplane orthogonal to ν. Then,*

$$\lim_{r \to 0} r^{n-1} \|D\chi_E\|[C(\varepsilon) \cap B(r)] = 0, \tag{5.6.11}$$

$$\lim_{r \to 0} r^{-n} |E \cap H^+ \cap B(r)| = 0, \text{ and} \tag{5.6.12}$$

$$\lim_{r \to 0} r^{-n} |(B(r) - E) \cap H^-| = 0. \tag{5.6.13}$$

Proof. Again we use the "blow-up" technique that was employed to obtain (5.6.1). Thus, let $T_r(x) = x/r$ and recall that

$$\|D\chi_{E_r}\|[B(1)] = r^{1-n} \|D\chi_E\|[B(r)].$$

Note that $T_r[C(\varepsilon) \cap B(r)] = C(\varepsilon) \cap B(1)$. Therefore

$$\|D\chi_{E_r}\|[C(\varepsilon) \cap B(1)] = r^{1-n} \|D\chi_E\|[C(\varepsilon) \cap B(r)],$$

and by Theorem 5.6.2,

$$r^{1-n} \|D\chi_E\|[C(\varepsilon) \cap B(r)] \to H^{n-1}[C(\varepsilon) \cap B(1) \cap \pi(0)] = 0.$$

This proves (5.6.11).

Similarly,

$$r^{-n} |E \cap B(r) \cap H^+| = |E_r \cap B(1) \cap H^+|$$

and since $\chi_{E_r} \to \chi_{H^-}$ in $L^1_{\text{loc}}(R^n)$ (Theorem 5.6.2),

$$\lim_{r \to 0} |E_r \cap B(1) \cap H^+| = |H^- \cap B(1) \cap H^+| = 0.$$

This establishes (5.6.12) and (5.6.13) is treated similarly. □

The following is an easy consequence of the relative isoperimetric inequality and complements (5.5.3).

5.6.6. Lemma. *There exists a constant $C = C(n)$ such that*

$$\liminf_{r \to 0} \frac{\|D\chi_E\|[B(x,r)]}{r^{n-1}} \geq C$$

whenever $x \in \partial^* E$.

Proof. Recall from Definition 5.6.4 that if $x \in \partial^* E$, then

$$\lim_{r \to 0} r^{-n} |B(x,r) \cap E \cap H^+(x)| = 0$$

and

$$\lim_{r\to 0} r^{-n}|[B(x,r)-E]\cap H^-(x)| = 0$$

where $H^+(x)$ and $H^-(x)$ are the half-spaces determined by the exterior normal, $n(x,E)$. Since $B(x,r)\cap H^-(x) = ([B(x,r)-E]\cap H^-(x))\cup(B(x,r)\cap E\cap H^-(x))$, the last equality implies that

$$\liminf_{r\to 0}\frac{|B(r)\cap E|}{|B(r)|} \geq \lim_{r\to 0}\frac{|B(x,r)\cap E\cap H^-(x)|}{|B(x,r)|} = \frac{1}{2}.$$

Similarly,

$$\liminf_{r\to 0}\frac{|B(r)-E|}{|B(r)|} \geq \frac{1}{2}$$

and consequently,

$$\lim_{r\to 0}\frac{|B(r)\cap E|}{|B(r)|} = \lim_{r\to 0}\frac{|B(r)-E|}{|B(r)|} = \frac{1}{2}.$$

The result now follows from the relative isoperimetric inequality (5.4.3). □

This result allows us to make our first comparison of the measures $\|D\chi_E\|$ and H^{n-1} restricted to $\partial^* E$.

5.6.7. Theorem. *There is a positive constant C such that if E is a set with locally finite perimeter, and $B\subset\partial^* E$ is a Borel set, then*

$$H^{n-1}(B) \leq C\|D\chi_E\|(B).$$

Proof. For each $x\in B$ we obtain from Lemma 5.6.6 that

$$\liminf_{r\to 0}\frac{\|D\chi_E\|[B(x,r)]}{r^{n-1}} \geq C.$$

Our conclusion thus follows from Lemma 3.2.1. □

5.6.8. Corollary. *If E is a set with locally finite perimeter, then*

$$H^{n-1}(\partial^* E - \partial^- E) = 0. \tag{5.6.14}$$

Moreover, $\|D\chi_E\|$ and the restriction of H^{n-1} to $\partial^* E$ have the same null sets.

Proof. From the discussion in Definition 5.5.1, we have that $\|D\chi_E\|(R^n - \partial^- E) = 0$ and therefore $\|D\chi_E\|(\partial^* E - \partial^- E) = 0$. Thus, (5.6.14) follows from the previous theorem. Moreover, if $B\subset\partial^- E$ with $H^{n-1}(B) = 0$, then $\|D\chi_E\|(B) = 0$ because of the second inequality in (5.5.3). This establishes the second assertion. □

5.7 Rectifiability of the Reduced Boundary

Thus far, we have shown that the measure-theoretic normal to a set E of locally finite perimeter exists whenever the generalized exterior normal exists (Theorem 5.6.5). Moreover, (5.6.11) states that the measure $\|D\chi_E\|$ has no mass inside the cone $C(\varepsilon)$, at least in the sense of measure density. This indicates that the reduced boundary may have some appealing tangential properties. Indeed, it will be shown that H^{n-1}-almost all of $\partial^- E$ can be decomposed into countably many sets each of which is contained within some C^1 manifold of dimension $(n-1)$.

5.7.1. Definition. A set $A \subset R^n$ is called countably $(n-1)$-rectificable if $A \subset A_0 \cup [\cup_{i=0}^{\infty} f_i(R^{n-1})]$ where $H^{n-1}(A_0) = 0$, and each $f_i\colon R^{n-1} \to R^n$ is Lipschitz, $i = 1, 2, \ldots$. Because a Lipschitz map defined on an arbitrary set in R^{n-1} can be extended to all of R^{n-1} (Theorem 3.6.2), countable $(n-1)$-rectifiability is equivalent to the statement that there exist sets $E_i \subset R^{n-1}$ and Lipschitz maps $f_i\colon E_i \to R^n$ such that $A \subset A_0 \cup [\cup_{i=1}^{\infty} f_i(E_i)]$.

The next result is an easy consequence of Rademacher's theorem and Theorem 3.6.2, concerning the approximation of Lipschitz functions.

5.7.2. Lemma. *A set $A \subset R^{n-1}$ is countably $(n-1)$-rectifiable if and only if $A \subset \cup_{i=0}^{\infty} A_i$ where $H^{n-1}(A_0) = 0$, and each A_i, $i \geq 1$, is an $(n-1)$-dimensional embedded C^1 submanifold of R^n.*

Proof. Obviously, only one direction requires proof. For this purpose, for each Lipschitz function f_i in the Definition 5.7.1, we may use Theorem 3.10.5 to find C^1 functions $g_{i,j}$, $j = 1, 2, \ldots$, such that

$$f_i(R^{n-1}) \subset N_i \cup \left[\bigcup_{j=1}^{\infty} g_{i,j}(R^{n-1})\right]$$

where $H^{n-1}(N_i) = 0$. Let $C_{i,j}$ denote the critical set of $g_{i,j}$:

$$C_{i,j} = R^{n-1} \cap \{y : Jg_{i,j}(y) = 0\},$$

where $Jg_{i,j}(y)$ denotes the Jacobian of $g_{i,j}$ at y. By an elementary area formula, see [F4, Theorem 3.2.3], $H^{n-1}[g_{i,j}(C_{i,j})] = 0$ and therefore the set

$$A_0 = \left(\bigcup_{i=1}^{\infty} N_i\right) \cup \left(\bigcup_{i,j=1}^{\infty} g_{i,j}(C_{i,j})\right) = 0$$

has zero H^{n-1} measure.

For each $y \in R^{n-1} - C_{i,j}$ an application of the implicit function theorem ensures the existence of an open set $U_{i,j}(y)$ containing y such that

$g_{i,j} \mid U_{i,j}(y)$ is univalent and that $g_{i,j}(U_{i,j}(y))$ is an $(n-1)$-dimensional C^1 submanifold of R^n. Clearly, there exists a sequence of points $y_1, y_2, \ldots$ in $R^{n-1} - C_{i,j}$ such that $\cup_{k=1}^{\infty} U_{i,j}(y_k) \supset R^{n-1} - C_{i,j}$ and

$$\bigcup_{k=1}^{\infty} g_{i,j}(U_{i,j}(y_k)) \supset g_{i,j}(R^{n-1} - C_{i,j}).$$

Therefore, for each i,

$$f_i(R^{n-1}) - A_0 \subset \bigcup_{j,k=1}^{\infty} g_{i,j}(U_{i,j}(y_k))$$

from which the result follows. □

5.7.3. Theorem. *If $E \subset R^n$ is of locally finite perimeter, then $\partial^- E$ is countably $(n-1)$-rectifiable.*

Proof. Clearly, in view of Corollary 5.5.3, we can reduce the argument to the case of E with finite perimeter. Now recall from the proof of Lemma 5.6.6, that if $x \in \partial^- E$, then

$$\lim_{r \to 0} r^{-n} |B(x,r) E \cap H^+(x)| = 0$$

and

$$\lim_{r \to 0} r^{-n} |[B(x,r) - E] \cap H^-(x)| = 0.$$

Since $B(x,r) \cap H^-(x) = ([B(x,r) - E] \cap H^-(x)) \cup (B(x,r) \cap E \cap H^-(x))$, the last equality implies that

$$\lim_{r \to 0} \frac{|B(x,r) \cap E \cap H^-(x)|}{|B(x,r)|} = \frac{1}{2}.$$

Therefore, with the aid of Egoroff's theorem, for each $0 < \varepsilon < 1$ and each positive integer i, there is a measurable set $F_i \subset \partial^- E$ and a positive number $r_i > 0$ such that $\|D\chi_E\|[(\partial^- E) - F_i] < 1/(2i)$ and

$$|E \cap H^+(x) \cap B(x,r)| < \frac{1}{4} \left(\frac{\varepsilon}{2}\right)^n |B(x,r)| \tag{5.7.1}$$

$$|E \cap H^-(x) \cap B(x,r)| > \frac{1}{4} |B(x,r)| \tag{5.7.2}$$

whenever $x \in F_i$ and $r < r_i$. Furthermore, by Lusin's theorem, there is a compact set $M_i \subset F_i$ such that $\|D\chi_E\|[F_i - M_i] < 1/(2i)$ and the restriction of $\nu(\cdot, E)$ to $M_i \cap \partial^- E$ is uniformly continuous. Since H^{n-1} restricted to $\partial^- E$ is absolutely continuous with respect to $\|D\chi_E\|$ (Theorem 5.6.7), our conclusion will follow if we can show that each M_i is countable $(n-1)$-rectifiable.

We will first prove that for each $x \in M_i$,

$$C(x,\varepsilon,\nu(x,E)) \cap M_i \cap B\left(x,\frac{r_i}{2}\right) = \emptyset \tag{5.7.3}$$

where $C(x,\varepsilon,\nu(x,E))$ is the cone introduced in Definition 5.6.3. Thus, we will show that $|\nu(x)\cdot(x-y)| \leq \varepsilon|x-y|$ whenever $x,y \in M_i$ and $|x-y| < (1/2)r_i$. If this were not true, first consider the consequences of $\nu(x)\cdot(y-x) > \varepsilon|x-y|$. Since the projection of the vector $y-x$ onto $\nu(x)$ satisfies $|\text{proj}_{\nu(x)}(y-x)| \geq \varepsilon|x-y|$, it would follow that $B(y,\varepsilon|x-y|) \subset H^+(x)$. Also, since $\varepsilon < 1$,

$$B(y,\varepsilon|x-y|) \subset B(x,2|x-y|)$$

and therefore

$$B(y,\varepsilon|x-y|) \subset H^+(x) \cap B(x,2|x-y|). \tag{5.7.4}$$

However, since $2|x-y| < r_i$, it follows from (5.7.1) and (5.7.2) that

$$\begin{aligned} |E \cap H^+(x) \cap B(x,2|x-y|)| &< \frac{1}{4}\left(\frac{\varepsilon}{2}\right)^n |B(x,2|x-y|)| \\ &\leq \frac{1}{4}\varepsilon^n |B(0,|x-y|)| \end{aligned} \tag{5.7.5}$$

and

$$\begin{aligned} |E \cap B(y,\varepsilon|x-y|)| &> |E \cap B(y,\varepsilon|x-y|) \cap H^-(y)| \\ &\geq \frac{1}{4}|B(y,\varepsilon|x-y|)| \\ &= \frac{1}{4}\varepsilon^n |B(0,|x-y|)|. \end{aligned} \tag{5.7.6}$$

Thus, from (5.7.4), a contradiction is reached because

$$\begin{aligned} \frac{1}{4}\varepsilon^n|B(0,|x-y|)| &< |E \cap B(y,\varepsilon|x-y|)| \\ &\leq |E \cap H^+(x) \cap B(x,2|x-y|)| < \frac{1}{4}\varepsilon^n|B(0,|x-y|)|. \end{aligned}$$

A similar contradiction is reached if $\nu(x)\cdot(y-x) < -\varepsilon|x-y|$ and thus, (5.7.3) is established.

We will now proceed to show that each M_i is countably $(n-1)$-rectifiable. In fact, we will show that M_i is finitely $(n-1)$-rectifiable. First, recall that M_i is compact and that $\nu(\cdot,E)$ is uniformly continuous on M_i. It will be shown that for each $x_0 \in M_i$ there exists a $t > 0$ such that $M_i \cap B(x_0,t)$ is the image of a set $A \subset R^{n-1}$ under a Lipschitz map. For this purpose, assume for notational simplicity that $\nu(x_0,E) = \nu(x_0)$ is the nth basis vector $(0,0,\ldots,1)$. Let $\pi(x_0)$ be the hyperplane orthogonal to $\nu(x_0)$ and let $p : M_i \to \pi(x_0)$ denote the orthogonal projection of M_i into $\pi(x_0)$. The conclusion will be established by showing that p is univalent on $B(x_0,t) \cap M_i$ and that $p^{-1}|p[B(x_0,t) \cap M_i]$ is Lipschitz.

To see that p is univalent, assume the contrary and suppose that $y, z \in M_i$ are points near x_0 with $|z-y| < \frac{1}{2}r_i$ and $p(y) = p(z)$. Let $u = z-y/|z-y|$ and note that $|\nu(x_0) \cdot u| = 1$. Since ν is continuous, it would follow that $|\nu(y) \cdot u| \geq \varepsilon$ if y were sufficiently close to x_0. However, (5.7.3) implies that $|\nu(y) \cdot u| < \varepsilon$, a contradiction. Thus, there exists $0 < t < \frac{1}{2}r_i$ such that p is univalent on $B(x_0,t) \cap M_i$.

Let L be the inverse of p restricted to $p[B(x_0,t) \cap M_i]$ and let $y, z \in p[B(x_0,t) \cap M_i]$. Then

$$\begin{aligned}\frac{|L(z) - L(y)|}{|z-y|} &= \frac{|L(z) - L(y)|}{(|L(z) - L(y)|^2 - |\mathrm{proj}_{\nu(x_0)}[L(z) - L(y)]|^2)^{1/2}} \\ &= \frac{1}{\left(1 - \frac{|\mathrm{proj}_{\nu(x_0)}(L(z)-L(y))|^2}{|L(z)-L(y)|^2}\right)^{1/2}}.\end{aligned}$$

Using again the continuity of ν, the last expression is close to

$$\frac{1}{\left(1 - \frac{|\mathrm{proj}_{\nu(y)}[L(z)-L(y)]|^2}{|L(x)-L(y)|^2}\right)^{1/2}} \tag{5.7.7}$$

provided that y is close to x_0. by (5.7.3), (5.7.7) is bounded above by $1/(1-\varepsilon^2)^{1/2}$, which proves that L is Lipschitz in some neighborhood of x_0. Since M_i is compact, this proves that M_i is finitely $(n-1)$-rectifiable. □

The following is an immediate consequence of Lemma 5.7.2 and the previous result.

5.7.4. Corollary. *If $E \subset R^n$ is of locally finite perimeter, then*

$$\partial^- E \subset \bigcup_{i=1}^{\infty} M_i \cup N$$

where $H^{n-1}(N) = 0$ and each M_i is an $(n-1)$-dimensional embedded C^1 submanifold of R^n.

5.8 The Gauss–Green Theorem

In this section it will be shown that the Gauss–Green formula is valid on sets of locally finite perimeter. The two main ingredients in the formulation of this result are the boundary of a set and the exterior normal. Since we are in the setting of sets of finite perimeter, it should not be surprising that the boundary of a set will be taken as the reduced boundary and the exterior normal as the measure-theoretic exterior normal.

In Definition 5.6.4, we introduced the notion of the measure-theoretic exterior normal and demonstrated (Theorem 5.6.5) that

$$\partial^- E \subset \partial^* E. \tag{5.8.1}$$

Moreover, from (5.6.14),

$$H^{n-1}(\partial^* E - \partial^- E) = 0. \tag{5.8.2}$$

One of the main objectives of this section is to strengthen this result by showing that if $B \subset \partial^* E$, then

$$H^{n-1}(B) = \|D\chi_E\|(B). \tag{5.8.3}$$

This is a crucial result needed for the proof of the Gauss–Green theorem.

5.8.1. Theorem. *If $E \subset R^n$ has locally finite perimeter, then*

$$H^{n-1}(B) = \|D\chi_E\|(B)$$

whenever $B \subset \partial^* E$ is a Borel set.

Proof. If $x \in \partial^- E$, it follows from Theorem 5.6.3 that

$$\begin{aligned} r^{1-n}\|D\chi_E\|[B(x,r)] &= \|D\chi_{E_r}\|[B(x,1)] \to \|D\chi_{H^-}\|[B(x,1)] \\ &= H^{n-1}[B(x,1) \cap \pi(x)] \\ &= \alpha(n-1) \end{aligned}$$

where $\pi(x)$ is the hyperplane orthogonal to $\nu(x,E)$. Therefore,

$$\lim_{r \to 0} \frac{\|D\chi_E\|[B(x,r)]}{\alpha(n-1)r^{n-1}} = 1, \quad x \in \partial^- E. \tag{5.8.4}$$

Since $H^{n-1}(\partial^* E - \partial^- E) = \|D\chi_E\|[\partial^* E - \partial^- E] = 0$ (Corollary 5.6.9) we may assume that $B \subset \partial^- E$ and $B \subset \cup_{i=1}^\infty M_i$, where each M_i is an $(n-1)$-manifold of class C^1 (Corollary 5.7.4). Fix i and let $\mu = H^{n-1}|M_i$. Since M_i is smooth,

$$\lim_{r \to 0} \frac{\mu[B(x,r)]}{\alpha(n-1)r^{n-1}} = 1, \quad x \in B \cap M_i,$$

and therefore, by (5.8.4),

$$\lim_{r \to 0} \frac{\mu[B(x,r)]}{\|D\chi_E\|[B(x,r)]} = 1, \quad x \in B \cap M_i.$$

By the Besicovitch Differentiation Theorem (Theorem 1.3.8 and Remark 1.3.9),

$$H^{n-1}(B \cap M_i) = \mu(B) = \|D\chi_E\|(B \cap M_i).$$

The result easily follows from this. □

We now are able to establish the Gauss–Green theorem in the context of sets of finite perimeter.

5.8.2. Theorem. *Let E be a set with locally finite perimeter. Then,*

$$\int_E \operatorname{div} V\, dx = \int_{\partial^* E} n(x, E) \cdot V(x) dH^{n-1}(x)$$

whenever $V \in C_0^1(R^n; R^n)$.

Proof. Choose a ball $B(r)$ containing spt V. Then

$$\begin{aligned}\int_E \operatorname{div} V\, dx &= -\int V \cdot d(D\chi_E) \quad \text{(from Lemma 5.5.2)}\\ &= \int_{\partial^- E} V(x) \cdot \nu(x, E) d\|D\chi_E\| \quad \text{(from Definition 5.5.1)}\\ &= \int_{\partial^* E} V(x) \cdot n(x, E) dH^{n-1}(x) \quad \text{(by the preceding theorem).}\end{aligned}$$

5.8.3. Remark. The Gauss–Green theorem is one of the basic results in analysis and therefore, the above result alone emphasizes the importance of sets of finite perimeter. Therefore, a question of critical importance is how large is the class of sets of finite perimeter. The definition alone does not allow easy identification of such sets. However, it is not difficult to see that a Lipschitz domain, Ω, is a set with locally finite perimeter. An outline of the proof will be given here while details are left as an exercise, for the reader. We may assume that Ω is locally of the form

$$\Omega = \{(w, y) : 0 \le y \le g(w)\}$$

where g is a non-negative Lipschitz function defined on an open cube $Q \subset R^{n-1}$. Since g admits a Lipschitz extension (Theorem (3.6.2) we may assume that g is defined on R^{n-1}. Let g_ε be a mollifier of g (Section 1.6) and recall that

$$\int_Q |Dg_\varepsilon| dx \le \int_Q |Dg| dx \tag{5.8.5}$$

for all $\varepsilon > 0$. Each set

$$\Omega_\varepsilon = \{(w, y) : 0 \le y \le g_\varepsilon(w),\ w \in Q\}$$

is obviously of finite perimeter because the classical Gauss–Green theorem applies to it (see Remark 5.4.2). Let χ_ε denote $\chi_{\Omega_\varepsilon}$ and observe that $\|D\chi_\varepsilon\|(R^n) = H^{n-1}(\partial\Omega_\varepsilon)$. Since

$$\int_Q \sqrt{1+|Dg_\varepsilon|^2}dx = H^{n-1}[\{(w,y) : y = g_\varepsilon(w),\ w \in Q\}],$$

it follows from (5.8.5) that $\|D\chi_\varepsilon\|(R^n) \le C$ where C is some constant independent of ε. We may apply the compactness property of BV functions (Corollary 5.3.4) to conclude that χ is BV in R^n, thus showing that Ω is locally of finite perimeter. Moreover, Rademacher's theorem on the almost everywhere total differentiability of Lipschitz functions (Theorem 2.2.1) implies that the measure-theoretic normal is H^{n-1}-almost everywhere given by

$$n(x,\Omega) = \frac{(Dg(w),1)}{\sqrt{1+|Dg(w)|^2}} \tag{5.8.6}$$

where $x = (w, g(w))$.

We conclude this section by stating without proof a useful characterization of sets of finite perimeter. This will be stated in terms of the measure-theoretic boundary.

5.8.4. Definition. If $E \subset R^n$ is a Lebesgue measurable set, the *measure-theoretic boundary of* E is defined by

$$\partial_M E = \{x : \overline{D}(E,x) > 0\} \cap \{x : \overline{D}(R^n - E, x) > 0\}.$$

If we agree to call the measure-theoretic interior (exterior) of E all points x for which $D(E,x) = 1$ ($D(E,x) = 0$), then $\partial_M E$ consists of those points that are in neither the measure-theoretic interior nor exterior of E. See Exercise 5.3 for more on this subject. In Lemma 5.9.5, we shall see that $\partial^* E$ and $\partial_M E$ differ by at most a set of H^{n-1}-measure 0.

5.8.5. Theorem. *Let $E \subset R^n$ be Lebesgue measurable. Then E has locally finite perimeter if and only if*

$$H^{n-1}(K \cap \partial_M E) < \infty$$

for every compact set $K \subset R^n$.

The reader is referred to [F4, Theorem 4.5.11] for the proof.

5.9 Pointwise Behavior of BV Functions

We now begin a treatment for BV functions analogous to that developed for Sobolev functions in the first three sections of Chapter 3. It will be shown that a BV function can be defined by means of its Lebesgue points everywhere except for a set of H^{n-1}-measure zero and a set that is analogous to the set of jump discontinuities in R^1.

In the definition below, the following notation will be used:

$$A_t = \{x : u(x) > t\},$$

$$B_t = \{x : u(x) < t\},$$

$$\overline{D}(E,x) = \limsup_{r \to 0} \frac{|E \cap B(x,r)|}{|B(x,r)|}$$

and

$$\underline{D}(E,x) = \liminf_{r \to 0} \frac{|E \cap B(x,r)|}{|B(x,r)|}.$$

In case the upper and lower limits are equal, we denote their common value by $D(E,x)$. Note that the sets A_t and B_t are defined up to sets of Lebesgue measure zero.

5.9.1. Definition. If u is a Lebesgue measurable function defined on R^n, the *upper* (*lower*) *approximate limit* of u at a point x is defined by

$$\operatorname{ap} \limsup_{y \to x} u(y) = \inf\{t : D(A_t, x) = 0\}$$

$$(\operatorname{ap} \liminf_{y \to x} u(y) = \sup\{t : D(B_t, x) = 0\}).$$

We speak of the approximate limit of u at x in case

$$\operatorname{ap} \limsup_{y \to x} u(y) = \operatorname{ap} \liminf_{y \to x} u(y).$$

u is said to be approximately continuous at x if

$$\operatorname{ap} \lim_{y \to x} u(y) = u(x).$$

5.9.2. Remark. If u is defined on an open set Ω, reference to the definitions imply that u is approximately continuous at x if for every open set U containing $u(x)$,

$$D[u^{-1}(U) \cap \Omega, x] = 1.$$

An equivalent and rather appealing formulation is the one used in Remark 3.3.5. It is as follows: u is approximately continuous at x if there exists a Lebesgue measurable set E containing x such that $D(E,x) = 1$ and $u \,|\, E$ is continuous at x. It is clear that this formulation implies the previous one. To see the validity of the opposite direction, let $r_1 > r_2 > r_3 > \dots$ be positive numbers tending to zero such that

$$\left| B(x,r) \cap \left\{ y : |u(y) - u(x)| > \frac{1}{k} \right\} \right| < \frac{|B(x,r)|}{2^k}, \quad \text{for } r \leq r_k.$$

Define

$$E = R^n - \bigcup_{k=1}^{\infty} \{B(x, r_k) - B(x, r_{k+1})\} \cap \left\{ y : |u(y) - u(x)| > \frac{1}{k} \right\}.$$

Clearly, $u \mid E$ is continuous at x. In order to complete the assertion, we will show that $D(\tilde{E}, x) = 0$. For this purpose, choose $\varepsilon > 0$ and let J be such that $\sum_{k=J}^{\infty} \frac{1}{2^k} < \varepsilon$. Let r be such that $0 < r < r_J$ and let $K \geq J$ be the integer such that $r_{K+1} \leq r < r_K$. Then,

$$\begin{aligned}
|(R^n - E) \cap B(x, r)| &\leq \sum_{k=K}^{\infty} \Bigg| \{B(x, r_k) - B(x, r_{k+1})\} \\
&\qquad \cap \left\{ y : |u(y) - u(x)| > \frac{1}{k} \right\} \Bigg| \\
&\leq \frac{|B(x, r)|}{2^K} + \sum_{k=K+1}^{\infty} \frac{|B(x, r_k)|}{2^k} \\
&\leq \frac{|B(x, r)|}{2^K} + \sum_{k=K+1}^{\infty} \frac{|B(x, r)|}{2^k} \\
&\leq |B(x, r)| \sum_{k=K}^{\infty} \frac{1}{2^k} \\
&\leq |B(x, r)| \varepsilon,
\end{aligned}$$

which yields the desired result since ε is arbitrary.

One of the main results of this section is that a BV function can be defined in terms of its approximate limits H^{n-1}-almost everywhere. For this, the following is needed.

5.9.3. Lemma. *Let $n > 1$ and $0 < \tau < 1/2$. Suppose E is a Lebesgue measurable set such that $\overline{D}(E, x) > \tau$ whenever $x \in E$. Then there exists a constant $C = C(\tau, n)$ and a sequence of closed balls $B(x_i, r_i)$ with $x_i \in E$ such that*

$$E \subset \bigcup_{i=1}^{\infty} B(x_i, r_i)$$

and

$$\sum_{i=1}^{\infty} (r_i)^{n-1} \leq C \|D\chi_E\|[R^n].$$

Proof. For each $x \in R^n$, the continuous function

$$f(r) = \frac{|B(x, r) \cap E|}{|B(x, r)|}$$

assumes the value τ for some $r_x > 0$ because it exceeds this value for some possibly different r and approaches zero as $r \to \infty$. Since $\tau < 1/2$, the relative isoperimetric inequality, (5.4.3), implies

$$[\tau\alpha(n)r_x^n]^{(n-1)/n} \leq C\|D\chi_E\|[B(x, r_x)].$$

Now apply Theorem 1.3.1 to the family of all such balls $B(x, r_x)$ to obtain a sequence of disjoint balls $B(x_i, r_i)$ such that $\cup_{i=1}^{\infty} B(x_i, 5r_i) \supset E$. Therefore,

$$\begin{aligned}[\tau\alpha(n)]^{(n-1)/n} \sum_{i=1}^{\infty}(5r_i)^{n-1} &\leq 5^{(n-1)}C \sum_{i=1}^{\infty} \|D\chi_E\|[B(x, r_i)] \\ &\leq 5^{(n-1)}C\|D\chi_E\|[R^n]. \end{aligned}$$

□

In addition to (5.5.3) concerning the $(n-1)$-density of the measure $\|D\chi_E\|$, we will need the following.

5.9.4. Lemma. *Let $E \subset R^n$ be a set with locally finite perimeter. Then, for H^{n-1}-almost every $x \in R^n - \partial^* E$,*

$$\limsup_{r \to 0} \frac{\|D\chi_E\|[B(x,r)]}{\alpha(n-1)r^{n-1}} = 0.$$

Proof. For each positive number λ let

$$A = (R^n - \partial^* E) \cap \left\{ x : \limsup_{r \to 0} \frac{\|D\chi_E\|[B(x,r)]}{\alpha(n-1)r^{n-1}} > \lambda \right\}.$$

It follows from Lemma 3.2.1 that

$$\|D\chi_E\|(A) \geq C\lambda H^{n-1}(A).$$

Therefore $H^{n-1}(A) = 0$ since $\|D\chi_E\|(A) = 0$, thus establishing the conclusion of the lemma. □

This leads directly to the next result which is needed to discuss the points of approximate continuity of BV functions. Recall the definition of the measure-theoretic boundary, $\partial_M E$, Definition 5.8.4. The next result, along with (5.8.2) shows that all of the boundaries associated with a set of finite perimeter, $\partial^- E$, $\partial^* E$, and $\partial_M E$, are the same except for a set of H^{n-1}-measure zero.

5.9.5. Lemma. *Let $E \subset R^n$ be a set with locally finite perimeter. Then $\partial^* E \subset \partial_M E$ and $H^{n-1}(\partial_M E - \partial^* E) = 0$.*

Proof. It follows immediately from Definition 5.6.4 that $\partial^* E \subset \partial_M E$. In order to prove the second assertion, consider a point $z \in \partial_M E$ such that

$\overline{D}(E,z) > \delta$ and $\overline{D}(R^n - E, z) > \delta$ where $0 < \delta < 1/2$ and define a continuous function f by

$$f(r) = \frac{|E \cap B(z,r)|}{|B(z,r)|} = 1 - \frac{|B(z,r) - E|}{|B(z,r)|}.$$

Thus,

$$\limsup_{r \to 0} f(r) = \overline{D}(E,z) > \delta$$

and

$$\liminf_{r \to 0} f(r) = 1 - \overline{D}(E,z) < 1 - \delta,$$

with $\delta < 1 - \delta$. Hence, there are arbitrarily small $r > 0$ such that $\delta < f(r) < 1 - \delta$ and for all such r, the relative isoperimetric inequality, (5.4.3), implies

$$[\delta\alpha(n)r^n]^{(n-1)/n} \leq C\|D\chi_E\|[B(z,r)].$$

Thus,

$$\limsup_{r \to 0} \frac{\|D\chi_E\|[B(z,r)]}{\alpha(n-1)r^{n-1}} > 0,$$

and reference to Lemma 5.9.4 now establishes the conclusion. □

In the next theorem, it is shown that a BV function is approximately continuous at all points except for a set of H^{n-1}-measure zero and a countably $(n-1)$-rectifiable set E which, roughly speaking, includes the points at which u has a jump discontinuity (in the sense of approximate limits). It is also shown that at H^{n-1}-almost all points of E, u has one-sided approximate limits. Later, these results will be refined and stated in terms of integral averages.

Recall from Definition 5.9.1 that $A_t = \{x : u(x) > t\}$.

5.9.6. Theorem. *Let $u \in BV(R^n)$. If*

$$\mu(x) = \operatorname{ap}\limsup_{y \to x} u(y),$$

$$\lambda(x) = \operatorname{ap}\liminf_{y \to x} u(y),$$

and

$$E = R^n \cap \{x : \lambda(x) < \mu(x)\},$$

then

(i) *E is countably $(n-1)$-rectifiable,*

(ii) *$-\infty < \lambda(x) \leq \mu(x) < \infty$ for H^{n-1}-almost all $x \in R^n$,*

(iii) *for H^{n-1}-almost all $z \in E$, there is a unit vector v such that $n(z, A_s) = v$ whenever $\lambda(z) < s < \mu(z)$.*

(iv) *For all* z *as in* (iii), *with* $-\infty < \lambda(x) < \mu(x) < \infty$, *let* $H^-(z) = \{y : (y-z)\cdot v < 0\}$ *and* $H^+(z) = \{y : (y-z)\cdot v > 0\}$. *Then, there are Lebesgue measurable sets* E^- *and* E^+ *such that*

$$\lim_{r\to 0} \frac{|E^- \cap H^-(z) \cap B(z,r)|}{|H^-(z) \cap B(z,r)|} = 1 = \lim_{r\to 0} \frac{|E^+ \cap H^+(z) \cap B(z,r)|}{|H^+(z) \cap B(z,r)|}$$

and

$$\lim_{\substack{x\to z \\ x\in E^-\cap H^-(z)}} u(x) = \mu(z), \qquad \lim_{\substack{x\to z \\ x\in E^+\cap H^+(z)}} u(x) = \lambda(z).$$

Proof. Applying Theorems 5.4.4 and 5.7.3, there exists a countable dense subset Q of R^1 such that $P(A_t) < \infty$ and $\partial^* A_t$ is countably $(n-1)$-rectifiable whenever $t \in Q$. From Remark 5.9.2 we see that

$$H^{n-1}[\{\cup(\partial_M A_t - \partial^* A_t) : t \in Q\}] = 0.$$

It follows immediately from definitions that

$$\{x : \lambda(x) < t < \mu(x)\} \subset \partial_M A_t \quad \text{for } t \in R^1, \tag{5.9.1}$$

and therefore $E \subset \{\cup \partial_M A_t : t \in Q\}$, $H^{n-1}[E - \{\cup\partial^* A_t : t \in Q\}] = 0$. This proves that E is $(n-1)$-countably rectifiable.

Let $I = \{x : \lambda(x) = -\infty\} \cup \{x : \mu(x) = \infty\}$. We will show that $H^{n-1}(I) = 0$. For this purpose it will be sufficient to assume that u has compact support. First, we will prove that $H^{n-1}[\{x : \lambda(x) = \infty\}] = 0$. Let $L_t = \{x : \lambda(x) > t\}$ and note that $D(L_t, x) = 1$ whenever $x \in L_t$. Now apply Lemma 5.9.3 to conclude that there is a sequence of balls $\{B(r_i)\}$ whose union contains L_t such that

$$\sum_{i=1}^{\infty} r_i^{n-1} \le C\|D\chi_{L_t}\|.$$

Since u has compact support, we may assume that $\operatorname{diam} B(r_i) < a$, for some positive number a. Therefore, Theorem 5.4.4 implies

$$\begin{aligned} H_a^{n-1}[\{x : \lambda(x) = \infty\}] &= H_a^{n-1}[\{\cap L_t : t \in R^1\}] \\ &\le C \liminf_{t\to\infty} \|D\chi_{L_t}\|(R^n) = 0. \end{aligned}$$

From this it easily follows that $H^{n-1}[\{x : \lambda(x) = \infty\}] = 0$. A similar proof yields $H^{n-1}[\{x : \mu(x) = -\infty\}] = 0$. Thus, the set $\{x : \mu(x) - \lambda(x)\}$ is well-defined for H^{n-1}-a.e. x and the proof of (ii) will be concluded by showing that $H^{n-1}[\{x : \mu(x) - \lambda(x) = +\infty\}] = 0$. Since E is countably $(n-1)$-rectifiable, it is σ-finite with respect to H^{n-1} restricted to E. Therefore, we may apply Lemma 1.5.1 to obtain

$$\begin{aligned}
\int_E (\mu - \lambda) dH^{n-1} &= \int_0^\infty H^{n-1}[\{x : \lambda(x) < t < \mu(x)\}]dt \\
&\leq \int_0^\infty H^{n-1}(\partial_M A_t)dt \quad \text{(by 5.9.1)} \\
&\leq \int_0^\infty H^{n-1}(\partial^* A_t)dt \quad \text{(by Lemma 5.9.5)} \\
&\leq C \int_0^\infty \|D\chi_{A_t}\| dt \quad \text{(by Theorem 5.6.7)} \\
&\leq C\|Du\|(R^n) \quad \text{(by Theorem 5.4.4)} \\
&< \infty, \quad \text{since spt } u \text{ is compact.}
\end{aligned}$$

We will prove that (iii) holds at each point

$$z \in E - \{\cup(\partial_M A_t - \partial^* A_t) : t \in Q\}.$$

If $t \in Q$ with $\lambda(z) < t < \mu(z)$, then $z \in \partial_M A_t$ and therefore $z \in \partial^* A_t$. Consequently, $n(z, A_t)$ exists. But is must be shown that $n(z, A_t) = n(z, A_s)$ whenever $\lambda(z) < s < \mu(z)$. It follows from the definition of the measure theoretic exterior normal (Definition 5.6.4) that

$$D(A_t, z) = 1/2 = D(A_s, z). \tag{5.9.2}$$

If $s < t$, then $A_s \supset A_t$ and therefore $D(A_s - A_t, z) = 0$, which implies that $n(z, A_t) = n(z, A_s)$.

For the proof of the first assertion of (iv), let $z \in E - I$ and choose $\varepsilon > 0$ such that $\lambda(z) < \mu(z) - \varepsilon < \mu(z)$. Observe that $D(A_{\mu(z)+\varepsilon}, z) = 0$ while

$$\lim_{r \to 0} \frac{|A_{\mu(z)-\varepsilon} \cap H^-(z) \cap B(z,r)|}{|H^-(z) \cap B(z,r)|} = 1,$$

from (5.9.2). Therefore

$$\lim_{r \to 0} \frac{|u^{-1}[\mu(z) - \varepsilon, \mu(z) + \varepsilon] \cap H^-(z) \cap B(z,r)|}{|H^-(z) \cap B(z,r)|} = 1.$$

By an argument similar to that in Remark 5.9.2, this implies that there is a set E^- with the desired properties. The second assertion is proved similarly. □

5.10 The Trace of a BV Function

For a given set $\Omega \subset R^n$ with suitably regular boundary and $u \in BV(\Omega)$, we will show that it is possible to assign values to u at H^{n-1}-almost all points of $\partial\Omega$ even though u, when considered as a member of $L^1(\Omega)$, is defined only as an element of an equivalence class of functions. Recall that two measurable functions are called equivalent if they differ at most on a set of Lebesgue measure zero. The difficulty with defining the trace of a function on the boundary is that $\partial\Omega$ may have zero Lebesgue measure, precisely where the function may be undefined. The theory requires further development in order for this difficulty to be circumvented. The approach we use for this is as follows. For a certain class of domains $\Omega \subset R^n$ (called admissible domains below), if $u \in BV(\Omega)$ is extended to all of R^n by defining $u \equiv 0$ on $R^n - \Omega$, then an easy application of the co-area formula shows that $u \in BV(R^n)$. By means of Theorem 5.9.6 we then are able to define u H^{n-1}-almost everywhere including E, the set of approximate jump discontinuities.

5.10.1. Definition. A bounded domain Ω of finite perimeter is said to be *admissible* if the following two conditions are satisfied:

(i) $H^{n-1}(\partial\Omega - \partial_M\Omega) = 0$,

(ii) There is a constant $M = M(\Omega)$ and for each $x \in \partial\Omega$ there is a ball $B(x,r)$ with

$$H^{n-1}[(\partial_M E) \cap (\partial_M\Omega)] \leq MH^{n-1}[(\partial_M E) \cap \Omega] \tag{5.10.1}$$

whenever $E \subset \overline{\Omega} \cap B(x,r)$ is a measurable set.

5.10.2. Remark. It is not difficult to see that a Lipschitz domain is admissible. For this purpose, we may assume that Ω is of the form

$$\Omega = \{(w,y) : 0 \leq y \leq g(w)\}$$

where g is a non-negative Lipschitz function defined on an open ball $B \subset R^{n-1}$. From Remark 5.8.3 we know that Ω is a set of finite perimeter. Let $E \subset \overline{\Omega}$ be a measurable set and we may as well assume that $H^{n-1}(\Omega \cap \partial_M E) < \infty$ for otherwise (5.10.1) is trivially satisfied. Since $\partial_M E = (\partial_M E \cap \partial\Omega) \cup (\Omega \cap \partial_M E)$ and $H^{n-1}(\partial\Omega) < \infty$, we conclude from Theorem 5.8.5 that E has finite perimeter. Hence, we may apply the Gauss–Green theorem (Theorem 5.8.2) with the constant vector field $V = (0,0,\ldots,1)$ and (5.8.6) to obtain

$$\int_{(\partial^* E)\cap(\partial\Omega)} V \cdot n(x,\Omega)dH^{n-1}(x) + \int_{(\partial^* E)\cap\Omega} V \cdot n(x,E)dH^{n-1}(x) = 0.$$

Therefore, if λ is the Lipschitz constant of g, we have

$$\frac{1}{\sqrt{1+|\lambda|^2}} H^{n-1}[(\partial^* E) \cap (\partial\Omega)] \leq H^{n-1}[(\partial^* E) \cap \Omega],$$

and reference to Lemma 5.9.5 establishes the desired conclusion.

5.10.3. Definition. Whenever u is a real valued Lebesgue measurable function defined on an open set Ω, we denote by u_0 the extension of u to R^n:

$$u_0(x) = \begin{cases} u(x) & x \in \Omega \\ 0 & x \in R^n - \Omega. \end{cases}$$

Observe that u_0 is merely a measurable function and is therefore defined only almost everywhere. Later in the development, we will consider $u \in BV(\Omega)$ where Ω is an admissible domain, and then we will be able to define u_0 everywhere except for an H^{n-1}-null set. If Ω is a smoothly bounded domain and $u \in BV(\Omega)$, it is intuitively clear that $u_0 \in BV(R^n)$ because the variation of u_0 is greater than that of u by only the amount contributed by $H^{n-1}(\partial\Omega)$. The next result makes this precise in the context of admissible domains.

5.10.4. Lemma. *If Ω is an admissible domain and $u \in BV(\Omega)$, then $u_0 \in BV(R^n)$ and $\|u_0\|_{BV(R^n)} \le C\|u\|_{BV(\Omega)}$ where $C = C(\Omega)$.*

Proof. It suffices to show that u_0 is BV in a neighborhood of each point of $\partial\Omega$ because $\partial\Omega$ is compact. If we write u in terms of its positive and negative parts, $u = u^+ - u^-$, it follows from Theorem 5.3.5 that $u \in BV(\Omega)$ if and only if $u^+(\Omega), u^-(\Omega) \in BV(\Omega)$. Therefore, we may as well assume that u is non-negative. For each $x \in \partial\Omega$, let $B(x,r)$ be the ball provided by condition (ii) of Definition 5.10.1. Let φ be a smooth function supported by $B(x,r)$ such that $0 \le \varphi \le 1$ and $\varphi \equiv 1$ on $B(x, r/2)$. Clearly, $\varphi u_0 \in BV(\Omega)$ and Theorem 5.4.4 and Lemma 5.9.5 implies

$$\int_0^\infty H^{n-1}[\Omega \cap \partial_M A_t]dt = \|D(\varphi u_0)\|(\Omega) < \infty \tag{5.10.2}$$

where $A_t = \{x : \varphi u_0(x) > t\}$. Since $|A_t - \overline{\Omega} \cap B(x,r)| = 0$ for $t > 0$, (5.10.1) and (5.10.2) imply

$$\int_0^\infty H^{n-1}(\partial_M A_t)dt \le C\|D(\varphi u_0)\|(\Omega) < \infty.$$

Hence, by Theorem 5.4.4, $\varphi u_0 \in BV(R^n)$ with

$$\|Du_0\|[B(x,r/2)]\| \le \|D(\varphi u_0)\|(R^n) \le C\|D(\varphi u_0)\|(\Omega).$$

However, by (5.1.2),

$$\|D(\varphi u_0)\|(\Omega) = \sup\left\{\int_\Omega u_0\varphi \operatorname{div} V\, dx : V \in C_0^1(\Omega; R^n), |V| \le 1\right\}$$

and

$$\int_{\Omega} u_0 \varphi \operatorname{div} V\, dx = \int_{\Omega} u \operatorname{div}(\varphi V) dx - \int_{\Omega} u\, D\varphi \cdot V\, dx.$$

Therefore

$$\begin{aligned} \|Du_0\|[B(x,r/2)]\| &\le C\|D(\varphi u_0)\|(\Omega) \le C\|Du\|(\Omega) + C(r)\|u\|_{1;\Omega} \\ &\le [C + C(r)]\|u\|_{BV(\Omega)}. \end{aligned}$$

This is sufficient to establish the result because $\partial\Omega$ is compact. □

We now are able to define the trace of u on the boundary of an admissible domain.

5.10.5. Definition. If Ω is an admissible domain and $u \in BV(\Omega)$, the *trace*, u^*, of u on $\partial\Omega$ is defined by

$$u^*(x) = \mu_{u_0}(x) + \lambda_{u_0}(x)$$

where $\mu_{u_0}(x)$ and $\lambda_{u_0}(x)$ are the upper and lower approximate limits of u_0 as discussed in Definition 5.9.1 and Theorem 5.9.6.

5.10.6. Remark. We will analyze some basic properties of the trace in light of Theorem 5.9.6. Let $E = \{x : \lambda_{u_0}(x) < \mu_{u_0}(x)\}$, $A_t = \{x : u_0(x) > t\}$, and select a point $x_0 \in E \cap \partial^*\Omega$ where (iii) of Theorem 5.9.6 applies. Thus, there is a unit vector v such that

$$n(x_0, A_t) = v \quad \text{whenever} \quad \lambda_{u_0}(x_0) < t < \mu_{u_0}(x_0).$$

We would like to conclude that

$$v = \pm n(x_0, \Omega). \tag{5.10.3}$$

For this purpose, note that $0 \in [\lambda_{u_0}(x_0), \mu_{u_0}(x_0)]$ and $A_t \subset \Omega$ for $t > 0$. If $t > 0$ and $v \neq \pm n(x_0, \Omega)$, then simple geometric considerations yield

$$\limsup_{r \to 0} \frac{|(A_t - \Omega) \cap B(x_0, r)|}{|B(x_0, r)|} > 0.$$

This is impossible since $A_t \subset \Omega$. On the other hand, if $t < 0$ and $v \neq \pm n(x_0, \Omega)$, then

$$\limsup_{r \to 0} \frac{|(B_t - \Omega) \cap B(x_0, r)|}{|B(x_0, r)|} > 0,$$

an impossibility since $B_t = \{x : u_0(x) < t\} \subset \Omega$. Hence, (5.10.3) is established.

Also, observe that

$$\text{if } v = n(x_0, \Omega), \text{ then } \lambda_{u_0}(x_0) = 0. \tag{5.10.4}$$

For, if $\lambda_{u_0}(x_0) < 0$ there would exist $t < 0$ such that $\lambda_{u_0}(x_0) < t < \mu_{u_0}(x_0)$. Because $v = n(x_0, A_t)$, it follows that

$$D(A_t \cap \{x : (x - x_0) \cdot v \geq 0\}, x_0) = 0.$$

But $t < 0$ implies $|(R^n - \Omega) - A_t| = 0$. This, along with the fact that $v = n(x_0, \Omega)$ yields

$$D(A_t \cap \{x : (x - x_0) \cdot v \geq 0\}, x_0) \geq 1/2,$$

a contradiction. Therefore, $\lambda_{u_0}(x_0) < 0$.

On the other hand, if $\lambda_{u_0}(x_0) > 0$, there would exist $t > 0$ such that $D(A_t, x_0) = 1$. This would imply that

$$D(A_t \cap \{x : (x - x_0) \cdot v \geq 0\}, x_0) = 1/2$$

which is impossible since $|A_t - \Omega| = 0$ and

$$D(\Omega \cap \{x : (x - x_0) \cdot v \geq 0\}, x_0) = 0.$$

Thus, (5.10.4) follows and a similar argument shows that

$$\text{if } v = -n(x_0, \Omega), \text{ then } \mu_{u_0}(x_0) = 0. \tag{5.10.5}$$

Later, in Section 5.12, after certain Poincaré-type inequalities have been established for BV functions we will be able to show that if Ω is admissible and $u \in BV(\Omega)$, then

$$\lim_{r \to 0} \fint_{B(x,r) \cap \Omega} |u(y) - u^*(x)|^{n/(n-1)} dy = 0 \tag{5.10.6}$$

for H^{n-1}-almost all $x \in \partial\Omega$.

We conclude this section with a result that ensures the integrability of u^* over $\partial\Omega$.

5.10.7. Theorem. *If Ω is an admissible domain, there is a constant $M = M(\Omega)$ such that*

$$\int_{\partial^* \Omega} |u^*| dH^{n-1} \leq M \|u\|_{BV(\Omega)}$$

whenever $u \in BV(\Omega)$.

Proof. Since by definition, $u^* = \lambda_{u_0} + \mu_{u_0}$, it suffices to establish the inequality for the non-negative function $\mu = \mu_{u_0}$, the case involving λ being treated in a similar manner. As in the proof of Lemma 5.10.4, we need only consider the case when μ is replaced by $\varphi\mu$, where φ is a smooth function with $0 \leq \varphi \leq 1$, $\varphi \equiv 1$ on $B(x, r/2)$, and $\operatorname{spt} u \subset B(x, r)$, where $B(x, r)$ is a ball satisfying the condition (5.10.1).

First, with $A_t = \{x : \varphi\mu(x) > t\}$, observe that $A_t \cap \partial_M \Omega \subset (\partial_M A_t) \cap (\partial_M \Omega)$, for $t > 0$. Indeed, let $x \in A_t \cap \partial_M \Omega$ and suppose $x \notin \partial_M A_t$. Then either $D(A_t, x) = 1$ or $D(A_t, x) = 0$. In the first case $D(\Omega, x) = 1$ since $|A_t - \Omega| = 0$ for $t > 0$. Hence, $x \notin \partial_M \Omega$, a contradiction. In the second case, a contradiction again is reached since the definition of $\varphi\mu(x)$ implies $x \notin A_t$. Therefore, we have

$$H^{n-1}[A_t \cap \partial_M \Omega] \leq H^{n-1}[(\partial_M A_t) \cap (\partial_M \Omega)], \quad t > 0. \tag{5.10.7}$$

Thus, we have

$$\begin{aligned}
\int_{B(x,r/2)\cap\partial_M\Omega} \mu \, dH^{n-1} &\leq \int_{B(x,r)\cap\partial_M\Omega} \varphi\mu \, dH^{n-1} \\
&\leq \int_0^\infty H^{n-1}(A_t \cap \partial_M \Omega) \quad \text{(by Lemma 1.5.1)} \\
&\leq \int_0^\infty H^{n-1}[(\partial_M A_t) \cap (\partial_M \Omega)] dt \quad \text{(by 5.10.7)} \\
&\leq M \int_0^\infty H^{n-1}[(\partial_M A_t) \cap \Omega] dt \quad \text{(by 5.10.1)} \\
&\leq M \|D(\varphi\mu)\|(\Omega) \quad \text{(by Theorem 5.4.4)} \\
&\leq M_1 \|u\|_{BV(\Omega)}.
\end{aligned}$$

Since $\mu = u$ almost everywhere, the last inequality follows as in the proof of Lemma 5.10.4. □

5.11 Poincaré-Type Inequalities for BV Functions

In this section we prove the main inequality (Theorem 5.11.1) from which essentially all Poincaré-type inequalities for BV functions will follow. This result is analogous to Theorem 4.2.1 which was established in the context of Sobolev spaces. In accordance with the previous section, throughout we will adopt the following conventions concerning the point-wise definition of BV functions. If $u \in BV(R^n)$ set

$$u(x) = \frac{1}{2}[\lambda_u(x) + \mu_u(x)] \tag{5.11.1}$$

at any point where the right side is defined. From Theorem 5.9.6(ii), we know that this occurs at H^{n-1}-almost all $x \in R^n$. If $u \in BV(\Omega)$, Ω admissible, then we know by Lemma 5.10.4 that $u_0 \in BV(R^n)$ and therefore u_0 is defined H^{n-1}-a.e. on R^n. Thus, we may define u on $\overline{\Omega}$ in terms of u_0 as follows:

$$u(x) = \begin{cases} u_0(x) & x \in \Omega \\ 2u_0(x) & x \in \partial\Omega. \end{cases} \tag{5.11.2}$$

At first glance, it may appear strange to define $u(x) = 2u_0(x)$ on $\partial\Omega$, but reference to (5.10.4) and (5.10.5) shows that this definition implies the intuitively satisfying fact that at H^{n-1}-a.e. $x \in \partial\Omega$, either $u(x) = \mu_{u_0}(x)$ or $u(x) = \lambda_{u_0}(x)$. Consequently, u is a Borel function defined H^{n-1}-a.e. on $\overline{\Omega}$. Note that we have for H^{n-1}-a.e. x,

$$au(x) + bu(x) = (au + bu)(x)$$

whenever $a, b \in R^1$.

5.11.1. Theorem. *Let Ω be a connected, admissible domain and suppose $u \in BV(\Omega)$. If $T \in [BV(\Omega)]^*$ and $T(\chi_\Omega) = 1$, then*

$$\|u - T(u)\|_{n/(n-1);\Omega} \le C\|T\|\,\|Du\|(\Omega), \tag{5.11.3}$$

where $\|T\|$ denotes the norm of T as an element of $[BV(\Omega)]^*$, and $C = C(\Omega, n)$.

Proof. It suffices to show that

$$\|u - T(u)\|_{1;\Omega} \le C\|T\|\,\|Du\|(\Omega), \tag{5.11.4}$$

for if we set $f = u - T(u)$, then by Sobolev's inequality and (5.11.4),

$$\|f\|_{n/(n-1);\Omega} = \|f_0\|_{n/(n-1);R^n} \le C_1\|f_0\|_{BV(R^n)} \le C\|f\|_{BV(\Omega)}.$$

The last inequality follows from Lemma 5.10.4. Also, note that the Sobolev inequality holds for f_0 because it holds for the regularizers of f_0, whose BV norms converge to $\|f_0\|_{BV(R^n)}$ by Corollary 5.2.4. Thus, in view of the fact that $\|T\| \ge |\Omega|^{-1}$, (5.11.3) follows from (5.11.4).

To prove (5.11.4), it is sufficient to assume $\int_\Omega u(x)dx = 0$ since the inequality is unchanged by adding a constant to u. With this assumption, (5.11.4) will follow if we can show

$$\|u\|_{1;\Omega} \le C\|Du\|(\Omega) \tag{5.11.5}$$

because

$$\begin{aligned} \|u - T(u)\|_{1;\Omega} &\le \|u\|_{1;\Omega} + |\Omega|\,\|T\|\,\|u\|_{BV(\Omega)} \\ &\le [1 + |\Omega|\,\|T\|](\|u\|_{1;\Omega} + \|Du\|(\Omega)) \\ &\le 2|\Omega|\,\|T\|(\|u\|_{1;\Omega} + \|Du\|(\Omega)). \end{aligned}$$

If (5.11.5) were not true for some constant C, there would exist a sequence $u_k \in BV(\Omega)$ such that

$$\int_\Omega u_k(x)dx = 0, \quad \|u_k\|_{1;\Omega} = 1, \text{ and } \|Du_k\|(\Omega) \to 0. \tag{5.11.6}$$

For each u_k, form the extension $u_{k,0}$ by setting $u = 0$ on $R^n - \Omega$. From Lemma 5.10.4, it follows that the sequence $\{\|u_{k,0}\|_{BV(R^n)}\}$ is bounded and therefore an application of Corollary 5.3.4 implies that there exist $u \in BV(R^n)$ and a subsequence of $\{u_{k,0}\}$ (which will still be denoted by the full sequence) such that $u_{k,0} \to u$ in $L^1(\Omega)$. Therefore $\|u\|_{1;\Omega} = 1$ from (5.11.6). But (5.11.6) also show that $\|Du\|(\Omega) = 0$ and therefore $u =$ constant on Ω since Ω is connected. Consequently, $u \equiv 0$ which contradicts $\|u\|_{1;\Omega} = 1$. Thus, (5.11.5) and therefore (5.11.4) is established. □

5.11.2. Corollary. *Let Ω be a connected, admissible domain. Let c and M be numbers such that*

$$|a_0 + a_1 c| \leq M \|a_0 + a_1 u\|_{BV(\Omega)}$$

for all $a_0, a_1 \in R^1$. Then there exists $C = C(\Omega)$ such that

$$\|u - c\|_{n/(n-1);\Omega} \leq CM\|Du\|(R^n). \tag{5.11.7}$$

Proof. Define a linear map T_0 on the subspace of $BV(\Omega)$ generated by χ_Ω and u by $T_0(\chi_\Omega) = 1$ and $T_0(u) = c$. From the hypotheses, the norm of T_0 is bounded by M and therefore an application of the Hahn–Banach theorem provides an extension, T, to $BV(\Omega)$ with the same norm. Now apply Theorem 5.11.1 to obtain the desired result. □

5.12 Inequalities Involving Capacity

We now will investigate the role that capacity plays in Sobolev-type inequalities by considering the implications of Theorem 5.11.1. Recall that the $BV(\Omega)$ is endowed with the norm

$$\|u\|_{BV(\Omega)} = \|u\|_{1;\Omega} + \|Du\|(\Omega).$$

However, for notational convenience, we will henceforth treat $BV(R^n)$ separately and its norm will be given by

$$\|u\|_{BV(R^n)} = \|Du\|(R^n).$$

In Section 2.6, Bessel capacity was introduced, developed, and subsequently applied to the theory of Sobolev spaces. Because of the irreflexivity of L^1, it was necessary to restrict our attention to $p > 1$. The case $p = 1$ is naturally associated with BV functions and the capacity in this case is defined as follows:

$$\gamma(E) = \inf\{\|Dv\|(R^n) : v \in Y(R^n),\ E \subset \operatorname{int}\{v \geq 1\}\},$$

where

$$Y(R^n) = L^{n/(n-1)}(R^n) \cap BV(R^n).$$

Note that $W^{1,1}(R^n) \subset Y(R^n)$ and by regularization (cf. Theorem 5.3.1), that

$$\|u\|_{n/(n-1)} \leq C\|Du\|(R^n) \tag{5.12.1}$$

for $u \in Y(R^n)$. A simple regularization argument also yields that in case E is compact,

$$\gamma(E) = \inf\{\|Dv\|_1 : v \in C_0^\infty(R^n),\ E \subset \operatorname{int}\{v \geq 1\}\}. \tag{5.12.2}$$

5.12.1. Lemma. *If $E \subset R^n$ is a Suslin set, then*

$$\gamma(E) = \sup\{\gamma(K) : K \subset E,\ K \text{ compact}\}.$$

Proof. Referring to Theorem 2.6.8, we see that it is only necessary to show that γ is left continuous on arbitrary sets since right continuity on compact sets follows directly from (5.12.2). Thus, it suffices to prove that if $E_1 \subset E_2 \subset \ldots$ are subsets of R^n, then

$$\gamma\left(\bigcup_{i=1}^{\infty} E_i\right) = \lim_{i\to\infty} \gamma(E_i).$$

For this purpose, suppose

$$\lambda = \lim_{i\to\infty} \gamma(E_i) < \infty, \quad \text{and} \quad \varepsilon > 0.$$

Choose non-negative $v_i \in Y(R^n)$ so that

$$E_i \subset \operatorname{int}\{x : v_i(x) \geq 1\} \quad \text{and} \quad \|Dv_i\|(R^n) \leq \gamma(E_i) + \varepsilon 2^{-i},$$

and let $h_i = \sup\{v_1, v_2, \ldots, v_i\}$. Note that $h_i \in Y(R^n)$ and

$$h_j = \sup\{h_{j-1}, v_j\},\ E_{j-1} \subset \operatorname{int}\{x : \inf\{h_{j-1}, v_j\}(x) \geq 1\}.$$

Therefore, letting $I_j = \inf\{h_{j-1}, v_j\}$, it follows from Theorem 5.4.4 (which remains valid for functions in $Y(R^n)$) and Lemma 5.9.5 that

$$\begin{aligned}\|Dh_j\|(R^n) + \gamma(E_{j-1}) &\leq \|Dh_j\|(R^n) + \|DI_j\|(R^n)\\ &= \int_{-\infty}^{+\infty} H^{n-1}(\partial_M\{h_j > t\})dt\\ &\quad + \int_{-\infty}^{+\infty} H^{n-1}(\partial_M\{I_j > t\})dt.\end{aligned}$$

It is an easy matter to verify that

$$\partial_M\{h_j > t\} \cup \partial_M\{I_j > t\} \subset \partial_M\{h_{j-1} > t\} \cup \partial_M\{v_j > t\}$$

$$\partial_M\{h_j > t\} \cap \partial_M\{I_j > t\} \subset \partial_M\{h_{j-1} > t\} \cap \partial_M\{v_j > t\}.$$

Consequently,

$$\begin{aligned}&H^{n-1}(\partial_M\{h_j > t\}) + H^{n-1}(\partial_M\{I_j > t\})\\ &\quad\le H^{n-1}(\partial_M\{h_{j-1} > t\}) + H^{n-1}(\partial_M\{v_j > t\})\end{aligned}$$

and therefore,

$$\begin{aligned}\|Dh_j\|(R^n) + \gamma(E_{j-1}) &\le \int_{-\infty}^{+\infty} H^{n-1}(\partial_M\{h_{j-1} > t\})dt\\ &\quad + \int_{-\infty}^{+\infty} H^{n-1}(\partial_M\{v_j > t\})dt\\ &= \|Dh_{j-1}\|(R^n) + \|Dv_j\|(R^n)\\ &\le \|Dh_{j-1}\|(R^n) + \gamma(E_j) + \varepsilon 2^{-j}.\end{aligned}$$

It follows by induction that

$$\|Dh_j\|(R^n) \le \gamma(E_j) + \sum_{i=1}^{j} \varepsilon 2^{-i}.$$

Therefore, letting $w = \lim_{j\to\infty} h_j$, (5.12.1) implies

$$\|w\|_{(n+1)/n} = \lim_{j\to\infty} \|h_j\|_{(n+1)/n} \le \limsup_{j\to\infty} C\|Dh_j\|(R^n) \le C(\lambda+\epsilon),$$

whereas the proof of Theorem 5.2.1 implies

$$\|Dw\|(R^n) \le \liminf_{j\to\infty} \|Dh_j\|(R^n) < \infty.$$

Thus, $w \in Y(R^n)$ and

$$\gamma\left(\bigcup_{i=1}^{\infty} E_i\right) \le \|Dw\|(R^n) \le \liminf_{j\to\infty} \|Dh_j\|(R^n) \le \lambda + \varepsilon. \qquad \square$$

In addition to the properties above, we will also need the following.

5.12.2. Lemma. *If $A \subset R^n$ is compact, then*

$$\gamma(A) = \inf\{P(U) : A \subset U, U \text{ open and } |U| < \infty\}. \tag{5.12.3}$$

Proof. Let $\gamma_1(A)$ denote the right side of (5.12.3).

$\gamma(A) \le \gamma_1(A)$: Choose $\eta > 0$ and let $A \subset U$ where U is bounded, open and

$$P(U) < \gamma_1(A) + \eta.$$

Let $\chi = \chi_U$ and $\chi_\varepsilon = \chi_U * \varphi_\varepsilon$, where φ_ε is a regularizer. Then $\chi_\varepsilon \geq 1$ on A for all sufficiently small $\varepsilon > 0$ and

$$\begin{aligned}\gamma(A) &\leq \int_{R^n} |D\chi_\varepsilon| dx \quad \text{(by (5.12.2))}\\ &\leq \|D\chi_U\|(R^n) \quad \text{(by Theorem 5.3.1)}\\ &< \gamma_1(A) + \eta.\end{aligned}$$

This establishes the desired inequality since η is arbitrary.

$\gamma_1(A) \leq \gamma(A)$: If $\eta > 0$, (5.12.2) yields $u \in C_0^\infty(R^n)$ such that $u \geq 1$ on A and

$$\int_{R^n} |Du| dx < \gamma(A) + \eta.$$

By the co-area formula,

$$\begin{aligned}\int_{R^n} |Du| dx &= \int_0^\infty H^{n-1}[u^{-1}(t)]dt\\ &\geq \int_0^1 H^{n-1}[u^{-1}(t)]dt\\ &\geq H^{n-1}[u^{-1}(t_0)]\end{aligned}$$

for some $0 < t_0 < 1$. Since $\partial\{u > t_0\} \subset u^{-1}(t_0)$, with the help of Lemma 5.9.5, it follows that

$$\begin{aligned}\gamma_1(A) \leq P(\{u > t_0\}) \leq H^{n-1}[\partial\{u > t_0\}] &\leq H^{n-1}[u^{-1}(t_0)]\\ &< \gamma(A) + \eta.\end{aligned}$$ □

We now are able to characterize the null sets of γ in terms of H^{n-1}.

5.12.3. Lemma. *If $E \subset R^n$ is a Suslin set, then*

$$\gamma(E) = 0 \quad \textit{if and only if} \quad H^{n-1}(E) = 0.$$

Proof. The sufficiency is immediate from the definition of H^{n-1} and the fact that $\gamma[B(r)] = Cr^{n-1}$. In fact, by a scaling argument involving $x \to rx$, it follows that $\gamma[B(r)] = \gamma[B(1)]r^{n-1}$.

To establish necessity, Lemma 5.12.1 along with the inner regularity of H^{n-1} [F4, Corollary 2.10.48] shows that it is sufficient to prove that if $A \subset R^n$ is compact with $\gamma(A) = 0$, then $H^{n-1}(A) = 0$. For $\varepsilon > 0$, the previous lemma implies the existence of an open set $U \supset A$ such that $P(U) < \varepsilon$. Lemma 5.9.3 provides a sequence of closed balls $\{B(r_i)\}$ such that $\cup_{i=1}^\infty B(r_i) \supset U \supset A$ and

$$\sum_{i=1}^\infty r_i^{n-1} \leq CP(U) < C\varepsilon.$$ □

We proceed with the following result which provides some information concerning the composition of $[BV(R^n)]^*$.

5.12.4. Theorem. *Let μ be a positive Radon measure on R^n. The following four statements are equivalent.*

(i) *$H^{n-1}(A) = 0$ implies that $\mu(A) = 0$ for all Borel sets $A \subset R^n$ and that there is a constant M such that $|\int u d\mu| \le M\|u\|_{BV(R^n)}$ for all $u \in BV(R^n)$.*

(ii) *There is a constant M such that $\mu(A) \le MP(A)$ for all Borel sets $A \subset R^n$ with $|A| < \infty$.*

(iii) *There is a constant M_1 such that $\mu(A) \le M_1\gamma(A)$ for all Borel sets $A \subset R^n$.*

(iv) *There is a constant M_2 such that $\mu[B(x,r)] \le M_2 r^{n-1}$ whenever $x \in R^n$ and $r \in R^1$.*

The ratios of the smallest constants M, M_1, and M_2, have upper bounds depending only on n.

Proof. By taking $u = \chi_A$, (ii) clearly follows from (i) since

$$\|u\|_{BV(R^n)} = \|Du\|(R^n) = P(A).$$

For the implication (ii) $\Rightarrow$ (iii), consider a compact set K and observe that from the regularity of μ and (ii),

$$\begin{aligned}\mu(K) &= \inf\{\mu(U) : K \subset U, U \text{ open and } |U| < \infty\} \\ &\le M \inf\{P(U) : K \subset U, U \text{ open and } |U| < \infty\}.\end{aligned}$$

Lemma 5.12.2 yields $\mu(K) \le M\gamma(K)$. The inner regularity of μ and Lemma (5.12.1) give (iii).

Since $\gamma[B(r)] = Cr^{n-1}$, (iii) implies (iv).

Clearly, (iv) implies that μ vanishes on sets of H^{n-1}-measure zero. Consequently, if $u \in BV(R^n)$, our convention (5.11.1) implies that u is defined μ-a.e. If u is also non-negative, we obtain from the co-area formula

$$\|Du\|(R^n) = \|u\|_{BV(R^n)} = \int_0^\infty P(A_t)dt \qquad (5.12.4)$$

where $A_t = \{x : u(x) > t\}$. In particular, this implies that for a.e. t, A_t has finite perimeter. For all such t, define

$$F_t = A_t \cap \{x : \overline{D}(A_t, x) \ge 1/2\}.$$

For $x \in A_t$, the upper approximate limit of u at x is greater than t (see 5.11.1), and therefore

$$A_t - F_t \subset \{x : 0 < \overline{D}(A_t, x) < 1/2\}. \qquad (5.12.5)$$

Therefore, $A_t - F_t \subset \partial_M A_t$. In fact, $A_t - F_t \subset \partial_M A_t - \partial^* A_t$ because $x \in \partial^* A_t$ implies that $D(A_t, x) = 1/2$. Therefore, $H^{n-1}(A_t - F_t) = 0$ by Lemma 5.9.5. Thus, we may apply Lemma 5.9.3 to F_t and obtain a sequence of balls $\{B(r_i)\}$ such that $F_t \subset \cup_{i=1}^{\infty} B(r_i)$ and

$$\sum_{i=1}^{\infty} r_i^{n-1} \leq CP(F_t).$$

Therefore (iv) yields

$$\mu(F_t) \leq CM_2 P(F_t). \tag{5.12.6}$$

Now, $P(A_t) = P(F_t)$ and since μ vanishes on sets of H^{n-1}-measure zero, we have $\mu(A_t) = \mu(F_t)$. Thus, Lemma 1.5.1, (5.12.6), and (5.12.4) imply

$$\int u \, d\mu = \int_0^{\infty} \mu(A_t) dt \leq CM_2 \|u\|_{BV(R^n)}.$$

If u is not non-negative, apply the above arguments to $|u|$ to obtain (i). □

5.12.5. Remark. A positive Radon measure μ satisfying one of the conditions of Theorem 5.12.4 can be identified with an element of $[BV(R^n)]^*$ and M can be chosen as its norm.

Suppose Ω is an admissible domain and μ a positive measure such that $\operatorname{spt} \mu \subset \overline{\Omega}$. In addition, if $\mu \in [BV(\Omega)]^*$, then there exists a constant $C = C(\Omega, \mu)$ such that

$$\left| \int u \, d\mu \right| \leq C\|u\|_{BV(\Omega)} \leq C\|u\|_{BV(R^n)}$$

whenever $u \in BV(R^n)$. Thus, $\mu \in [BV(R^n)]^*$ and Theorem 5.12.4 applies. On the other hand, if $\operatorname{spt} \mu \subset \overline{\Omega}$ and one of the conditions of Theorem 5.12.4 holds, then $\mu \in [BV(\Omega)]^*$ because of Lemma 5.10.4. Thus, for measures μ supported by $\overline{\Omega}$, $\mu \in [BV(\Omega)]^*$ if and only if one of the conditions of Theorem 5.12.4 holds and in this case there is a constant $C = C(\Omega)$ such that

$$C^{-1}\|\mu\|_{[BV(R^n)]^*} \leq \|\mu\|_{[BV(\Omega)]^*} \leq C\|\mu\|_{[BV(R^n)]^*}. \tag{5.12.7}$$

For the applications that follow, it will be necessary to have yet another formulation for the capacity γ.

5.12.6. Lemma. *If $A \subset R^n$ is a Suslin set, then*

$$\gamma(A) = \sup\{\mu(A)\} \tag{5.12.8}$$

where the supremum is taken over the set of positive Radon measures $\mu \in [BV(R^n)]^*$ with $\|\mu\|_{[BV(R^n)]^*} \leq 1$.

Proof. Because of the inner regularity of γ (Lemma 5.12.1) it suffices to consider the case when A is compact. Referring to the Minimax theorem stated in Section 2.6, let X denote the set of all positive Radon measures μ with $\operatorname{spt}\mu \subset A$ and $\mu(R^n) = 1$. Let Y be the set of all non-negative functions $f \in C_0^\infty(R^n)$ such that $\|Df\|_1 \leq 1$. From the Minimax Theorem, we have

$$\sup_{f\in Y}\inf_{\mu\in X}\int f\,d\mu = \inf_{\mu\in X}\sup_{f\in Y}\int f\,d\mu.$$

It is easily seen that the left side is equal to the reciprocal of $\gamma(A)$ whereas the right side is the reciprocal of the right side of (5.12.8). □

One of the most frequently used Poincaré-type inequalities is

$$\|u\|_{p;\Omega} \leq C(\Omega)\|Du\|_{p;\Omega}$$

where $p > 1$, $u \in W^{1,p}(\Omega)$, and $\int_\Omega u(x)dx = 0$. Inequalities of this type were treated from a general perspective in Section 4.2. In the next theorem, we will obtain a Poincaré-type inequality for BV functions normalized so that their integral with respect to a measure in $[BV(\Omega)]^*$ is zero. That is, the measures under consideration are those with the property that $\mu[B(x,r)] \leq Mr^{n-1}$ for all balls $B(x,r)$. For example, this includes Lebesgue measure restricted to a bounded domain or $(n-1)$-Hausdorff measure restricted to a compact smooth hypersurface in R^n.

5.12.7. Theorem. *Let Ω be a connected admissible domain in R^n and let μ be a non-trivial positive Radon measure such that* $\operatorname{spt}\mu \subset \overline{\Omega}$ *and for some constant $M > 0$ that*

$$\mu[B(x,r)] \leq Mr^{n-1}$$

for all balls $B(x,r)$ in R^n. Then, there exists a constant $C = C(\Omega)$ such that for each $u \in BV(\Omega)$,

$$\|u - \overline{u}(\mu)\|_{n/(n-1);\Omega} \leq C\frac{M}{\mu(\overline{\Omega})}\|Du\|(\Omega)$$

where $\overline{u}(\mu) = \frac{1}{\mu(\overline{\Omega})}\int u(x)\,d\mu(x)$.

Proof. Theorem 5.12.4 states that $\mu \in [BV(R^n)]^*$ and because Ω is admissible, (5.12.7) shows that μ may be regarded as an element of $[BV(\Omega)]^*$. Therefore, Theorem 5.11.1 is applicable and the result follows immediately. □

This leads directly to the Poincaré inequality for BV functions.

5.12.8. Corollary. *Let Ω be a connected, admissible domain and let $A \subset \overline{\Omega}$ be a Suslin set with $H^{n-1}(A) > 0$. Then for $u \in BV(\Omega)$ with the property* *that* $u(x) = 0$ *for* H^{n-1}*-almost all* $x \in A$, *there exists a constant* $C = C(\Omega)$ *such that*

$$\|u\|_{n/(n-1);\Omega} \leq \frac{C}{\gamma(A)} \|Du\|(\Omega).$$

Proof. From Lemma 5.12.6 we find that A supports a positive Radon measure $\mu \in [BV(\Omega)]^*$ such that $\mu(A) \geq 2^{-1}\gamma(A)$ and $\|\mu\|_{[BV(R^n)]^*} \leq 1$. Thus, for any $u \in BV(\Omega)$ with the property in the statement of the corollary, $\int u\, d\mu = 0$. Our conclusion now follows from the preceding theorem. □

We now consider inequalities involving the median of a function rather than the mean. The definition of the median is given below.

5.12.9. Definition. If $u \in BV(\Omega)$ and μ a positive Radon measure in $[BV(\Omega)]^*$, we define $\operatorname{med}(u, \mu)$ as the set of real numbers t such that

$$\mu[\overline{\Omega} \cap \{x : u(x) > t\}] \leq \frac{1}{2}\mu(\overline{\Omega})$$

$$\mu[\overline{\Omega} \cap \{x : u(x) < t\}] \leq \frac{1}{2}\mu(\overline{\Omega}).$$

It is easily seen that $\operatorname{med}(u, \mu)$ is a non-empty compact interval and that if a_0 and a_1 are constants, then

$$\operatorname{med}(a_0 + a_1 u, \mu) = a_0 + a_1 \operatorname{med}(u, \mu). \tag{5.12.9}$$

If $c \in \operatorname{med}(u, \mu)$, then $\mu(A_c) \geq \frac{1}{2}\mu(\overline{\Omega})$ where $A_c = \overline{\Omega} \cap \{x : u(x) \geq c\}]$. Consequently,

$$\frac{c}{2}\mu(\overline{\Omega}) \leq c\mu(A_c) \leq \int_{\overline{\Omega}} |u(x)| d\mu(x).$$

Similarly, if $c \leq 0$, then $\mu(B_c) \geq \frac{1}{2}\mu(\overline{\Omega})$ where $B_c = \overline{\Omega} \cap \{x : u(x) \leq c\}]$ and

$$-\frac{c}{2}\mu(\overline{\Omega}) \leq -c\mu(B_c) \leq \int_{B_c} -u(x) d\mu(x) \leq \int_{\overline{\Omega}} |u(x)| d\mu(x).$$

Therefore,

$$|c| \leq \frac{2}{\mu(\overline{\Omega})} \int_{\overline{\Omega}} |u| d\mu, \tag{5.12.10}$$

and (5.12.9) thus implies

$$|a_0 + a_1 c| \leq \frac{2}{\mu(\overline{\Omega})} \|\mu\|_{[BV(\Omega)]^*} \|a_0 + a_1 u\|_{BV(\Omega)}.$$

The following is now a direct consequence of Corollary 5.11.2.

5.12.10. Theorem. *Let* Ω *be a connected admissible domain in* R^n *and let* μ *be a positive Radon measure such that* $\operatorname{spt} \mu \subset \overline{\Omega}$ *and for some constant* $M > 0$ *that*

$$\mu[B(x, r)] \leq M r^{n-1}$$

for all balls $B(x, r)$ *in* R^n. *Then there exists a constant* $C = C(\Omega)$ *such that for* $u \in BV(\Omega)$ *and* $c \in \operatorname{med}(u, \mu)$,

$$\|u - c\|_{n/(n-1);\Omega} \leq C\frac{M}{\mu(\overline{\Omega})}\|Du\|(\Omega).$$

5.12.11. Corollary. *Let* Ω *be a connected, admissible domain and let* A *and* B *be disjoint Suslin subsets of* $\overline{\Omega}$ *of positive* H^{n-1}*-measure. Then there exists a constant* $C = C(\Omega)$ *such that for each* $u \in BV(\Omega)$ *with* $u > 0$ H^{n-1}-a.e. *on* A *and* $u < 0$ H^{n-1}-a.e. *on* B,

$$\|u\|_{n/(n-1);\Omega} \leq C[\gamma(A)^{-1} + \gamma(B)^{-1}]\|Du\|(\Omega).$$

Proof. Lemma 5.12.6 yields measures μ and ν supported by A and B respectively such that

$$\gamma(A) \geq \mu(A) \geq \frac{1}{2}\gamma(A), \quad \|\mu\|_{[BV(R^n)]^*} \leq 1,$$

$$\gamma(B) \geq \nu(B) \geq \frac{1}{2}\gamma(B), \quad \|\nu\|_{[BV(R^n)]^*} \leq 1.$$

Define

$$\lambda = \mu(A)\nu + \nu(B)\mu$$

and observe that $0 \in \operatorname{med}(u, \lambda)$ for $u \in BV(\Omega)$. Since $\lambda(\overline{\Omega}) = 2\mu(A)\nu(B)$ and $\|\lambda\|_{[BV(R^n)]^*} \leq \mu(A) + \nu(B)$, the conclusion follows from Theorems 5.12.4 and 5.12.10. □

5.13 Generalizations to the Case $p > 1$

Since a BV function u is defined H^{n-1}-almost everywhere by means of (5.11.1), an obvious question arises whether the results of the previous section can be extended by replacing $\|u\|_{n/(n-1);\Omega}$ that appears on the left side of the inequalities by the appropriate L^p-norm of u defined relative to a measure that is absolutely continuous with respect to H^{n-1}. We will show that this can be accomplished by establishing Poincaré-type inequalities that involve $\|u\|_{n/(n-1),\lambda}$ where λ is a positive measure that satisfies one of the conditions of Theorem 5.12.2.

5.13.1. Theorem. *Let* λ *be a positive Radon measure on* R^n. *The following two conditions are equivalent:*

(i) $H^{n-1}(A) = 0$ *implies* $\lambda(A) = 0$ *for all Borel sets* A *and for* $1 \leq p \leq n/(n-1)$, *there exists a constant* $C = C(p, n, \lambda)$ *such that*

$$\|u\|_{p,\lambda} \leq C\|Du\|[R^n]$$

for all $u \in BV(R^n)$.

(ii) *There is a constant* C_1 *such that*

$$\lambda[B(x,r)] \leq C_1^p r^{p(n-1)}$$

for all balls $B(x,r)$.

The ratios of the smallest constants C *and* C_1 *have upper bounds depending only on* n.

Proof. The case $p = 1$ is covered by Theorem 5.12.2, so we may assume that $p > 1$. Suppose (ii) holds and let $f \in L^{p'}(\lambda)$, $f \geq 0$. Then, by Holder's inequality,

$$\begin{aligned}\int_{B(x,r)} f(x)d\lambda &\leq \left(\int_{B(x,r)} f^{p'} d\lambda\right)^{1/p'} \lambda[B(x,r)]^{1/p} \\ &\leq C_1 \|f\|_{p',\lambda} r^{n-1}.\end{aligned}$$

Thus, the measure $f\lambda$ defined by

$$f\lambda(E) = \int_E f(x)d\lambda(x)$$

satisfies condition (ii) with $p = 1$. Therefore, by Theorem 5.12.2,

$$\left|\int uf\, d\lambda\right| \leq MC_1 \|f\|_{p',\lambda} \|u\|_{BV(R^n)}$$

for all $u \in BV(R^n)$. From the definition of Hausdorff measure, it is clear that $H^{n-1}(A) = 0$ implies $\lambda(A) = 0$. Thus, (i) is established.

Now assume that (i) holds. For each Borel set $A \subset R^n$ and each $\varepsilon > 0$, reference to Lemma 5.12.1 supplies an open set $U \supset A$ such that $P(U) \leq \gamma(A) + \varepsilon$. Therefore, from (i) with $u = \chi_A$,

$$\begin{aligned}\lambda(A)^{1/p} \leq \lambda(U)^{1/p} &\leq C\|D\chi_U\|(R^n) \\ &= CP(U) \\ &\leq C\gamma(A) + C\varepsilon.\end{aligned}$$

In view of the fact that $\gamma[B(x,r)] = Cr^{n-1}$, (ii) is established. □

With the help of the preceding theorem, results analogous to those of the Section 5.12 are easily obtained. For example, we have the following.

5.13.2. Theorem. *Let* Ω *be a connected, admissible domain in* R^n. *Let* μ *and* λ *be positive Radon measures supported by* $\overline{\Omega}$ *such that*

$$\mu[B(x,r)] \leq C_1 r^{n-1},$$

$$\lambda[B(x,r)] \le C_2^p r^{p(n-1)}, \quad 1 \le p \le n/(n-1),$$

for all balls $B(x,r)$. *Then, there exists a constant* $M = M(\Omega)$ *such that*

$$\|u - \overline{u}(\mu)\|_{p,\lambda} \le M \frac{C_1 C_2}{\mu(\overline{\Omega})} \|Du\|(\Omega)$$

for all $u \in BV(\Omega)$.

Proof. From Theorem 5.13.1 and Lemma 5.10.3 we have

$$\begin{aligned} \|u\|_{p,\lambda;\Omega} &= \|u_0\|_{p,\lambda} \\ &\le C\|u_0\|_{BV(R^n)} \\ &\le C\|u\|_{BV(\Omega)}. \end{aligned}$$

Applying this inequality to $u - \overline{u}(\mu)$, we obtain

$$\begin{aligned} \|u - \overline{u}(\mu)\|_{p,\lambda;\Omega} &\le C\|u - \overline{u}(\mu)\|_{BV(\Omega)} \\ &\le C\|u - \overline{u}(\mu)\|_{1;\Omega} + C\|Du\|(\Omega) \\ &\le C\|Du\|(\Omega). \end{aligned}$$

□

Other results analogous to those in the preceding section are established in a similar way and are stated without proof.

5.13.3. Theorem. *If* $c \in \operatorname{med}(u,\mu)$, *then*

$$\|u - c\|_{p,\lambda} \le M \frac{C_1 C_2}{\mu(\overline{\Omega})} \|Du\|(\Omega).$$

Also,

If $u(x) = 0$ *on* A *where* A *is a Suslin set of positive* H^{n-1}*-measure, then*

$$\|u\|_{p,\lambda} \le M \frac{C_2}{\gamma(A)} \|Du\|(\Omega).$$

5.14 The Trace Defined in Terms of Integral Averages

For $u \in BV(R^n)$, recall the following facts established in Theorem 5.9.6:

(i) $E = \{x : \lambda(x) < \mu(x)\}$ is countably $(n-1)$-rectifiable,

(ii) $-\infty < \lambda(x) \le \mu(x) < \infty$ for H^{n-1}-almost all $x \in R^n$,

(iii) For H^{n-1}-almost every $z \in E$, there exists a unit vector v such that $n(z, A_s) = v$ whenever $\lambda(z) < s < \mu(z)$.

Although our convention (5.11.1) of setting $u(x) = \frac{1}{2}[\lambda_u(x) + \mu_u(x)]$ allows a meaningful pointwise definition of u at H^{n-1}-almost all points, the simple example of u as the characteristic function of a ball shows that it is not possible to define u in terms of its Lebesgue points H^{n-1}-almost everywhere. This is merely one of the ways that the BV theory differs from the Sobolev theory developed in Chapters 3 and 4. However, in this section we will show that a slightly weaker result holds:

$$\lim_{r\to 0} \fint_{B(x,r)} u(y)dy = \frac{1}{2}[\lambda_u(x) + \mu_u(x)]$$

for H^{n-1}-almost all $x \in \Omega$. If Ω is admissible, then a similar result will be shown to hold for the trace u^*, the only difference being that the ball $B(x,r)$ in the above expression will be replaced by $B(x,r)\cap\Omega$. Briefly stated then, a BV function can be defined pointwise H^{n-1}-almost everywhere on $\overline{\Omega}$ as the limit of its integral averages.

5.14.1. Remark. If $u \in BV(R^n)$ is bounded, then it is easily seen that (with convention (5.11.1) in force),

$$\lim_{r\to 0} \fint_{B(x_0,r)} |u(x) - u(x_0)|^\sigma dx = 0 \tag{5.14.1}$$

for $x_0 \notin E$ and that for all $z \in E$ for which (iii) above holds,

$$\lim_{r\to 0} \fint_{B^+(z,r)} |u(x) - \lambda_u(z)|^\sigma dz = 0, \tag{5.14.2}$$

$$\lim_{r\to 0} \fint_{B^-(z,r)} |u(x) - \mu_u(z)|^\sigma dz = 0, \tag{5.14.3}$$

where

$$B^+(z,r) = B(z,r)\cap\{y : (y-z)\cdot v > 0\},$$
$$B^-(z,r) = B(z,r)\cap\{y : (y-z)\cdot v < 0\},$$

and

$$\sigma = \frac{n}{n-1}.$$

To verify (5.14.1), use Remark 5.9.2 to conclude that there is a Lebesgue measurable set A such that $D(A, x_0) = 1$ and

$$\lim_{\substack{x\to x_0\\ x\in A}} = u(x_0).$$

Then

$$\lim_{r\to 0} r^{-n} \int_{B(x_0,r)} |u(x) - u(x_0)|^\sigma dx = \lim_{r\to 0} r^{-n} \int_{B(x_0,r)\cap A} |u(x) - u(x_0)|^\sigma dx$$
$$+ \lim_{r\to 0} r^{-n} \int_{B(x_0,r)\cap \tilde{A}} |u(x) - u(x_0)|^\sigma dx.$$

The first term tends to 0 by the continuity of $u \mid A$. The second term also tends to 0 because u is bounded and $D(\tilde{A}, x_0) = 0$, where $\tilde{A} = R^n - A$.

Now consider (5.14.3), the proof of (5.14.2) being similar. From the definition of $\mu(z)$, we have that $D(A_t, z) = 0$ for each $t > \mu(z)$. From (iii) above, $D(\tilde{A}_s \cap \{y : (y - z) \cdot v < 0\}) = 0$ for $\lambda(z) < s < \mu(z)$. Thus, if $\varepsilon > 0$, $t - s < \varepsilon$, and $s < \mu(z) < t$, we have

$$\limsup_{r \to 0} r^{-n} \int_{B^-(z,r)} |u(x) - \mu_u(z)|^\sigma dx \le \limsup_{r \to 0} r^{-n} \int_{B^-(z,r) \cap (A_s - A_t)} \varepsilon^\sigma$$

$$+ \sup_{x \in R^n} |u(x) - \mu_u(z)|^\sigma \cdot \limsup_{r \to 0} \frac{|B^-(z,r) \cap (A_t \cup \tilde{A}_s)|}{r^n}.$$

The last term is zero since u is bounded and therefore the conclusion follows since ε is arbitrary.

Our task now is to prove (5.14.1), (5.14.2), and (5.14.3) without the assumption that u is bounded. For this we need the following lemma.

5.14.2. Lemma. *If $u \in BV(R^n)$ and $\lambda_u(x_0) = \mu_u(x_0)$, then there is a constant $C = C(n)$ such that*

$$\limsup_{r \to 0} \left(\fint_{B(x_0,r)} |u(x) - u(x_0)|^\sigma dx \right)^{1/\sigma} \le C \limsup_{r \to 0} r^{1-n} \|Du\|[B(x_0, r)].$$

Proof. For each $r > 0$, consider the median of u in $B(x_0, r)$,

$$t_r = \inf\{t : |B(x_0, r) \cap \{x : u(x) > t\}| \le \frac{1}{2}|B(x_0, r)|\},$$

and apply Theorem 5.12.10 and Minkowski's inequality to conclude

$$\limsup_{r \to 0} \left(\fint_{B(x_0,r)} |u(x) - u(x_0)|^\sigma dx \right)^{1/\sigma}$$

$$\le C \limsup_{r \to 0} r^{1-n} \|Du\|[B(x_0, r)] + C'|t_r - u(x_0)|.$$

Moreover, $t_r \to u(x_0)$ since $\lambda_u(x_0) = \mu_u(x_0)$. □

5.14.3. Theorem. *If $u \in BV(R^n)$, then* (5.14.1) *holds for H^{n-1}-almost all $x_0 \in R^n - E$, whereas* (5.14.2) *and* (5.14.3) *hold for H^{n-1}-almost all $x_0 \in E$.*

Proof. For each positive integer i, let

$$u_i(x) = \begin{cases} i & \text{if } u(x) > i \\ u(x) & \text{if } |u(x)| \le i \\ -i & \text{if } u(x) < -i, \end{cases}$$

$$W_i = \{x : -i \le \lambda_u(x) \le \mu_u(x) \le i\},$$

and observe that

$$H^{n-1}\left(R^n - \bigcup_{i=1}^{\infty} W_i\right) = 0 \tag{5.14.4}$$

by Theorem 5.9.6(ii).

For each $\varepsilon > 0$, let

$$Z_i = \left\{x_0 : \limsup_{r\to 0} \frac{\|D(u-u_i)\|[B(x_0,r)]}{\alpha(n-1)r^{n-1}} \le \varepsilon\right\} \tag{5.14.5}$$

and refer to Lemma 3.2.1 to obtain

$$\varepsilon H^{n-1}(U - Z_i) \le \|D(u-u_i)\|(U) \tag{5.14.6}$$

whenever $U \subset R^n$ is open. By Theorem 5.4.4,

$$\begin{aligned} \|Du(u-u_i)\|(U) &\le \int_0^{\infty} P[\{x : (u-u_i)(x) \ge s\}, U]ds \\ &\quad + \int_{\infty}^{0} P[\{x : (u-u_i)(x) < s\}, U]ds \\ &= \int_0^{\infty} P[\{x : u(x) \ge i + s\}, U]ds \\ &\quad + \int_{\infty}^{0} P[\{x : u(x) < -i + s\}, U]ds \\ &= \int_{|s|>i} P[\{x : u(x) > s\}, U]ds. \end{aligned} \tag{5.14.7}$$

If U is bounded, then

$$\int_{-\infty}^{\infty} P[\{x : u(x) > s\}, U]ds = \|Du\|(U) < \infty,$$

and therefore the last integral in (5.14.7) tends to zero as $i \to \infty$. Hence, we obtain from (5.14.6) that $H^{n-1}(U - Z_i) \to 0$ as $i \to \infty$. Then,

$$H^{n-1}\left[\bigcup_{j=1}^{\infty}\bigcap_{i=j}^{\infty}(R^n - Z_i)\right] = 0,$$

and

$$H^{n-1}\left[R^n - \bigcap_{j=1}^{\infty}\bigcup_{i=j}^{\infty} Z_i\right] = 0. \tag{5.14.8}$$

To prove (5.14.1), it suffices by (5.14.4) to consider $x_0 \in (\cup_{i=1}^{\infty} W_i) - E$. Because $u - u_i$ has 0 as an approximate limit at each such point x_0 of W_i, it follows from Lemma 5.14.2 that

$$\limsup_{r \to 0} \left(\fint_{B(x_0,r)} |u(x) - u_i(x)|^{\sigma} dx \right)^{1/\sigma}$$
$$\leq C \limsup_{r \to 0} r^{1-n} \|D(u - u_i)\|[B(x_0, r)]. \tag{5.14.9}$$

From (5.14.8) we may as well assume that $x_0 \in \cap_{j=1}^{\infty} \cup_{i=j}^{\infty} Z_i$. For i sufficiently large, reference to Remark 5.14.1 yields

$$\lim_{r \to 0} \left(\fint_{B(x_0,r)} |u_i(x) - u(x)|^{\sigma} dx \right)^{1/\sigma} = 0.$$

From (5.14.9), there exists i sufficiently large such that

$$\lim_{r \to 0} \left(\fint_{B(x_0,r)} |u(x) - u_i(x)|^{\sigma} dx \right)^{1/\sigma} \leq C\varepsilon.$$

Consequently, (5.14.1) follows for unbounded $u \in BV(R^n)$ by Minkowski's inequality and the fact that ε is arbitrary.

Essentially the same argument establishes (5.14.2) and (5.14.3). □

As an immediate consequence of the above result, we obtain the following.

5.14.4. Theorem. *If $\Omega \subset R^n$ is open and $u \in BV(\Omega)$, then*

$$\lim_{r \to 0} \fint_{B(x_0,r)} |u(x) - u(x_0)|^{n/(n-1)} dx = 0$$

for H^{n-1}-almost all $x_0 \in \Omega - E$, and

$$\lim_{r \to 0} \fint_{B(x_0,r)} u(x) dx = u(x_0)$$

for H^{n-1}-almost all $x_0 \in \Omega$. If Ω is admissible, then the trace u^* satisfies

$$\lim_{r \to \infty} \int_{B(x_0,r) \cap \Omega} |u(x) - u^*(x_0)|^{n/(n-1)} dx = 0$$

for H^{n-1}-almost every $x_0 \in \partial\Omega$.

Proof. The statements concerning the integral averages of u follow immediately from (5.14.2) and (5.14.3). Also, referring to Remark 5.10.6 leads to the last part of the theorem. □

Exercises

5.1. Suppose u is a function of a single variable and that $u \in BV(a,b)$, $a \leq b$. Prove that $\|Du\|(a,b) = \text{ess } V_a^b(u)$. Hint: Use regularizers of u.

5.2. Suppose $\Omega \subset R^n$ is an open set and $u \in BV(\Omega)$. Prove that there exists a sequence of polyhedral regions $\{P_k\}$ invading Ω and piecewise linear maps $L_k : P_k \to R^1$ such that

$$\lim_{k\to\infty} \int_{P_k} |L_k - u| = 0,$$

$$\lim_{k\to\infty} \int_{P_k} |DL_k| = \|Du\|(\Omega).$$

Hint: By Theorem 5.3.3, it suffices to consider the case when $u \in C^\infty(\Omega) \cap BV(\Omega)$. Let $P_1 \subset P_2 \subset \ldots \subset \Omega$ be polyhedral regions such that $|\Omega - P_k| \to 0$. Choose each P_k as a simplicial complex, so that it is composed of n-dimensional simplices. Since $\|Du\|_{1;\Omega} < \infty$, we may choose k so as to make $\int_{\Omega - P_k} |Du|$ arbitrarily small. Moreover, we may assume without loss of generality that each simplex, σ, in P_k has its diameter small enough to ensure that the oscillation of $|u|$ and $|Du|$ over σ is small, uniformly with respect to all σ. Suppose σ is spanned by the unit vectors $v_1, v_2, \ldots, v_n$. Define the linear map L_k on σ so that it agrees with u at the $n+1$ vertices of σ. Clearly, $\|L_k\|_{1;P_k} \to \|u\|_{1;\Omega}$. To see that the L^1-norm of the gradients also converge, note that

$$\int_\sigma |Du| = |Du(p)|\,|\sigma|$$

for some $p \in \sigma$. On each of the edges of σ determined by the vectors v_i, $i = 1, 2, \ldots, n$, there is a point p_i such that $[Du(p_i) - DL_k(p_i)] \cdot v_i = 0$ (by the Mean Value theorem). But $DL_k(p_i) = DL_k(p)$ since L_k is linear and $|Du(p_i)|$ is close to $|Du(p)|$ because of the small oscillation of $|Du|$ over σ. Therefore, $|Du(p) - DL_k(p)|$ is small and consequently, $\int_\sigma |DL_k|$ is close to $\int_\sigma |Du|$, uniformly with respect to all σ.

5.3. If $E \subset R^n$ is a measurable set, let us say that E is open in the density topology if $D(E,x) = 1$ for each $x \in E$. Prove that the sets open in this sense actually produce a topology. In order to show this, it must be established that an uncountable union of density open sets is open; in particular, it must be shown that it is measurable. Hint: Use the Vitali covering theorem. If we agree to call the exterior of E all points x such that $D(E,x) = 0$, we see that $\partial_M E$ is the boundary of E in the density topology.

5.4. In the setting of metric spaces, there is an inequality, called the *Eilenberg Inequality,* that vaguely has the form of the co-area formula. It states that if X is a separable metric space and $u : X \to R^n$ is a Lipschitz map, then for any $E \subset X$ and all integers $0 \le k \le n$

$$\int_{R^n}^{*} H^{n-k}(E \cap u^{-1}(y))dH^k(y) \le 2^n \frac{\alpha(n-k)\alpha(n)}{\alpha(n)} L^k H^n(E).$$

Here, L is the Lipschitz constant of u and $\int^*$ denotes the upper Lebesgue integral. Also, H^m denotes m-dimensional Hausdorff measure which has a meaningful definition in a metric space.

STEP 1. By the definition of H^n, for every integer $s > 0$ there exists a countable covering of E in X by sets $E_{i,s}$, $i = 1, 2, \ldots$, such that $\operatorname{diam} E_{i,s} < 1/s$ and

$$H^n(E) \ge \frac{\alpha(n)}{2^n} \sum_{i=1}^{\infty} (\operatorname{diam} E_{i,s})^n - \frac{1}{s}.$$

Hence,

$$H^{n-k}(E \cap u^{-1}(y)) \le \frac{\alpha(n-k)}{2^{n-k}} \liminf_{s\to\infty} \sum_{i=1}^{\infty} [\operatorname{diam}(E_{i,s} \cap u^{-1}(y)]^{n-k}.$$

STEP 2. Consider the characteristic function of the set $\overline{u(A)}$, $\chi(\overline{u(A)})$, where $\overline{u(A)}$ denotes the closure of $u(A)$. Then

$$[\operatorname{diam}(A \cap u^{-1}(y))]^{n-k} \le (\operatorname{diam} A)^{n-k} \chi(\overline{u(A)}, y).$$

Hence, from Step 1,

$$H^{n-k}(E \cap u^{-1}(y)) \le \frac{\alpha(n-k)}{2^{n-k}} \liminf_{s\to\infty} \sum_{i=1}^{\infty} (\operatorname{diam} E_{i,s})^{n-k} \chi(\overline{u(E)}, y).$$

STEP 3. Apply Fatou's lemma (which is valid with the upper integral) to obtain

$$\int_{R^n}^{*} H^{n-k}(E \cap u^{-1}(y))dH^n(y) \le \frac{\alpha(n-k)}{2^{n-k}} \liminf_{s\to\infty} \sum_{i=1}^{\infty} (\operatorname{diam} E_{i,s})^{n-k} \cdot \int_{R^n} \chi(\overline{u(E_{i,s})}, y)dH^n(y).$$

However,

$$\int_{R^n} \chi(\overline{u(E_{i,s})}, y)dH^n(y) = H^n(u(E_{i,s})) \le \alpha(n)[\operatorname{diam} u(E_{i,s})]^n.$$

Now use Step 1 to reach the desired conclusion.

5.5. Let $u \in BV(\Omega)$ where $\Omega \subset R^n$ is an open set. For each real number t, let $A_t = \{x : u(x) \geq t\}$. As usual, let χ_A denote the characteristic function of the set A. For any Borel set E, prove that

$$Du(E) = \int_{-\infty}^{+\infty} D\chi_{A_t}(E)dt.$$

5.6. Prove the following version of the Gauss–Green theorem for BV vector fields. Let Ω be a bounded Lipschitz domain and suppose $u : \Omega \to R^n$ is a vector field such that each of its components is an element of $BV(\Omega)$. Then the trace, u^*, of u on $\partial\Omega$ is defined and

$$\operatorname{div} u(\Omega) = \int_{\partial\Omega} u^*(x)\nu(x,\Omega)dH^{n-1}(x).$$

For this it is sufficient to prove that

$$D_i u(\Omega) = \int_{\partial\Omega} u^*(x)\nu_i(x,\Omega)dH^{n-1}(x)$$

where $u \in BV(\Omega)$, $1 \leq i \leq n$, and $D_i = \partial/\partial x_i$. With the notation of the previous exercise, observe that

$$\begin{aligned}
D_i\chi_{A_t}(\Omega) &= -\int_{\Omega\cap\partial^* A_t} \nu_i(x, A_t)dH^{n-1}(x) \\
&= -\int_{\partial\Omega\cap\partial^* A_t} \nu_i(x, A_t)dH^{n-1}(x) \\
&\qquad (\text{since } D_i\chi_{A_t}(R^n) = 0) \\
&= -\int_{\partial\Omega\cap\partial^* A_t} \nu_i(x, \Omega)dH^{n-1}(x) \\
&= -\int_{A_t\cap\partial^*\Omega} \nu_i(x, \Omega)dH^{n-1}(x) \\
&= D_i\chi_\Omega(\partial^* A_t).
\end{aligned}$$

With the help of the previous exercise, conclude that

$$\begin{aligned}
D_i u(\Omega) &= \int_{-\infty}^{+\infty} D_i\chi_{A_t}(\Omega)dt \\
&= \int_{-\infty}^{+\infty} D_i\chi_\Omega(\partial^* A_t)dt \\
&= \int_{-\infty}^{0} D_i\chi_\Omega(\partial^* A_t)dt + \int_{0}^{+\infty} D_i\chi_\Omega(\partial^* A_t)dt \\
&= \int_{0}^{+\infty} D_i\chi_\Omega(\partial^* A_t)dt - \int_{-\infty}^{0} D_i\chi_\Omega(\partial^*\Omega - \partial^* A_t)dt
\end{aligned}$$

$$= \int_0^\infty D_i\chi_\Omega(\{x : u^* \geq t\})dt - \int_{-\infty}^0 D_i\chi_\Omega(\{x : u^* \leq t\})dt$$
$$= \int_{\partial\Omega} u^* dD_i\chi_\Omega$$
$$= \int_{\partial\Omega} u^*(x)\nu_i(x,\Omega)dH^{n-1}(x).$$

5.7. Give a description of the result in the previous exercise on the line, i.e., when $n = 1$.

5.8. Under the conditions of Corollary 5.2.4, prove that $\{Du_i\} \to Du$ weakly as measures as $i \to \infty$.

5.9. Suppose Ω is an open subset of R^{n-1} and that $u\colon \Omega \to R^1$ is differentiable at $x_0 \in \Omega$ in the sense defined by (2.2.2). Also, let $M = \{(x, u(x)) : x \in \Omega\}$. Show that there exists a vector, ν, that satisfies (5.6.10) at $(x_0, u(x_0))$.

5.10. Let $\Omega \subset R^n$ be an open set with the property that $\partial\Omega$ has a tangent plane at $x_0 \in \partial\Omega$. That is, assume for each $\varepsilon > 0$, that

$$C(\varepsilon) \cap \partial\Omega \cap B(x_0, r) = \emptyset$$

for all small $r > 0$, where $C(\varepsilon)$ is introduced in Definition 5.6.3. Assume also that

$$\limsup_{r\to 0} \frac{|\Omega \cap B(x_0,r)|}{|B(x_0,r)|} > 0$$

and

$$\limsup_{r\to 0} \frac{|(R^n - \Omega) \cap B(x_0,r)|}{|B(x_0,r)|} > 0.$$

Prove that $x_0 \in \partial^*\Omega$.

Historical Notes

5.1. BV functions were employed in several areas such as area theory and the calculus of variations before the formal introduction of distributions, cf., [CE], [TO]. However, the definition employed at that time was in the spirit of Theorem 5.3.5.

5.3. Theorem 5.3.3 is a result adopted from Krickerberg [KK]. This result is analogous to the one obtained by Meyers and Serrin [MSE] for Sobolev functions, Theorem 2.3.2. Serrin [SE] and Hughs [HS] independently discovered Theorem 5.3.5.

5.4. The theory of sets of finite perimeter was initiated by Caccioppoli [C] and DeGiorgi [DG1], [GD2] and subsequently developed by many contributors including [KK], [FL], [F1], and [F2]. Sets of finite perimeter can be regarded as n-dimensional integral currents in R^n and therefore they can be developed within the context of geometric measure theory. The isoperimetric inequality with sets of finite perimeter is due to DeGiorgi [D1], [D2] and the co-area formula for BV functions, Theorem 5.4.4 was proved by Fleming and Rishel [FR].

5.5. The notion of generalized exterior normal is due to DeGiorgi [D2] and is basic to the development of sets of finite perimeter. Essentially all of the results in this section are adapted from DeGiorgi's theory.

5.6. The concept of the measure-theoretic normal was introduced by Federer [F1] who proved [F2] that it was essentially the same as the generalized exterior normal of DeGiorgi.

Definition 5.6.3 implicitly invokes the notion of an approximate tangent plane to an arbitrary set which is of fundamental importance in geometric measure theory, cf. [F4]. Theorem 5.6.5 states that the plane orthogonal to the generalized exterior normal at a point x_0 is the approximate tangent plane to the reduced boundary at x_0.

5.7. Countably k-rectifiable sets and approximate tangent planes are closely related concepts. Indeed, from the definition and Rademacher's theorem, it follows that a countably k-rectificable set has an approximate tangent k-plane at H^k almost all of its points. Lemma 5.7.2 is one of the important results of the theory developed in [F4]. Theorem 5.7.3 is due to DeGiorgi [D1], [D2], although his formulation and proof are not the same.

5.8. In his earlier work Federer, [F1], was able to establish a version of the Gauss–Green theorem which employs the measure-theoretic normal for all every open subset of R^n whose boundary has finite H^{n-1}-measure. After DeGiorgi had established the regularity of the reduced boundary (Theorems 5.7.3 and 5.8.1), Federer proved Theorem 5.8.7 [F2].

5.9. A different version of Lemma 5.9.3 was first proved by William Gustin [GU]. This version is due to Federer and appears in [F4, Section 4.5.4]. Theorem 5.9.6 is only a part of the development of BV functions that appears in [F4, Section 4.5.9]. Other contributors to the pointwise behavior of BV functions include Goffman [GO] and Vol'pert [VO]. In particular, Vol'pert proved that the measure-theoretic boundary $\partial_M E$ is equivalent to the reduced boundary.

5.10–5.11. The trace of a BV function on the boundary of a regular domain as developed in this section is taken from [MZ]. Alternate developments can be found in [GI1] and [MA3]. This treatment of Poincaré-type inequalities was developed in [MZ].

5.12. Lemma 5.12.3 was first proved by Fleming [FL]. The proof of the lemma depends critically on Lemma 5.9.3. Fleming publicly conjectured that the claim of the lemma (his statement had a slightly different form) was true and Gustin proved it in [GU]. Theorems 5.12.2, 5.12.8, and the material in Section 5.13 were proved in [MZ].

Bibliography

[AD1]
D.R. ADAMS, *Traces of potentials arising from translation invariant operators,* Ann. Sc. Norm. Super. Pisa., **25** (1971), 203–217.

[AD2]
D.R. ADAMS, *A trace inequality for generalized potentials,* Stud. Math., **48** (1973), 99–105.

[AD3]
D.R. ADAMS, *Traces of potentials,* Ind. U. Math. J.,**22** (1973), 907–918.

[AD4]
D.R. ADAMS, *On the existence of capacitary strong type estimates in R^n*, Arkiv for Matematik, **14** (1976), 125–140.

[AD5]
D.R. ADAMS, *Quasi-additivity and sets of finite L^p-capacity,* Pacific J. Math., **79** (1978), 283–291.

[AD6]
D.R. ADAMS, *Lectures on L^p-potential theory,* University of Umea, Sweden, Lecture Notes 1981.

[AD7]
D.R. ADAMS, *Weighted nonlinear potential theory,* Trans. Amer. Math. Soc., **297** (1986), 73–94.

[AD8]
D.R. ADAMS, *A sharp inequality of J.Moser for higher order derivatives,* Ann. Math., **128** (1988), 385–398.

[AM]
D.R. ADAMS & N. MEYERS, *Thinness and Wiener criteria for nonlinear potentials,* Indiana Univ. Math. J., **22** (1972), 169–197.

[AR1]
R.A. ADAMS, *Capacity and compact imbeddings,* J. Math. Mech., **19** (1978), 923–929.

[AR2]
R.A. ADAMS, *Sobolev spaces,* Academic Press, 1975.

[ARS1]
N. ARONSZAJN & K.T. SMITH, *Functional spaces and functional completion,* Ann. Inst. Fourier (Grenoble), **6** (1956), 125–185.

[ARS2]
N. ARONSZAJN & K.T. SMITH, *Theory of Bessel potentials I, Studies in eigenvalue problems,* Technical Report No. 22, University of Kansas, 1959.

[AMS]
N. ARONSZAJN, P. MULLA & P. SZEPTYCKI, *On spaces of potentials connected with L^p-classes,* Ann. Inst. Fourier **13** (1963), 211–306.

[AV]
A. AVANTAGGIATI, *On compact embedding theorems in weighted Sobolev spaces,* Czech. Math. J., **29** (1979), 635–647.

[AS]
S. AXLER & A.L. SHIELDS, *Univalent multipliers of the Dirichlet space,* Mich. Math. J., **32** (1985), 65–80.

[BAZ]
T. BAGBY & W. ZIEMER, *Pointwise differentiability and absolute continuity,* Trans. Amer. Math. Soc., **194** (1974), 129–148.

[BB]
H. BERESTYCKI & H. BREZIS, *On a free boundary problem arising in plasma physics,* Nonlinear Anal. The., Met., and Appl., **4** (1980), 415–436.

[BE1]
A.S. BESICOVITCH, *A general form of the covering principle and relative differentiation of additive functions,* (I), Proc. Camb. Phil. Soc., **41** (1945), 103–110.

[BE2]
A.S. BESICOVITCH, *A general form of the covering principle and relative differentiation of additive functions,* (II), Proc. Camb. Phil. Soc., **42** (1946), 1–10.

[BL]
G.A. BLISS, *An integral inequality,* J. London Math. Soc., **5** (1930), 40–46.

[BS]
H.P. BOAS & E.J. STRAUBE, *Integral inequalities of Hardy and Poincaré-type,* Preprint, 1986.

[BRT]
M. BRELOT, *Lectures on potential theory,* Tata Institute, Bombay, 1960.

[BR1]
H. BREZIS, *Laser beams and limiting cases of Sobolev inequalities,* Nonlinear partial differential equations, College de France Seminar; Research Notes in Math., Pitman Press, (1982).

[BR2]
H. BREZIS, *Large harmonic maps in two dimensions,* preprint, Workshop Isola d'Elba, Sept. 1983.

[BW]
H. BREZIS & S. WAINGER, *A note on limiting cases of Sobolev embeddings and convolution inequalities,* Comm. Partial Diff. Eq. **5** (1980), 773–789.

[BL]
H. BREZIS & E.H. LIEB, *Sobolev inequalities with remainder terms,* J. Funct. Analysis, **62** (1985), 73–86.

[BZ]
J. BROTHERS & W. ZIEMER, *Minimal rearrangements of Sobolev functions,* J. Für die Reine und Angewandte Math., **384** (1988), 153–179.

[BUZ]
Y. BURAGO & V. ZALGALLER, *Geometric inequalities,* Grundlehren (285), Springer-Verlag, 1988.

[C]
R. CACCIOPPOLI, *Misure e integrazione sugli insiemi dimensionalmente orientati,* Rend. Accad. Naz. dei Lincei., **12** (1953), 3–11, 137–146.

[CKR]
L. CAFFARELLI, R. KOHN & L. NIRENBERG, *First order interpolation inequalities with weights,* Comp. Math., **53** (1984), 259–275.

[CA1]
A.P. CALDERÓN, *Lebesgue spaces of differentiable functions and distributions,* Proc. Symp. Pure Math., **IV** (1961), 33–49.

[CA2]
A.P. CALDERÓN, *Intermediate spaces and interpolation,* Studia Math., Seria Spec., **1** (1960), 31–34.

[CA3]
A.P. CALDERÓN, *Intermediate spaces and interpolation, the complex method,* Studia Math., **24** (1964), 113–190.

[CA4]
A.P. CALDERÓN, *Uniqueness of distributions,* Unión Mate. Argentina, **25** (1970), 37–55.

[CZ]
A.P. CALDERÓN & A. ZYGMUND, *Local properties of solutions of elliptic partial differential equations,* Studia Math., **20** (1961), 171–225.

[CFR]
C. CALDERÓN, E. FABES & N. RIVIERE, *Maximal smoothing operators,* Ind. Univ. Math. J., **23** (1973), 889–898.

[CA]
J.W. CALKIN, *Functions of several variables and absolute continuity, I.*, Duke Math. J., **6** (1970), 170–185.

[CAR1]
H. CARTAN, *La théorie général du potentiel dans les espace homogènes,* Bull. Sci. Math. **66** (1942), 126–132, 136–144.

[CAR2]
H. CARTAN, *Théorie générale du balayage en potentiel newtonien,* Ann. Univ. Grenoble, **22** (1946), 221–280.

[CAY]
C. CARATHÉODORY, *Über das lineare Mass von Punktmengen eine Verallgemeinerung des Längenbegriffs,* Nach. Ges. Wiss. Gottingen, (1914), 404–426.

[CE]
L. CESARI, *Sulle funzioni a variazione limitata,* Ann. Scuola Norm, Sup. Pisa, **5** (1936), 299–313.

[CW]
S. CHANILLO & R.L. WHEEDEN, *L^p-estimates for fractional integrals and Sobolev inequalities with applications to Schrodinger operators,* Comm. Partial Diff. Eqns., **10** (1985), 1077–1116.

[CF]
F. CHIARENZA & M. FRASCA, *Noncoercive and degenerate elliptic equations in unbounded domains,* Unione Mat. Ital. Boll., **1** (1987), 57–68.

[CH]
G. CHOQUET, *Theory of capacities,* Ann. Inst. Fourier, **5** (1955), 131–295.

[DA1]
B. DAHLBERG, *A note on Sobolev spaces,* Proc. Symp. Pure Math., AMS, **35** (1979), 183–185.

[DA2]
B. DAHLBERG, *Regularity properties of Riesz potentials,* Ind. Univ. Math. J., **28** (1979), 257–268.

[DZ]
D. DEIGNAN & W.P. ZIEMER, *Strong differentiability properties of Sobolev functions,* Trans. Amer. Math. Soc., **225** (1977), 113–122.

[DE]
J. DENY, *Les potentiels d'énergie finie,* Acta Math. **82** (1950), 107–183.

[DL]
J. DENY & J.L. LIONS, *Les espaces du type de Beppo Levi,* Ann. Inst. Fourier, Grenoble, **5** (1953–54), 305–370.

[D1]
E. DeGIORGI, *Su una teoria generale della misure* $(r-1)$-*dimensionale in uno spazio ad r dimensioni,* Ann. Mat. Pura Appl., IV. Ser., **36** (1954), 191–213.

[D2]
E. DeGIORGI, *Nuovi teoremi relative alle misure* $(r-1)$-*dimensionale in spazio ad r dimensioni,* Ric. Mat., **4** (1955), 95–113.

[DG]
N. DeGUZMAN, *Differentiation of integrals in* R^n, Lecture Notes in Mathematics, Springer-Verlag, 481(1975).

[DT]
T.K. DONALDSON & N.S. TRUDINGER, *Orlicz-Sobolev spaces and imbedding theorems,* J. Funct. Analysis, **8** (1971), 52–75.

[DO]
WILLIAM F. DONOGHUE, *Distributions and Fourier transforms,* Academic Press, 1969.

[E]
A. EHRHARD, *Inegalités isoperimetriques et integrales de Dirichlet Gaussiennes,* Ann. Scient. Ec. Norm. Sup., **17** (1984), 317–332.

[EH]
G. EHRLING, *On a type of eigenvalue problem for certain elliptic differential operators,* Math. Scand., **2** (1954), 267–285.

[FA]
K. FAN, *Minimax theorems,* Proc. Nat. Acad. Sci., **39** (1953), 42–47.

[F1]
H. FEDERER, *The Gauss–Green theorem,* Trans. Amer. Math. Soc., **58** (1945), 44–76.

[F2]
H. FEDERER, *A note on the Gauss–Green theorem,* Proc. Am. Math. Soc., **9** (1958), 447–451.

[F3]
H. FEDERER, *Curvature measures,* Trans. Amer. Math. Soc., **93** (1959), 418–491.

[F4]
H. FEDERER, *Geometric measure theory,* Springer-Verlag, New York, Heidelberg, 1969.

[FF]
H. FEDERER & W.H. FLEMING, *Normal and integral currents,* Ann. Math., **72** (1960), 458–520.

[FZ]
H. FEDERER & W. ZIEMER, *The Lebesgue set of a function whose distribution derivatives are p-th power summable,* Ind. Univ. Math. J., **22** (1972), 139–158

[FL]
W. FLEMING, *Functions whose partial derivatives are measures,* Ill. J. Math., **4** (1960), 452–478.

[FR]
W.H. FLEMING & R. RISHEL, *An integral formula for total gradient variation,* Arch. Math., **11** (1960), 218–222.

[FRE]
JENS FREHSE, *A refinement of Rellich's theorem,* Bonn Lecture Notes, 1984.

[FU]
B. FUGLEDE, *Extremal length and functional completion,* Acta Math., **98** (1957), 171–219.

[GA1]
E. GAGLIARDO, *Proprieta di alcune classi di funzioni in piu variabili,* Ricerche di Mat. Napli, **7** (1958), 102–137.

[GA2]
E. GALIARDO, *Proprieta di alcune classi di funzioni in piu variabili,* Ric. Mat., **7** (1958), 102–137.

[GA3]
E. GALIARDO, *Ulteriori proprieta di alcune classi di funzioni in piu variabili,* Ric. Mat., **8** (1958), 102–137.

[GE1]
I. GELFAND, M. GRAEV & N. VILENKIN, *Obobščennye funkcii, Integral'naja geometrija i svjazannye s neí voprosy teorii predstavolenií,* (Generalized functions, Integral geometry and related questions in the theory of representations), Gosudarstvennoe izdatel'stvo fiziko–matematičeslpoí literatury, Moscow, 1962.

[GE2]
I. GELFAND & G.E. ŠILOV, *Generalized functions, Properties and operations,* Vol. I, Academic Press, New York-London, 1964.

[GE3]
I. GELFAND & ŠILOV, *Les distributions, Espaces fondamentaux,* Tome 2, Collection Universitaire de Mathématiques, 15, Dunod, Paris, 1964.

[GE4]
I. GELFAND & G.E. ŠILOV, *Verallgemeinerte Funktionen (Distributionen.), Einige Fragen zur Theorie der Differentialgleichungen,* III, VEB Deutscher Verlag der Wissenschaften, Berlin, 1964.

[GE5]
I. GELFAND & N. VILENKIN, *Applications to Harmonic Analysis*, Academic Press, New York-London, 1964.

[GT]
D. GILBARG & N. TRUDINGER, *Elliptic Partial Differential Equations of Second Order*, Springer-Verlag, New York, 1983, Second Ed.

[GI]
E. GIUSTI, *Precisazione delle funzioni $H^{1,p}$ e singolarità delle soluzioni deboli di sistemi ellittici non lineari*, Boll. UMI, **2** (1969), 71–76.

[GI2]
E. GIUSTI, *Minimal surfaces and functions of bounded variation*, Birkhäuser, 1985.

[GO]
C. GOFFMAN, *A characterization of linearly continuous functions whose partial derivatives are measures*, Acta Math., **117** (1967), 165–190.

[GU]
W. GUSTIN, *Boxing Inequalities*, J. Math. Mech., **9** (1960), 229–239.

[HA]
K. HANSSON, *Imbedding theorems of Sobolev type in potential theory*, Math. Scand., **45** (1979), 77–102.

[H]
R. HARDT, *An introduction to Geometric Measure Theory*, Lecture Notes, Melbourne University, (1979).

[HL]
G.G. HARDY & J.E. LITTLEWOOD, *A maximal theorem with function-theoretic applications*, Acta Math., **54** (1930), 81–86.

[HAU]
F. HAUSDORFF, *Dimension und äusseres Mass*, Math. Ann., **79** (1919), 157–179.

[HM]
V. HAVIN & V. MAZ'YA, *Nonlinear potential theory*, Russian Math. Surveys, **27** (1972), 71–148.

[HE1]
L. HEDBERG, *On certain convolution inequalities*, Proc. Amer. Math. Soc., **36** (1972), 505–510.

[HE2]
L. HEDBERG, *Spectral synthesis in Sobolev spaces, and uniqueness of solutions of the Dirichlet problem*, Acta Math., **147** (1981), 237–264.

[HW]
L. HEDBERG & T. WOLFF, *Thin sets in nonlinear potential theory*, Ann. Inst. Fourier (Grenoble), **23** (1983), 161–187.

[HMT]
J.G.R. HEMPEL, J.G.R. MORRIS & N.S. TRUDINGER, *On the sharpness of a limiting case of the Sobolev imbedding theorem,* Bull. Australian Math. Soc., **3** (1970), 369–373.

[HS]
R.E. HUGHS, *Functions of BVC type,* Proc. Amer. Math. Soc., **12** (1961), 698–701.

[HU]
R. HUNT, *On $L(p,q)$ spaces,* L'enseignement Math., **12** (1966), 249–276.

[JMZ]
B. JESSEN, J. MARCINKIEWICZ & A. ZYGMUND, *Note on the differentiability of multiple integrals,* Fund. Math., **25** (1934), 217–234.

[JO]
P. JONES, *Quasiconformal mappings and extendability of functions in Sobolev spaces,* Acta Math., **147** (1981), 71–88.

[KO]
V.I. KONDRACHOV, *Sur certaines propriétés fonctions dans l'espace L^p*, C. R. (Doklady) Acad. Sci. URSS (N.S.) **48** (1945),535–538.

[KK]
K. KRICKERBERG, *Distributionen, Funktionen beschränkter Variation und Lebesguescher Inhalt nichtparametrischer Flächen,* Ann. Mat. Pura Appl., IV Ser., **44** (1957), 105–134.

[KR]
A.S. KRONROD, *On functions of two variables,* Usp. Mat. Nauk, **5** (1950), 24–134, (Russian).

[KU]
A. KUFNER, *Weighted Sobolev spaces,* John Wiley & Sons, 1983.

[LU]
O.A. LADYZHENSKAYA & N.N. URAL'TSEVA, *Linear and quasilinear elliptic equations,* Academic Press, New York, (1968).

[LE1]
H. LEBESGUE, *Leçons sur l'integration et la recherche des fonctions primitives,* Gauthier–Villars, Paris (1904).

[LE2]
H. LEBESGUE, *Sur l'integration des fonctions discontinues,* Ann. Ecole Norm., **27** (1910), 361–450.

[LB]
E.H. LIEB, *Sharp constants in the Hardy–Littlewood–Sobolev and related inequalities,* Ann. of Math. **118** (1963), 349–374.

[LI]
FON-CHE LIU, *A Lusin type property of Sobolev functions,* Ind. Univ. Math. J., **26** (1977). 645–651.

[LM]
J.L. LIONS & E. MAGENES, *Non-homogeneous boundary value problems and application,* Springer-Verlag, New York, Heidelberg, 1972.

[LP]
J.L. LIONS & J. PEETRE, *Sur une classe d'espaces d'interpolation,* Inst. Hautes Etudes Sci. Publ. Math., **19** (1964), 5–68.

[LO1]
G.G. LORENTZ, *Some new functional spaces,* Ann. of Math., **51** (1950), 37–55.

[LO2]
G.G. LORENTZ, *On the theory of spaces,* Pacific J. Math., **1** (1951), 411–429.

[MM1]
M. MARCUS & V.J. MIZEL, *Complete characterization of functions which act, via superposition, on Sobolev spaces,* Trans. Amer. Math. Soc., **251** (1979), 187–218.

[MM2]
M. MARCUS & V.J. MIZEL, *Absolute continuity on tracks and mappings of Sobolev spaces,* Arch. Rat. Mech. Anal., **45** (1972), 294–320.

[MA1]
V.G. MAZ'YA, *On some integral inequalities for functions of several variables,* Problems in Math. Analysis, Leningrad Univ.,(Russian) **3** (1972).

[MA2]
V.G. MAZ'YA, *Imbedding theorems and their applications,* Proc. Symp. Baku. 1966, Isdat Nauka., Moscow, 1970.

[MA3]
V.G. MAZ'YA, V.G., *Sobolev Spaces,* Springer-Verlag, Springer Series in Soviet Mathematics, 1985.

[ME1]
N. MEYERS, *A theory of capacities for potentials of functions in Lebesgue spaces,* Math. Scand., **26** (1970), 255–292.

[ME2]
N. MEYERS, *Taylor expansion of Bessel potentials,* Ind. U. Math. J., **23** (1974), 1043–1049.

[ME3]
N. MEYERS, *Continuity properties of potentials,* Duke Math. J., **42** (1975), 157–166.

[ME4]
N. MEYERS, *Integral inequalities of Poincaré and Wirtinger Type,* Arch. Rat. Mech. Analysis, **68** (1978), 113–120.

[MSE]
N. MEYERS & J. SERRIN, $H = W$, Proc. Nat. Acad. Sci. USA, **51** (1964), 1055–1056.

[MZ]
N.G. MEYERS & W.P. ZIEMER, *Integral inequalities of Poincaré and Wirtinger type for BV functions,* Amer. J. Math., **99** (1977), 1345–1360.

[MI]
J. MICHAEL, *The equivalence of two areas for nonparametric discontinuous surfaces,* Ill. J. Math., **7** (1963), 255–292.

[MS]
J. MICHAEL & L. SIMON, *Sobolev and mean-value inequalities on generalized submanifolds of* R^n, Comm. Pure Appl. Math., **26** (1973), 361–379.

[MIZ]
J. MICHAEL & W.P. ZIEMER, *A Lusin type approximation of Sobolev functions by smooth functions,* Contemporary Mathematics, AMS, **42** (1985), 135–167.

[MSE1]
A.P. MORSE, *The behavior of a function on its critical set,* Ann. Math., **40** (1939), 62–70.

[MSE2]
A.P. MORSE, *Perfect blankets,* Trans. Amer. Math. Soc., **6** (1947), 418–442.

[MO1]
C.B. MORREY, *Functions of several variables and absolute continuity,* II, Duke Math. J., **6** (1940), 187–215.

[MO2]
C.B. MORREY, *Multiple integrals in the calculus of variations,* Springer-Verlag, Heidelberg, New York, (1966).

[MOS]
J. MOSER, *A sharp form of an inequality by N. Trudinger,* Ind. Univ. Math. J., **20** (1971), 1077–1092.

[MW]
B. MUCKENHOUPT & R. WHEEDEN, *Weighted norm inequalities for fractional integrals,* Trans. Amer. Math. Soc., **192** (1974), 261–274,

[NI1]
L. NIRENBERG, *An extended interpolation inequality,* Ann. Scuola Norm. Sup. Pisa, **20** (1966), 733–737.

[NI2]
L. NIRENBERG, *On elliptic partial differential equations,* Ann. Scuola Norm. Pisa(III), **13** (1959), 1–48.

[O]
R. O'NEIL, *Convolution operators and $L(p, q)$ spaces,* Duke Math. J., **30** (1963), 129–142.

[PE]
J. PEETRE, *Nouvelles proprietés d'espaces d'interpolation,* C. R. Acad. Sci., **256** (1963), 1424–1426.

[P]
J. POLKING, *Approximation in L^p by solutions of elliptic partial differential equations,* Amer. J. Math., **94** (1972), 1231–1234.

[PS]
G. POLYA & G. SZEGO, *Isoperimetric inequalities in mathematical physics,* Annals of Math. Studies, Princeton, **27** (1951).

[PO]
Ch.J. de la VALLEE POUSSIN, *Sur l'integrale de Lebesgue,* Trans. Amer. Math. Soc., **16** (1915), 435–501.

[RA]
H. RADEMACHER, *Über partielle und totale Differenzierbarkeit* I., Math. Ann., **79** (1919), 340–359.

[RR]
T. RADO & P. REICHELDERFER, *Continuous transformations in analysis,* Grundlehren der Math., Springer-Verlag, Heidelberg, New York, 1955.

[RE]
R. RELLICH, *Ein Satz über mittlere Konvergenz.,* Nachr. Akad. Wiss. Göttingen Math.–Phys. K1., (1930), 30–35.

[RM]
M. RIESZ, *Integrales de Riemann-Liouville et potentiels,* Acta Sci. Math. Szeged, **9** (1938), 1–42.

[RES]
Ju.G. RESETNJAK, *On the concept of capacity in the theory of functions with generalized derivatives,* Sib. Mat. Zh. **10** (1969), 1109–1138, (Russian). English translation: Siberian Mat. J., **10** (1969), 818–842.

[RO]
E. RODEMICH, *The Sobolev inequalities with best possible constants,* Analysis Seminar at California Institute of Technology, 1966.

[SK]
S. SAKS, *Theory of the integral,* Warsaw, 1937.

[SA]
A. SARD, *The measure of the critical values of differentiable maps,* Bull. Amer. Math. Soc., **48** (1942), 883–890.

[SC]
M. SCHECTER, *Weighted norm estimates for Sobolev spaces,* Trans. Amer. Math. Soc., **304** (1987), 669–687.

[SCH]
L. SCHWARTZ, *Théorie des distributions,* I, II, Act. Sci. Ind., 1091, 1122, Hermann et Cie., Paris (1951).

[SE]
J. SERRIN, *On the differentiability of functions of several variables,* Arch. Rat. Mech. Anal., **7** (1961), 359–372.

[SG]
V.G. SIGILLITO, *Explicit a priori inequalities with applications to boundary value problems,* Research Notes in Mathematics, Pittman Publishing, 13(1977).

[S]
L. SIMON, *Lectures on geometric measure theory,* Proc. Centre Math. Analysis, ANU, **3** (1983).

[SO1]
S.L. SOBOLEV, *On some estimates relating to families of functions having derivatives that are square integrable,* Dokl. Adak. Nauk SSSR, **1** (1936), 267–270, (Russian).

[SO2]
S.L. SOBOLEV, *On a theorem of functional analysis,* Mat. Sb. **46** (1938), 471–497, (Russian). English translation: Am. Math. Soc. Translations, (2)**34** (1963), 39–68.

[SO3]
S.L. SOBOLEV, *Applications of functional analysis in mathematical physics,* Leningrad: Izd. LGU im. A.A. Zdanova 1950 (Russian), English Translation: Amer. Math. Soc. Translations, **7** (1963).

[ST]
E.M. STEIN, *Singular integrals and differentiability properties of functions,* Princeton University Press, 1970.

[SW]
E. STEIN & G. WEISS, *Introduction to Fourier analysis on euclidean spaces,* Princeton Univ. Press, (1971).

[STR]
R.S. STRICHARTZ, *A note on Trudinger's extension of Sobolev's inequalities,* Ind. Univ. Math. J., **21** (1972), 841–842.

[TA]
G. TALENTI, *Best constant in Sobolev inequality,* Ann. Mat. Pura Appl., **110** (1976), 353–372.

[TO]
L. TONELLI, *Sulla quadratura delle superficie,* Atti Reale Accad. Lincei, **6** (1926), 633–638.

[TR]
N.S. TRUDINGER, *On imbeddings into Orlicz spaces and some applications,* J. Math. Mech., **17** (1967), 473–483.

[U]
ALEJANDRO URIBE, *Minima of the Dirichlet and Toeplitz operators,* preprint, 1985.

[PO]
Ch.J. de la VALLEE POUSSIN, *Sur l'integrale de Lebesgue,* Trans. Amer. Math. Soc., **16** (1915), 435–501.

[VI]
G. VITALI, *Sui gruppi di punti e sulle funzioni di variabili reali,* Atti Accad. Sci. Torino, **43** (1908), 75–92.

[VO]
A.I. VOL'PERT, *The spaces BV and quasi-linear equations,* Mat. Sb. **73** (1967), 255–302 (Russian); English translation: Math. USSR Sb., **2** (1967) 225–267.

[WI]
N. WIENER, *The ergodic theorem,* Duke Math. J., **5** (1939), 1–18.

[WH]
H. WHITNEY, *Analytic extensions of differentiable functions defined in closed sets,* Trans. Amer. Math. Soc., **36** (1934), 63–89.

[Y]
L.C. YOUNG, *Partial area,* Rivista Mat., **10** (1959).

[Z1]
W.P. ZIEMER, *Extremal length and conformal capacity,* Trans. Amer. Math. Soc., **126** (1967), 460–473.

[Z2]
W.P. ZIEMER, *Change of variables for absolutely continuous functions,* Duke Math. J., **36** (1969), 171–178.

[Z3]
W.P. ZIEMER, *Extremal length and p-capacity,* Mich. Math. J., **17** (1969), 43–51.

[Z4]
W.P. ZIEMER, *Extremal length as a capacity,* Mich. Math. J., **17** (1970), 117–128.

[Z5]
W.P. ZIEMER, *Mean values of subsolutions of elliptic and parabolic equations,* Trans. Amer. Math. Soc., **279** (1983), 555–568.

[Z6]
W.P. ZIEMER, *Poincaré inequalities for solutions of elliptic equations,* Proc. Amer. Math. Soc., **97** (1986), 286–292.

List of Symbols

1.1

$\emptyset$	empty set	1
χ_E	characteristic function of E	2
R^n	Euclidean n-space	1
x	$x = (x_1, \ldots, x_n)$, point in R^n	1
$x \cdot y$	inner product	1
$\lvert x \rvert$	norm of x	1
$\overline{S}$	closure of S	1
∂S	boundary of S	1
spt u	support of u	1
$d(x, E)$	distance from x to E	1
$\text{diam}(E)$	diameter of E	1
$B(x, r)$	open ball center x, radius r	2
$\overline{B}(x, r)$	closed ball center x, radius r	2
$\alpha(n)$	volume of unit ball	2
α	$\alpha = (\alpha_1, \cdots, \alpha_n)$, multi-index	2
$\lvert \alpha \rvert$	length of multi-index	2
x^α	$x^\alpha = x_1^{\alpha_1} \cdot x_2^{\alpha_2} \cdots x_n^{\alpha_n}$	2
$\alpha!$	$\alpha! = \alpha_1! \alpha_2! \cdots \alpha_n!$	2
$D_i = \partial/\partial x_i$	i^{th} partial derivative	2
D^α	$D^\alpha = D_1^{\alpha_1} \cdots D_n^{\alpha_n} = \frac{\partial^{\lvert \alpha \rvert}}{\partial x_1^{\alpha_1} \cdots \partial x_n^{\alpha_n}}$	2
Du	gradient of u	2
$D^k u$	$D^k u = \{D^\alpha u\}_{\lvert \alpha \rvert = k}$	2
$C^0(\Omega)$	space of continuous functions on Ω	2
$C^k(\Omega)$	$C^k(\Omega) = \{u : D^\alpha u \in C^0(\Omega)\}$	2
$C_0^k(\Omega)$	$C_0^k(\Omega) = C^k(\Omega) \cap \{u : \text{spt } u \text{ compact, spt } u \subset \Omega\}$	2
$C^k(\overline{\Omega})$	$\{u : D^\alpha u \text{ uniformly continuous on } \Omega, \lvert \alpha \rvert \leq k\}$	2
$C^k(\Omega; R^m)$	R^m-valued functions	2
$C^{0,\alpha}(\overline{\Omega})$	Hölder continuous functions on Ω	3
$C^{k,\alpha}(\overline{\Omega})$	$C^{k,\alpha}(\overline{\Omega}) = \{u : D^\beta u \in C^{0,\alpha}(\overline{\Omega}), 0 \leq \lvert \beta \rvert \leq k\}$	3

1.2

$\lvert E \rvert$	Lebesgue measure of E	3

$d(A, B)$	distance between sets	4

1.3

$\hat{B}$	ball concentric about B with 5 times its radius	7
$\fint_{B(x,r)} f\, d\mu$	integral average	14

1.4

H^γ	γ-dimensional Hausdorff measure	15

1.5

$L^p(\Omega)$	functions p^{th}-power integrable on Ω	18
$L^p_{\text{loc}}(\Omega)$	$u \in L^p(K)$ for each compact $K \subset \Omega$	18
$\|u\|_{p;\Omega}$	L^p-norm on Ω	18
$\|u\|_{\infty;\Omega}$	sup norm on Ω	18
$L^p(\Omega; \mu)$	underlying measure is μ	18
$L^p_{\text{loc}}(\Omega; \mu)$	underlying measure is μ	18
$\|u\|_{p;\Omega;\mu}$	underlying measure is μ	19
$\|u\|_{\infty;\Omega;\mu}$	underlying measure is μ	19
p'	$p' = \frac{p}{p-1}$ if $p > 1$, $p' = \infty$, if $p = 1$	20

1.6

φ_ε	regularizer (or mollifier)	21
u_ε	$u_\varepsilon = \varphi_\varepsilon * u$, convolution of u with φ_ε	21

1.7

T	Schwartz distribution	24

1.8

E^f_s	$E^f_s = \{x : |f(x)| > s\}$	26
$\alpha_f(s)$	$\alpha_f(s) = |E^f_s|$	26
f^*	non-increasing rearrangement of f	26
f^{**}	$f^{**}(t) = \frac{1}{t}\int_0^t f^*(r)\,dr$	28

$L(p,q)$	Lorentz spaces	28
$\|f\|_{p,q}$	Lorentz norm	28
$L(p,\infty)$	weak L^p	28

2.1

$W^{k,p}(\Omega)$	Sobolev space on Ω	43
$\|u\|_{k,p;\Omega}$	Sobolev norm on Ω	43
$W_0^{k,p}(\Omega)$	norm closure of $C_0^\infty(\Omega)$	43
$BV(\Omega)$	BV functions on $\Omega \subset R^n$	43
u^+	$u^+ = \max\{u,0\}$	46
u^-	$u^- = \min\{u,0\}$	46

2.2

$\Omega' \subset\subset \Omega$	$\overline{\Omega}' \subset \Omega$, $\overline{\Omega}'$ compact	52

2.4

$\omega(n-1)$	area of unit sphere in R^n	58
p^*	Sobolev exponent, $p^* = np/(n-kp)$	58

2.6

I_α	Riesz kernel of order α	64
γ_α	Riesz constant $\gamma_\alpha = \frac{\pi^{n/2} 2^\alpha \Gamma(\frac{\alpha}{2})}{\Gamma(\frac{n}{2} - \frac{\alpha}{2})}$	64
g_α	Bessel kernel of order α	65
$a(x) \sim b(x)$	$a(x)/b(x)$ bounded away from 0 and ∞	65
$L^{\alpha,p}$	space of Bessel potentials	66
$B_{\alpha,p}$	Bessel capacity	66
$R_{\alpha,p}$	Riesz capacity	66
$\gamma_{\alpha,p}$	variational capacity, Exer. 2.8	104
$b_{\alpha,p}$	$b_{\alpha,p}^p = B_{\alpha,p}$	71

2.7

div V	divergence of the vector field V	77

2.8

$M(f)(x)$	maximal function of f at x	84

3.4

$P_x^{(m)}$	Taylor polynomial of degree m at x	126

3.5

$T^k(E)$	bounded k^{th}-order difference quotients on E	130
$t^k(E)$	formal Taylor expansions of degree k on E	131
$T^{k,p}(x)$	bounded k^{th}-order difference quotients in L^p	132
$t^{k,p}(x)$	k^{th}-order differentiability in L^p	132

3.10

$M_{p,R}u(x)$	L^p maximal function	154

4.1

$(W^{m,p}(\Omega))^*$	dual of $W^{m,p}(\Omega)$	179

4.7

$\mathcal{M}_k\mu(x)$	fractional maximal operator	204
$\mathcal{E}_{k,p}(\mu)$	(k,p)-energy of μ	204

5.1

$\|u\|_{BV}$	norm of BV function	221
$\|Du\|$	total variation measure of BV function u	221

5.3

$\text{ess } V_a^b(u)$	essential variation of u on $[a,b]$	227

5.4

$P(E,\Omega)$	perimeter of E in Ω	229

5.5

$\partial^- E$	reduced boundary of E	233
$\nu(x,E)$	generalized exterior normal of E at x	233

5.6

$C(x,\varepsilon,\nu)$	$C(x,\varepsilon,\nu) = R^n \cap \{y : \|(y-x)\cdot\nu\| > \varepsilon\|y-x\|\}$	240
$n(x,E)$	measure-theoretic normal to E at x	240
$\partial^* E$	$\partial^* E = \{x : n(x,E) \text{ exists}\}$	240

5.8

$\partial_M E$	measure-theoretic boundary of E	249

5.9

$\overline{D}(E,x)$	upper density of E at x	250
$\underline{D}(E,x)$	lower density of E at x	250
$D(E,x)$	density of E at x	250
$\operatorname{ap}\limsup_{y\to x} u(y)$	upper approximate limit of u at x	250
$\operatorname{ap}\limsup_{y\to x} u(y)$	lower approximate limit of u at x	250

5.12

$\gamma(E)$	BV-capacity of E	262

Index

Page numbers are enclosed in parentheses; section numbers precede page numbers.

Graduate Texts in Mathematics

continued from page ii

93 DUBROVIN/FOMENKO/NOVIKOV. Modern Geometry—Methods and Applications Vol. I.
94 WARNER. Foundations of Differentiable Manifolds and Lie Groups.
95 SHIRYAYEV. Probability, Statistics, and Random Processes.
96 CONWAY. A Course in Functional Analysis.
97 KOBLITZ. Introduction to Elliptic Curves and Modular Forms.
98 BRÖCKER/TOM DIECK. Representations of Compact Lie Groups.
99 GROVE/BENSON. Finite Reflection Groups. 2nd ed.
100 BERG/CHRISTENSEN/RESSEL. Harmonic Anaylsis on Semigroups: Theory of Positive Definite and Related Functions.
101 EDWARDS. Galois Theory.
102 VARADARAJAN. Lie Groups, Lie Algebras and Their Representations.
103 LANG. Complex Analysis. 2nd ed.
104 DUBROVIN/FOMENKO/NOVIKOV. Modern Geometry—Methods and Applications Vol. II.
105 LANG. $SL_2(\mathbf{R})$.
106 SILVERMAN. The Arithmetic of Elliptic Curves.
107 OLVER. Applications of Lie Groups to Differential Equations.
108 RANGE. Holomorphic Functions and Integral Representations in Several Complex Variables.
109 LEHTO. Univalent Functions and Teichmüller Spaces.
110 LANG. Algebraic Number Theory.
111 HUSEMÖLLER. Elliptic Curves.
112 LANG. Elliptic Functions.
113 KARATZAS/SHREVE. Brownian Motion and Stochastic Calculus.
114 KOBLITZ. A Course in Number Theory and Cryptography.
115 BERGER/GOSTIAUX. Differential Geometry: Manifolds, Curves, and Surfaces.
116 KELLEY/SRINIVASAN. Measure and Integral, Volume 1.
117 SERRE. Algebraic Groups and Class Fields.
118 PEDERSEN. Analysis Now.
119 ROTMAN. An Introduction to Algebraic Topology.
120 ZIEMER. Weakly Differentiable Functions: Sobolev Spaces and Functions of Bounded Variation.

MIX
Papier aus verantwortungsvollen Quellen
Paper from responsible sources
FSC® C105338

If you have any concerns about our products,
you can contact us on
ProductSafety@springernature.com

In case Publisher is established outside the EU,
the EU authorized representative is:
Springer Nature Customer Service Center GmbH
Europaplatz 3, 69115 Heidelberg, Germany

Printed by Libri Plureos GmbH
in Hamburg, Germany